Lecture Notes in Business Information Processing 597

Series Editors

LNBIP reports state-of-the-art results in areas related to business information systems and industrial application software development – timely, at a high level, and in both printed and electronic form.

The type of material published includes

- Proceedings (published in time for the respective event)
- Postproceedings (consisting of thoroughly revised and/or extended final papers)
- Other edited monographs (such as, for example, project reports or invited volumes)
- Tutorials (coherently integrated collections of lectures given at advanced courses, seminars, schools, etc.)
- Award-winning or exceptional theses

LNBIP is abstracted/indexed in DBLP, EI and Scopus. LNBIP volumes are also submitted for the inclusion in ISI Proceedings.

Anna Bernasconi · Mario Cortès-Cornax ·
Jose Luis de la Vara · Elsa Negre ·
Thomas Polacsek · Jolita Ralyté · Marcela Ruiz
Editors

Research Challenges in Information Science

RCIS 2026 Forum and Doctoral Consortium

Toulouse, France, May 26–29, 2026
Revised Selected Papers

Springer

Editors
Anna Bernasconi
Politecnico di Milano
Milan, Italy

Mario Cortès-Cornax
Université Grenoble Alpes
Saint-Martin-d'Hères, France

Jose Luis de la Vara
Universidad de Castilla-La Mancha
Albacete, Spain

Elsa Negre
Paris-Dauphine University
Paris, France

Thomas Polacsek
ONERA
Toulouse, France

Jolita Ralyté
University of Geneva
Carouge, Switzerland

Marcela Ruiz
ZHAW Zurich University of Applied Sciences
Winterthur, Switzerland

ISSN 1865-1348 ISSN 1865-1356 (electronic)
Lecture Notes in Business Information Processing
ISBN 978-3-032-35042-8 ISBN 978-3-032-35043-5 (eBook)
https://doi.org/10.1007/978-3-032-35043-5

This Springer imprint is published by the registered company Springer Nature Switzerland AG
The registered company address is: Gewerbestrasse 11, 6330 Cham, Switzerland

Preface RCIS 2026 Forum

The RCIS 2026 Forum track provided a platform for researchers and practitioners to present innovative research ideas, work in progress, and emerging prototypes addressing current challenges in Information Science. In line with the theme of RCIS 2026, "Bridging the Gap: Enhancing Understandability in Information Science", the Forum track encouraged contributions that explore understandable, explainable, and trustworthy information systems across diverse domains.

The 20th edition of the RCIS conference was held in Toulouse, France, May 26–29, 2026. In alignment with the aims and scope of the conference, the Forum considered two types of submissions:

- Research idea papers, intended to present new ideas, challenges, and work in progress where at least some preliminary results are available;
- Tool demonstrations, intended to describe tools or prototypes that are relevant for facing challenges related to information science and topics of interest for the RCIS conference.

This year, the Forum track attracted a diverse set of 8 submissions covering areas such as digital transformation, cybersecurity & privacy, health information systems, smart environments, process redesign, and artificial intelligence, among others. Fourteen papers were ultimately selected. Nine of them were invited through the RCIS main conference call, while the remaining five were selected through direct submissions to the Forum track.

Each submission followed a single-blind peer-review process involving at least three members of either the RCIS main conference committee or the dedicated Forum committee. Authors received constructive feedback aimed at strengthening their contributions and supporting future developments. The accepted papers reflect the interdisciplinary nature of the RCIS community. They address both theoretical and practical challenges related to Information Systems, with two contributions demonstrating tools and the other twelve presenting research ideas.

During the conference, Forum papers were presented through a short two-minute teaser presentation, followed by interactive poster and demonstration sessions designed to encourage discussion, collaboration, and feedback among participants. These exchanges reflect the RCIS spirit of bringing together researchers and practitioners to address emerging challenges in Information Science.

We would like to thank the authors for their contributions and the reviewers for their commitment to maintaining the quality of the research presented at the RCIS Forum. We extend our gratitude to the General Chairs and Program Chairs of the main conference for their support in coordinating the Forum organization.

June 2026

Anna Bernasconi
Mario Cortès-Cornax

Organization

Chairs

Anna Bernasconi	Politecnico di Milano, Italy
Mario Cortès-Cornax	Université Grenoble Alpes, France

Program Committee

Greta Adamo	Free University of Bozen-Bolzano, Italy
Clara Ayora	Universidad de Castilla-La Mancha, Spain
Nelly Barret	INSA Lyon, France
Judith Barrios Albornoz	Universidad de Los Andes, Venezuela
Dominik Bork	TU Wien, Austria
Francesco Cambria	Politecnico di Milano, Italy
Cinzia Cappiello	Politecnico di Milano, Italy
Riccardo Coppola	Politecnico di Torino, Italy
Istvan David	McMaster University, Canada
Victoria Döller	University of Vienna, Austria
Alberto García S.	Universitat Politècnica de València, Spain
Giovanni Giachetti	Universitat Politècnica de València, Spain
Francesco Invernici	Politecnico di Milano, Italy
Abdelaziz Khadraoui	University of Geneva, Switzerland
Elena Kornyshova	CNAM, France
Georgios Koutsopoulos	Stockholm University, Sweden
Sotirios Liaskos	York University, Canada
Giovanni Meroni	Technical University of Denmark, Denmark
Christoforos Ntantogian	Ionian University, Greece
Michalis Pavlidis	University of Brighton, UK
Francisca Pérez	Universitat Politècnica de València, Spain
Jose Luis de la Vara	University of Castilla-La Mancha, Spain

Preface RCIS 2026 Doctoral Consortium

The RCIS 2026 Doctoral Consortium (DC) provided a dedicated forum for PhD students to present and discuss their ongoing research in Information Science and Information Systems. It offered a supportive environment aimed at fostering constructive feedback, methodological guidance, and scientific exchange between doctoral researchers and experienced mentors.

In line with the theme of RCIS 2026, "Bridging the Gap: Enhancing Understandability in Information Science", the DC encouraged PhD contributions addressing complex and emerging challenges in areas such as data governance, artificial intelligence, human-centered systems, cybersecurity, and digital transformation.

The Doctoral Consortium attracted 11 submissions, out of which 8 papers were accepted after a rigorous single-blind peer-review process conducted by two members of the dedicated DC program committee per submission.

Each doctoral student benefited from a structured and interactive presentation format, including a 20-minute presentation of their research work, followed by a 10-minute peer discussion with another PhD student from the consortium, and a 15-minute in-depth discussion with two assigned mentors. The sessions were further enriched by open exchanges with the audience, fostering a collaborative and constructive environment.

The research topics covered in the DC reflect the diversity and richness of the RCIS community. They include, among others, formal modeling of data protection and consent, data governance and compliance with large language models, hybrid human-AI collaborative agile practices, scenario analysis for future systems, IoT-based human activity and workload assessment, as well as the use of NLP and LLMs for market research and customer journey analysis. Representative contributions include work on knowledge graph- and LLM-assisted curriculum engineering, as well as cyber-resilience frameworks for critical infrastructure systems.

Following the consortium, authors were invited to revise and extend their papers based on feedback received from peers and mentors. These improved versions form the post-proceedings of the Doctoral Consortium.

We would like to warmly thank all doctoral students for their high-quality contributions, as well as the members of the DC program committee and mentors for their commitment and insightful guidance. We also acknowledge the support of the RCIS 2026 organizing committee for making this event possible.

June 2026

Elsa Negre
Jose Luis de la Vara

Organization

Chairs

Elsa Negre	Université Paris Dauphine-PSL, France
Jose Luis de la Vara	University of Castilla-La Mancha, Spain

Program Committee

Lylia Abrouk	Université de Bourgogne, France
Carina Alves	Universidade Federal de Pernambuco, Brazil
Pierre-Emmanuel Arduin	Université Paris Dauphine-PSL, France
Saïd Assar	Institut Mines-Télécom Business School, France
Jean-Michel Bruel	IRIT, France
Nelly Condori Fernández	University of Santiago de Compostela, Spain
Sergio Firmenich	National University of La Plata and CONICET, Argentina
Xavier Franch	Universitat Politècnica de Catalunya, Spain
Jänis Grabis	Riga Technical University, Latvia
Patricia Martin-Rodilla	Spanish National Research Council (CSIC), Spain
Imen Megdiche	IRIT, France
Selmin Nurcan	Université Paris 1 Panthéon-Sorbonne, France
Jolita Ralyté	University of Geneva, Switzerland
Maribel Yasmina Santos	University of Minho, Portugal
Sagar Sen	Simula Research Laboratory, Norway
Nathalie Vallès-Parlangeau	Université de Pau et des Pays de l'Adour, France
Tanja E. J. Vos	Open Universiteit and Universitat Politècnica de València, Netherlands/Spain
Jinfeng Zhong	Université Paris 8 Vincennes-Saint-Denis, France

Mentors

Saïd Assar	Institut Mines-Télécom Business School, France
Jänis Grabis	Riga Technical University, Latvia
Selmin Nurcan	Université Paris 1 Panthéon-Sorbonne, France
Jolita Ralyté	University of Geneva, Switzerland
Marcela Ruiz	Zurich University of Applied Sciences (ZHAW), Switzerland

Contents

Forum

Data Chameleon: Biological-Inspired Contextual Privacy for Automated Compliance Evaluation in Trustworthy Data Sharing 3
Shudan Yang and Pierluigi Plebani

Towards an IoT Plan for Smart Buildings: Lessons from a Real-World Construction Project 16
Lasse Harjumaa, Petri Jyrkkä, Ilkka Kivelä, and Ismo Hakala

When Fake News Sounds Right: Assessing Message Credibility in AI-Generated Misinformation 29
Paulina Wojciechowska-Dzięcielak, Pierre-Emmanuel Arduin, and Myriam Merad

From Data Governance Policies to Policy-as-Code: An LLM-Based Framework 41
Matteo Falconi, Valeria Maria Fortina, Marco Tonnarelli, Indika Kumara, and Pierluigi Plebani

MAPS-AI a Model-driven AI-Assisted Tool for IT Project Plan and Scope: Case Study and Evaluation 54
Megija Krista Milune and Oksana Nikiforova

Low-Code Platforms Adoption: A Motivational Adoption Process Perspective 66
Elisa De Coster, Anthony Simonofski, and Monique Snoeck

Quantifying ICT Third-Party Concentration Risk Using DORA Reporting Data 78
Beāte Krauze and Laura Kruglova

Probabilistic Dempster–Shafer Fusion for Multimodal Breast Cancer Analysis 90
Abir Baâzaoui, Malek Baâzaoui, Ikram Bellara, and Walid Barhoumi

Adoption of AI-Based Recruitment Systems: Benefits, Challenges, and User Trust 102
Hao Yue Liu and Rébecca Deneckère

Querying Structured Data Through Natural Language Using Language Models ... 115
Valentin Micu-Hontan, Andrei-Alexandru Bunea, Nikolaos Dimitrios Tantaroudas, and Dan-Matei Popovici

Towards a Meta-model for Complex Performance Indicators: A Systematic Literature Review and an Illustration in the Cardiology Domain ... 127
Benito Giunta, Corentin Burnay, and Alexandre Faulkner

Mapping Capabilities with LLMs: Requirements and Feasibility Assessment of Capability Broker ... 140
Adrian Olbricht, Jennifer Skopac, Jan Lundberg, Jelena Zdravkovic, and Janis Stirna

Insights from Triage Processes for Fairness-Aware Process Redesign ... 153
Rachmadita Andreswari, Stephan A. Fahrenkrog-Petersen, and Jan Mendling

Can We Typify Health Information Seekers? Identification of Personas Among Laypeople ... 165
Mariana Dias and Carla Teixeira Lopes

Doctoral Consortium

Toward Hybrid Agile Teams: Enhancing Collaboration in Agile Projects by Exploring Traditional and AI-Based Tools for Process Optimization and Best Practices ... 179
Haoyue Liu

Towards a Conceptual Model for Evaluating Prospective Scenarios - A Case Study in Future Air Transport ... 191
Brieuc Danet

A Bounded-Autonomy Multi-agent Architecture for Policy-Constrained Data Governance ... 202
Yasaman Hemmati

Assessing Occupational Physical Workload Using IoT and Human Activity Recognition ... 213
Djalal Bouchekif

Cyber Resilience by Design for Critical-Infrastructure Information Systems: A Three-Phase Framework for Systemic Risk Assessment, Management, and Recovery ... 223
Fatimah Faraji

DIAPROD: A Framework for Formal Consent Modelling, Dynamic Storage Limitation, Data Minimisation, and LLM-Assisted Compliance for Data Protection by Design 235
Samuel Opoku Daniels

Combining Large Language Models with Named Entity Recognition to Automate Market Research Interview Transcription and Customer Journey Mapping – Empirical Feedback by Business Professionals 247
Tarmo Koppel

Towards a Knowledge Graph-Based and LLM-Assisted Support System for University Curriculum Engineering 261
Aritha Kumarasinghe

Author Index 273

Forum

Data Chameleon: Biological-Inspired Contextual Privacy for Automated Compliance Evaluation in Trustworthy Data Sharing

Shudan Yang(✉) and Pierluigi Plebani

Politecnico di Milano, Piazza Leonardo Da Vinci, 32, 20133 Milan, Italy
{shudan.yang,pierluigi.plebani}@polimi.it

Abstract. Data sharing has become a key factor in data analytics, as it allows organizations to operate on others' data for secondary usages. At the same time, when sharing datasets, due to privacy reasons, some transformations are required to the shared datasets to limit the visibility according to an agreement between a consumer and the data provider. Usually, upon this agreement a specific access control setting is established to ensure the data privacy protection. However, when data sharing occurs at scale, reconciling the conceptual discrepancy between data sharing and privacy protection requires automated mechanisms to evaluate whether the shared data is compliant with the data sharing agreement. In this work, inspired by the biological mechanism by which chameleons adapt their coloration in response to environmental stimuli, we propose an automated, context-aware compliance evaluation mechanism for data sharing. This mechanism characterizes how the privacy level of data can vary across interaction contexts, such as access roles and query content, thereby providing interpretable decision support for compliance evaluation and fine-grained governance throughout the data-sharing process.

Keywords: Data Sharing · Privacy Evaluation · Context-aware Compliance

1 Introduction

Data have become a core organizational asset, and sharing them across teams and organizational boundaries can improve analytics, support decision-making, and foster collaboration. In response, approaches such as Data Mesh [2] and Data Spaces [1] have been proposed to enable governed data exchange at scale. However, cross-domain sharing remains difficult to operationalize in practice [5]. The core tension is clear: data consumers seek broader and more integrated views, whereas data providers need to retain continuous control over downstream access, usage, and disclosure consequences once data are exposed beyond organizational boundaries. At the same time, regulatory requirements such as purpose

A. Bernasconi et al. (Eds.): RCIS 2026, LNBIP 597, pp. 3–15, 2027.
https://doi.org/10.1007/978-3-032-35043-5_1

limitation, traceability, and auditability introduce substantial governance and engineering overhead. Together, these pressures motivate the demand for data sovereignty, namely, that providers should remain able to constrain external access and usage in a continuous, fine-grained, and verifiable manner [6].

To support this demand, cross-domain sharing is shifting from copy-based exchange to consortium-style federated sharing delivered through API-mediated services [8]. Under this paradigm, authorized consumers interact with provider-side query interfaces, while providers preserve governance controls and accountability evidence after sharing. In practice, sharing agreements specify who may access what, for which purpose, and in what form outputs may be returned. However, such contractual clauses do not automatically translate into runtime behavior. The system must assess the exposure level of an ongoing session and align it with the limits permitted by the agreement for different roles and purposes. When exposure approaches or exceeds those limits, the system should enable a downstream enforcement layer to apply more conservative output strategies.

The challenge is that multi-round interactions can weaken the effectiveness of per-request compliance checks [3]. In realistic query–answer workflows, requests are often refined step by step, moving from population-level statistics to subpopulation analyses and eventually toward near-individual inference. Risk escalation therefore arises not only from a single obviously out-of-policy request, but also from the accumulation of semantically related refinements. As the current query becomes more intrusive and prior disclosures amplify the sensitivity of later outputs, even a sequence of individually compliant queries may gradually strengthen inferential power and push the session toward over-exposure. These observations suggest the need for a runtime mechanism that can track cumulative exposure across the session rather than judging each request in isolation.

To address this need, inspired by the biological mechanism by which chameleons adapt their coloration in response to environmental stimuli, this paper proposes a dynamic privacy-state evaluation tool that characterizes how the privacy state of the same data evolves across interaction contexts. The proposed mechanism integrates field-level sensitivity, scenario-level baseline context, query intrusiveness, history-driven amplification, and role-dependent perspective into an auditable item-level privacy state. When multiple fields are implicated, it further derives a conservative session-level signal to support compliance alignment and auditing.

The remainder of this paper is organized as follows. Section 2 reviews related work. Section 3 introduces the system definition. Section 4 presents the dynamic privacy-state assessment method. Section 5 reports the validation. Section 6 concludes the paper.

2 Related Work

Recent conceptual blueprints increasingly frame cross-domain sharing as API-mediated, federated provision rather than data replication, enabling sustained consumption relationships while preserving post-sharing governance and accountability [1]. In cross-organizational settings, data spaces further foreground identity, trust, and policy enforcement to operationalize data sovereignty [10]. Building on this service-based sharing premise, we target a capability that remains under-specified at runtime: an interpretable and auditable signal that tracks session-level privacy exposure under multi-turn interactions.

In privacy-aware systems, privacy state has been used to characterize an individual's privacy posture in context, including quantified states and state-transition formulations [7]. Related work also models privacy state as a system-level variable for reasoning about dependencies among entities and behaviors, and stage-based views have been adopted to describe how privacy state evolves as requirements are progressively satisfied. These studies motivate our formulation of privacy as a state variable, but they do not address how a data item's state should evolve within a governed multi-turn sharing session.

A complementary body of work emphasizes that privacy and access decisions are context-dependent. Contextual Integrity formalizes privacy as norm-conformant information flow in context [9], while Zero Trust and related context-aware access-control mechanisms support continuous evaluation using contextual signals at runtime [11]. Recent work on contextual sensitive-data detection further shows that sensitivity judgments may also depend on context [12]. However, these approaches mainly support per-request authorization, policy evaluation, or pre-release detection. They do not provide an auditable session-level signal that summarizes query semantics, requester context, and disclosure history to indicate whether cumulative exposure is increasing.

Our work addresses this gap by introducing such a runtime signal for governed multi-turn data sharing. We model privacy risk as a dynamic session-level process driven by query semantics, disclosure history, and requester role.

3 System Definition

We consider a consortium-style federated sharing setting consistent with data mesh and data spaces: organizations collaborate under a shared governance framework, while raw data remain within each provider's administrative boundary and are exposed only through governed interfaces rather than centralized copying. Within this setting, we distinguish two roles. A Data Provider collects, stores, and governs data assets and exposes consumable access endpoints. A Data Consumer issues requests through agreed interfaces under legitimate credentials in order to obtain governed data views or statistical results.

We focus on a practically important threat actor in cross-organizational sharing: the authorized-but-curious consumer. Such a consumer does not exceed formal privileges, yet may gradually strengthen inferential capability through semantically related query refinements. Therefore, per-request compliance checking is insufficient: even if each query appears acceptable in isolation, a sequence of related interactions may accumulate into session-level over-exposure.

Under this setting, the core runtime governance question is: how can the system produce an auditable compliance-assessment signal at each interaction

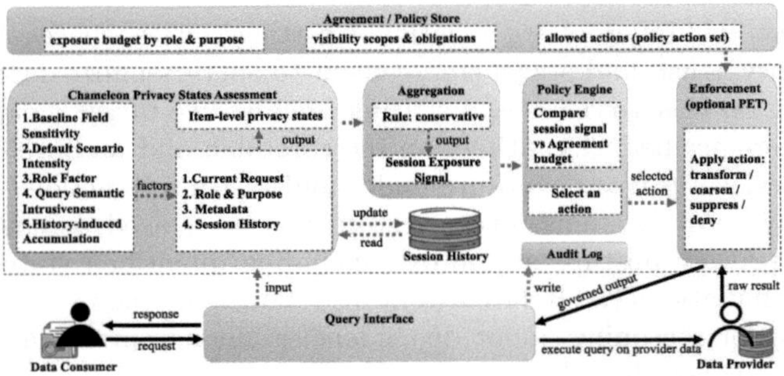

Fig. 1. System overview.

round, and how can this signal be used to constrain API outputs so that session exposure remains within the sharing agreement? To answer this question, we abstract the runtime system into three coupled elements: (i) protected information and interaction context, including protected items, query semantics, requester role-purpose context, and session history; (ii) the provider-side query pipeline, which executes requests and returns governed outputs; and (iii) a regulation loop that evaluates exposure and drives policy-aligned actions.

At interaction round t, a request (q_t, u, p) enters through a query interface under role-purpose context (u, p). The provider executes the query on in-boundary data assets and obtains a raw result y_t. A policy-controlled enforcement stage may then transform this result, through coarsening, masking, suppression, or other PET-based operations, to produce the governed response $\tilde{y}_t$. The interaction is written to an audit log for traceability. Figure 1 summarizes this flow.

The central regulatory component in this paper is the privacy-state assessment module (Chameleon). For each protected data item x implicated by the interaction, the module outputs an auditable item-level privacy state $\sigma_x(t) \in [0, 1)$, where $\sigma_x(t)$ is the bounded privacy state defined in Sect. 4. When a query implicates multiple protected items, such as multiple fields or a composite view, the system further derives a conservative session-level governance signal

$$s_t = \mathrm{Agg}\Big(\big\{\sigma_x(t)\big\}_{x \in X(q_t)}\Big) \in [0, 1), \tag{1}$$

where $X(q_t)$ denotes the set of items implicated by q_t, either directly requested or made inferable, and $\mathrm{Agg}(\cdot)$ is a monotone aggregation rule. In this paper, we use a conservative choice such as max for system-level alignment, and the item-level state $\sigma_x(t)$ is the primary state variable studied. s_t is a derived session-level signal used only for interface-level governance to summarize the current session exposure into a single decision input for downstream policy enforcement.

The assessment output does not itself determine which PET or output strategy should be applied. Instead, the policy engine consumes the session-level signal s_t together with agreement metadata and selects an action $a_t \in \mathcal{A}$, such as *release*, *coarsen*, or *suppress*. The selected action is then executed by a downstream enforcement stage. Accordingly, our focus in this paper is on modeling and computing privacy states, namely the item-level state $\sigma_x(t)$ and its conservative aggregation s_t, while enforcement strategies and PET instantiations are treated as pluggable downstream components. A risk signal is actionable only when it is compared against a normative bound specific to role and purpose. To make data sovereignty operational, providers and consumers establish a sharing agreement specifying who may access what, under which conditions, and in what output form. Such agreements typically encode purpose limitation, visibility scope, output granularity, required transformations, and traceability obligations. At runtime, these clauses must be translated into a decision boundary against which exposure can be checked.

Formally, for each consumer role $u \in \mathcal{U}$ and purpose or scenario $p \in \mathcal{P}$, the agreement induces a maximum exposure budget $\tau : \mathcal{U} \times \mathcal{P} \to [0, 1)$. The threshold $\tau(u, p)$ is configured as an auditable policy table during agreement operationalization and is read at runtime by the policy engine. It is not learned autonomously by the model. Instead, it represents the maximum session-level exposure under the given role-purpose combination. Runtime compliance therefore reduces to an alignment problem between an assessed session signal and an agreement budget. At round t, the policy engine compares s_t against $\tau(u, p)$ and selects an action $a_t \in \mathcal{A}$. If $s_t \leq \tau(u, p)$, the current interaction remains within the permitted exposure boundary. If $s_t > \tau(u, p)$, the session is deemed to approach or exceed the agreement's acceptable exposure level, and the enforcement layer may apply more conservative output actions at the API boundary. An example illustrates why session-level signaling is necessary. Consider a medical data-sharing interface over a COVID-19 cohort, where approved researchers are allowed to access only aggregate statistics and individual records are prohibited. A researcher may begin with broad cohort summaries and then iteratively refine the population through increasingly specific constraints. Although each individual query may satisfy per-round checks, overlapping disclosure history can amplify inferential power through refinement, differencing, or intersection effects. The resulting risk is therefore not purely query-local, but session-dependent.

For this reason, privacy risk in interactive sharing should be modeled as a state variable evolving over the session rather than as a static property of a single query. Its transitions are driven jointly by the instantaneous semantic stimulus of the current query, the cumulative amplification induced by prior history, and the consumer's role-dependent perspective. The concrete evaluation of this dynamic privacy state is presented in Section 4.

4 The "Chameleon Effect" of Data: Dynamic Assessment of Privacy States

In biological systems, chameleons can rapidly change their skin coloration in response to changes in external conditions such as illumination, threat, and social context. A common explanation is structural coloration: an ordered nanocrystal lattice selectively reflects incident light, producing the observed color shift [14]. Under the first-order Bragg reflection approximation ($m = 1$), the reflected wavelength satisfies: $m\lambda = 2n_{\text{eff}} d \sin\theta$, where n_{eff} is the effective refractive index, d is the lattice spacing, and θ is the incidence/observation angle. A key property is that the lattice spacing is not fixed. Prior studies show that reversible shifts of the reflection peak can be induced by changes in lattice spacing [13]. To keep the analogy at an abstract level, we summarize these influences using an external stimulus $P(t)$ and an internal state signal $S(t)$. Let d_0 denote the resting lattice spacing and let $\Delta d(P(t), S(t))$ denote the time-dependent deformation induced jointly by stimulus and internal state. Then

$$\lambda(t) = 2n_{\text{eff}} \big[d_0 + \Delta d\big(P(t), S(t)\big)\big] \sin\theta. \tag{2}$$

This equation provides an abstract separable structure: a stable substrate, an external stimulus, and an internal state jointly determine the observable state of the same individual over time.

Motivated by this separable form, we use the same structural idea to characterize privacy dynamics in interactive data sharing. Our goal is not to claim a physical equivalence, but to borrow the decomposition "substrate + external stimulus + internal state + observation condition" as a modeling template. Intuitively, the privacy level of the same data item need not remain fixed across a multi-round query–answer session [3,4]; rather, it may be progressively "recolored" by the semantics of the current query, the historical interaction trajectory, and the requester role.

We therefore map the key quantities in the structural-color model to a privacy system (Table 1) and model the privacy state of the same data item x at time t in two stages. First, we define an interpretable unnormalized risk driver:

$$\tilde{\sigma}_x(t) = B(x)\Big[C_0 + \Delta c\big(C_{\text{local}}(t), H_{\text{history}}(t)\big)\Big] R(u), \tag{3}$$

where $\tilde{\sigma}_x(t)$ preserves a linear decomposition and is therefore convenient for attribution and ablation. Second, we map it into a bounded privacy state:

$$\sigma_x(t) = \psi\big(\tilde{\sigma}_x(t)\big) \in [0, 1), \qquad \psi(z) = 1 - \exp(-\beta_\sigma z), \ \beta_\sigma > 0, \tag{4}$$

which ensures stability and comparability across sessions and roles by preventing unbounded growth caused by $R(u)$ or other weighting coefficients. Unless otherwise stated, privacy state refers to the bounded state $\sigma_x(t)$.

(1) Structural Substrate: $B(x)$ and C_0 The terms $B(x)$ and C_0 represent the baseline substrate of the privacy state. Here $B(x) \in (0, 1]$ denotes the intrinsic sensitivity of data item x. For example, direct identifiers are typically highly sensitive; in medical settings, some health attributes may also be highly sensitive, whereas general demographic fields are often less sensitive, subject to governance policy and intended use. The quantity $C_0 \in [0, 1]$ denotes a scenario-level baseline context intensity, such as routine clinical use versus third-party commercial analytics. When both semantic stimulus and historical accumulation are weak, the dominant contribution to the risk driver is $B(x)\, C_0\, R(u)$.

Table 1. Mapping between the chameleon color model and the privacy-state model

Chameleon	Biological meaning	Privacy	Privacy meaning
$\lambda(t)$	Reflected wavelength	$\sigma_x(t)$	Dynamic privacy state at time t
n_{eff}	Refractive index	$B(x)$	Intrinsic sensitivity of data item x
d_0	Resting lattice spacing	C_0	Default context intensity
$\Delta d(\cdot)$	Lattice deformation	$\Delta c(\cdot)$	Context deformation driven by semantics and history
$P(t)$	External stimulus	$C_{\text{local}}(t)$	Semantic intrusiveness of the current query
$S(t)$	Internal state signal	$H_{\text{history}}(t)$	History-induced cumulative risk
θ	Observation angle	$R(u)$	Consumer role factor

(2) Instantaneous Stimulus: $C_{\text{local}}(t)$ The term $C_{\text{local}}(t)$ is analogous to the external stimulus $P(t)$ in the structural-color model and captures the instantaneous semantic intrusiveness of the current query q_t. Throughout this section, we fix a data item x within a session and track how its state evolves over time. Even when a query does not explicitly request x, it may still increase the inferability of x by shrinking the candidate set, narrowing a target individual, or revealing correlated clues. Accordingly, we treat each query as a stimulus acting on a fixed x. Formally, C_{local} is a bounded real-valued function $C_{\text{local}} : \mathcal{Q} \to [0, 1]$, where $\mathcal{Q}$ denotes the space of natural-language queries. This function satisfies three requirements: (i)Boundedness: $C_{\text{local}}(t) \in [0, 1]$; (ii)Monotone comparability: if query q_a is semantically more individual-targeted or identifiable than q_b, then $C_{\text{local}}(q_a) \geq C_{\text{local}}(q_b)$; (iii)Locality: $C_{\text{local}}(t)$ depends only on the current query q_t and does not explicitly encode history or requester role.

We interpret $C_{\text{local}}(t)$ as the probability that the current query triggers an individual-level access intent:

$$C_{\text{local}}(t) = \Pr(\mathsf{individual} \mid q_t), \tag{5}$$

where the event $\mathsf{individual}$ indicates that the query semantically targets individual-level or individually inferable information access. In our instantiated implementation, this probability is estimated by a calibrated semantic classifier, whose output is directly used as $C_{\text{local}}(t)$. Details of label definition, training, and calibration are given in the experimental section.

(3) Historical Accumulation: $H_{\text{history}}(t)$ The term $H_{\text{history}}(t)$ is analogous to the internal state signal $S(t)$ in the structural-color model and captures the surplus risk induced by prior interaction history. Unlike the instantaneous stimulus $C_{\text{local}}(t)$, it captures session-level semantic dependency: when a requester progressively refines queries around the same target, prior disclosures can amplify the sensitivity of subsequent queries, even if each round appears acceptable in isolation. Formally, H_{history} is a bounded functional over history, $H_{\text{history}} : \mathcal{Q}^{t-1} \times \mathcal{Q} \to [0, 1)$, taking as input the historical query sequence $(q_1, \ldots, q_{t-1})$ and the current query q_t. This functional should satisfy: (i) temporal decay, so that distant history contributes less; (ii) semantic relatedness, so that accu-

mulation is substantial only when history and the current query are close; (iii) bounded saturation, so that cumulative risk grows but does not diverge.

We therefore define an unsaturated accumulation quantity $z(t) = \sum_{i=1}^{t-1} w(t-i) \cdot \mathrm{sim}(q_i, q_t) \cdot C_{\mathrm{local}}(i)$,where $w(\cdot)$ is a lag-dependent decay function and $\mathrm{sim}(\cdot,\cdot) \in [0,1]$ is a semantic similarity metric between historical and current queries. We then compress $z(t)$ into $[0,1)$ through a bounded monotone mapping:

$$H_{\mathrm{history}}(t) = \phi\big(z(t)\big), \qquad H_{\mathrm{history}}(t) \in [0,1), \tag{6}$$

where $\phi(\cdot)$ can be instantiated as a saturating function such as $1 - \exp(-\beta_h z)$.

In the implementation used in this paper, we adopt exponential decay $w(\Delta) = \exp(-\lambda\Delta)$ and choose $\phi(z) = 1 - \exp(-\beta_h z)$. The specific realization of $\mathrm{sim}(\cdot,\cdot)$ can be obtained from any semantic representation method; the experimental section reports the realization and parameter settings used here.

Operationally, $H_{\mathrm{history}}(t)$ is session-scoped. It is reset when the session ends or when governance scope changes (e.g., role switch or dataset switch), unless explicit policy specifies otherwise. This avoids carrying a permanently elevated risk state across unrelated interaction contexts.

(4) Role Perspective: $R(u)$ The observation angle θ affects the observed color; analogously, privacy risk depends not only on what is queried, but also on who is querying. To capture differences in privileges, responsibilities, and misuse risk across requesters, we introduce a role factor: $R : \mathcal{U} \to \mathbb{R}_{>0}$, where $\mathcal{U}$ denotes the set of requester roles (e.g., public, researcher, clinician, admin). The factor $R(u)$ scales the overall risk driver: roles with weaker privileges or higher potential misuse risk correspond to larger $R(u)$, whereas roles operating under stronger professional norms and auditable governance typically correspond to smaller $R(u)$. For example, one may set: $R(\mathsf{clinician}) < R(\mathsf{public})$, reflecting that the effective sensitivity of the same field may differ across requester roles.

(5) Context Deformation Function: $\Delta c(\cdot)$ The function $\Delta c(\cdot)$ corresponds to the lattice deformation $\Delta d(\cdot)$ in the structural-color model and integrates instantaneous semantic stimulus and history-induced cumulative risk into a unified context deformation quantity. Formally, we define: $\Delta c : [0,1] \times [0,1) \to \mathbb{R}_{\geq 0}$, and require: (i) monotonicity in both inputs; (ii) stability, so that privacy states do not grow without bound; and (iii) interpretability, so that instantaneous and historical contributions remain distinguishable.

To obtain a transparent decomposition and support ablation analysis, we adopt a linear instantiation:

$$\Delta c\big(C_{\mathrm{local}}(t), H_{\mathrm{history}}(t)\big) = \alpha_1 C_{\mathrm{local}}(t) + \alpha_2 H_{\mathrm{history}}(t), \qquad \alpha_1, \alpha_2 \geq 0. \tag{7}$$

Here α_1 controls the contribution of instantaneous semantic shifts, and α_2 controls the contribution of historical accumulation. Substituting Eq. (7) into Eq. (3) yields an explicit form of $\tilde{\sigma}_x(t)$, and applying Eq. (4) produces the final bounded state $\sigma_x(t) \in [0,1)$ used in subsequent visualization and trajectory analysis.

To make the dynamic interpretation explicit, we further examine how the risk driver changes over time. Under a continuous-time approximation, if $\tilde{\sigma}_x(t)$ is differentiable, then: $\frac{d\tilde{\sigma}_x(t)}{dt} = B(x)\,R(u)\Big[\alpha_1 \frac{dC_{\text{local}}(t)}{dt} + \alpha_2 \frac{dH_{\text{history}}(t)}{dt}\Big]$. Since $\sigma_x(t) = \psi(\tilde{\sigma}_x(t))$, we obtain: $\frac{d\sigma_x(t)}{dt} = \psi'(\tilde{\sigma}_x(t)) \cdot \frac{d\tilde{\sigma}_x(t)}{dt}, \qquad \psi'(z) = \beta_\sigma \exp(-\beta_\sigma z)$. This shows that the evolution of the bounded privacy state is jointly driven by two sources: (i) changes in the instantaneous semantic stimulus, and (ii) changes in the accumulated history term. Because $\psi'(\cdot)$ is positive and decays with the magnitude of the risk driver, the model preserves monotonicity while damping excessive growth. Large values of $\left|\frac{d\sigma_x(t)}{dt}\right|$ correspond to sharp privacy transitions, whereas smaller values indicate smoother state evolution.

In practical interactive systems, however, privacy state is naturally indexed by discrete query rounds $t = 1, 2, \ldots$ rather than continuous time. We therefore introduce the round-to-round difference: $\Delta\sigma_x(t) = \sigma_x(t) - \sigma_x(t-1)$, where larger $|\Delta\sigma_x(t)|$ indicates a pronounced privacy jump between adjacent rounds, and smaller $|\Delta\sigma_x(t)|$ indicates smoother evolution over the session. In the experiments, we use the trajectory of $\sigma_x(t)$ together with changes in $\Delta\sigma_x(t)$ to illustrate the chameleon-like privacy dynamics of the same data item under different query semantics, historical paths, and role configurations.

5 Validation

5.1 Assumptions and Setup

We perform a controlled, mechanism-level validation of the proposed privacy-state assessment model. The objective is to test whether its main components behave as intended in multi-round governance settings.

Experiments are conducted on a synthetic COVID-19 patient registry stored as a `.csv` file with 100 records. The dataset contains demographic, clinical, and test-related attributes. Queries are generated automatically using a template-and-slot-filling procedure to reduce manual subjectivity. The resulting query pool contains two categories: (i) aggregate queries, requesting population statistics, distributions, or grouped summaries; and (ii) targeted queries, requesting individual-level outputs or identifier-revealing information.

All components are implemented in Python 3.9. We validate whether (i) $C_{\text{local}}(t)$ increases as query semantics become more intrusive, (ii) $H_{\text{history}}(t)$ differentiates sessions with different refinement paths under the same terminal query, and (iii) $R(u)$ acts as a contextual modulator on top of the semantic/history-driven trajectory. Unless otherwise specified, $B(x)$, C_0, and $R(u)$ are fixed to default settings, and each experiment varies only the component under study. A larger $\sigma_x(t)$ indicates higher sensitivity and stronger governance demand.

Instantaneous Term $C_{\text{local}}(t)$. We instantiate $C_{\text{local}}(t) = \Pr(\mathsf{targeted} \mid q_t)$, where a query is labeled targeted if its expected output (i) contains a single patient identifier, (ii) can be directly attributed to one patient through attribute values or an individual record, or (iii) narrows the candidate identifier set close to

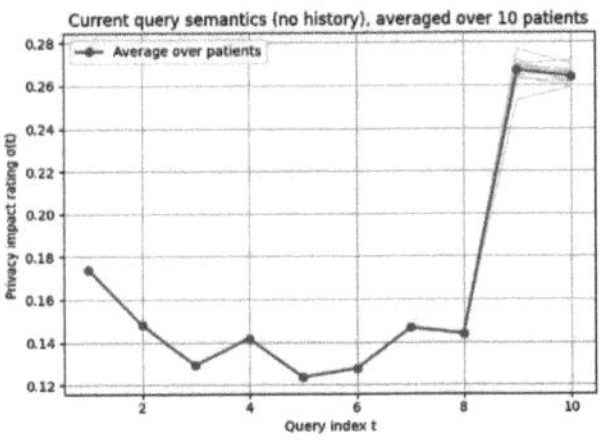

Fig. 2. Semantic refinement with history disabled.

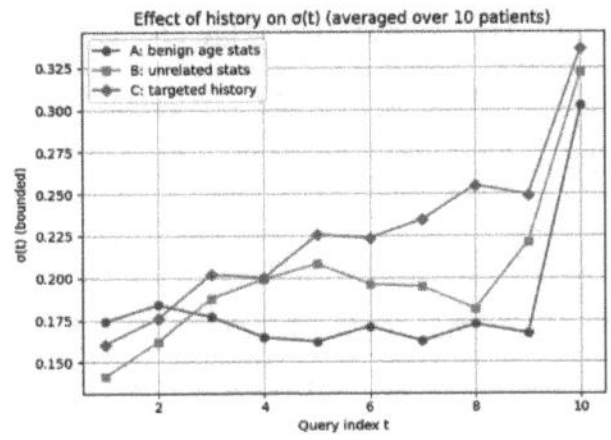

Fig. 3. History effect under a fixed terminal query.

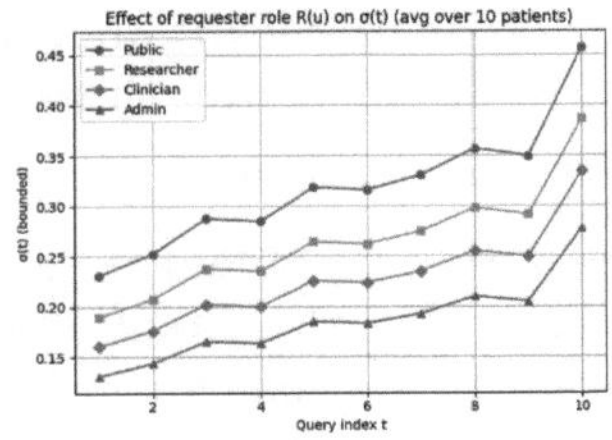

Fig. 4. Role-based scaling via $R(u)$.

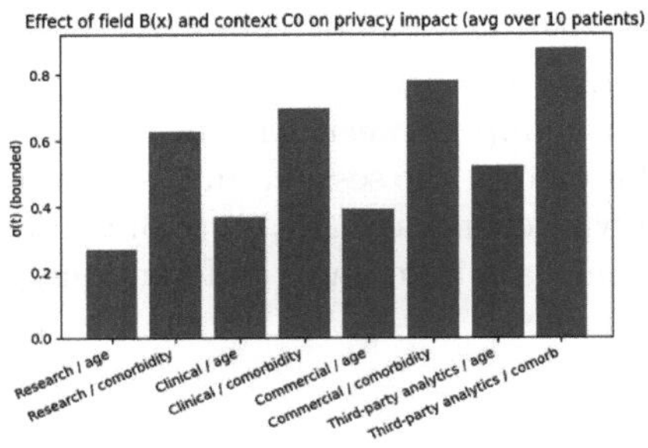

Fig. 5. Effect of $(B(x), C_0)$ on $\sigma(t)$.

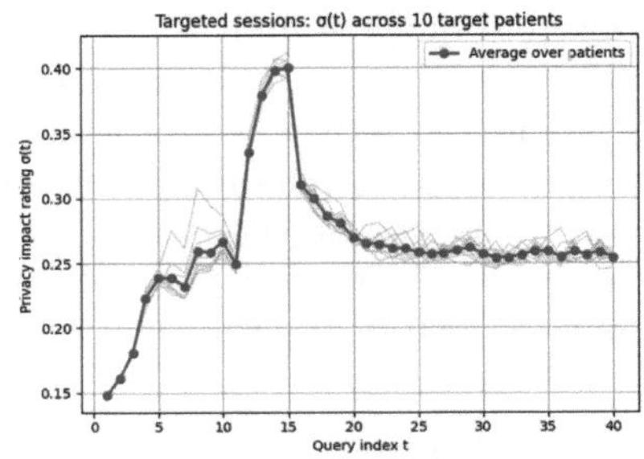

Fig. 6. Cross-target consistency under fixed refinement.

individual-level disclosure. Otherwise, it is labeled non-targeted. Each query is encoded using Sentence-BERT, ℓ_2-normalized, and classified by logistic regression; the predicted probability is used as $C_{\text{local}}(t)$. To reduce superficial template memorization, evaluation adopts a semantic-family holdout protocol, where validation templates come from unseen semantic families.

History Term $H_{\text{history}}(t)$. Given a session history $(q_1, \ldots, q_{t-1})$, we define $H_{\text{history}}(t)$ as a bounded accumulation of semantically related past stimuli. Let $\mathbf{e}_i$ denote the ℓ_2-normalized embedding of q_i, and let $C_{\text{local}}(i)$ denote its instantaneous strength. For the current query q_t, we compute $\text{sim}_{01}(q_i, q_t) = \frac{\langle \mathbf{e}_i, \mathbf{e}_t \rangle + 1}{2}$, apply temporal decay $w(\Delta) = \exp(-\lambda\Delta)$ with default $\lambda = 0.35$, and obtain the accumulated and saturated history term using Eqs. (4)–(6). History is session-scoped and reset at the beginning of each new session.

5.2 Component-Wise Validation

Single-Round Semantic Intrusiveness (Isolating C_{local}.) This experiment disables history accumulation to test whether $C_{\text{local}}(t)$ separates aggregate queries from individual-targeted ones. We construct a length-10 query sequence whose semantics progressively move from population-level statistics to individual-level requests. To isolate the instantaneous stimulus, we set $H_{\text{history}}(t) = 0$

throughout. The same query structure is repeated for 10 target patients with $\alpha_1 = 0.6$, $\alpha_2 = 0.4$, $\beta_\sigma = 2.0$. We use the *age* field as an example with $B(age) = 0.3$ and $C_0 = 0.2$. Representative queries and their $C_{\text{local}}(t)$ outputs are:

0.026: What is the average age of all patients?
0.152: How many patients live in Bologna?
0.591: What is the exact age of patient P0007 in the registry?
0.976: Show me the full medical record of patient P0007.

As shown in Fig. 2, $\sigma_x(t)$ stays low in the aggregate phase and rises sharply in the targeted phase. Since history is disabled, the increase is attributable to $C_{\text{local}}(t)$ alone. This supports the intended response of the instantaneous term to increasingly intrusive semantics.

History-Path Effect (*Validating* H_{history}.) This experiment tests whether different session histories produce different state responses under the same terminal query. We fix the terminal query as "What is the exact age of patient {id}?" and construct three histories of equal length (9 history queries + 1 terminal query): A: mild history (repeated age-related aggregate statistics); B: unrelated history (statistics weakly related or unrelated to age, such as vaccination, symptoms, or test type); C: targeted refinement history (progressive narrowing via city, sex, retirement status, and comorbidity).

Each path is executed as an independent session and averaged over 10 target patients. As shown in Fig. 3, the final $\sigma_x(t)$ under path C is higher than under paths B and A. This indicates that targeted refinement yields a larger $H_{\text{history}}(t)$ and maps the same terminal query to a higher session-level state. The result supports the intended accumulation effect of the history term.

Role-Factor Effect (*Validating Contextual Scaling via* $R(u)$). This experiment changes only the requester role factor $R(u)$ to test whether it acts as a stable contextual scaling on the same underlying trajectory. We assume the ordinal relation $R(\mathsf{Public}) > R(\mathsf{Researcher}) > R(\mathsf{Clinician}) > R(\mathsf{Admin})$, with $R(\mathsf{Clinician}) = 1$.

As shown in Fig. 4, the role-specific curves largely preserve the same shape while their levels vary monotonically with $R(u)$. This indicates that $R(u)$ mainly scales the semantic/history-driven trajectory: roles with weaker privileges correspond to higher $\sigma_x(t)$ and thus stronger governance demand.

Stratification by Field and Context Baselines ($B(x)$ *and* C_0). This experiment tests whether the static terms provide an interpretable baseline level for the state. We compare two fields (*age* and *comorbidity*) under four preset values of C_0, yielding eight $(B(x), C_0)$ configurations.

As shown in Fig. 5, larger $B(x)$ and larger C_0 both lead to higher $\sigma_x(t)$. This is consistent with the model decomposition: static terms set the reference level, while dynamic terms drive within-session variation.

Consistency Under a Fixed Refinement Skeleton. Finally, we test whether the trajectory remains structurally stable when only the patient changes under the same refinement skeleton. We randomly sample 10 target patients and generate a length-40 targeted refinement session for each. All sessions share the same query structure, while slot values are filled from the corresponding patient record.

As shown in Fig. 6, the trajectories exhibit consistent phase patterns, with only small inter-patient amplitude differences. This suggests that, under controlled target substitution, the model responds primarily to the refinement skeleton and accumulation path rather than to incidental textual variations.

6 Conclusion

Existing governed data-sharing solutions cannot adequately capture how privacy risk evolves under multi-round interactive access, where exposure may accumulate through semantic refinement, disclosure history, and role-dependent usage contexts. In this paper, we presented a biologically inspired dynamic privacy-state assessment method for governed data sharing. The method models the privacy condition of the same data item as an evolving state and derives an interpretable item-level privacy state together with a conservative session-level signal for runtime compliance alignment and audit-oriented governance. Although the current study is limited to preliminary mechanism-level validation, these results provide initial evidence that dynamic privacy-state tracking is a feasible way to characterize session-level exposure beyond isolated per-request checks. Future work will focus on integration with downstream privacy-enhancing and enforcement mechanisms, and on validation in real-world data-sharing scenarios.

References

1. Bacco, M., Kocian, A., Chessa, S., Crivello, A., Barsocchi, P.: What are data spaces? Systematic survey and future outlook. Data Brief **57**, 110969 (2024)
2. Dehghani, Z.: Data Mesh. O'Reilly Media, Inc., Sebastopol, CA (2022)
3. Dinur, I., Nissim, K.: Revealing information while preserving privacy. In: Proceedings of the twenty-second ACM SIGMOD-SIGACT-SIGART symposium on Principles of database systems, pp. 202–210 (2003)
4. Dwork, C., Roth, A.: The algorithmic foundations of differential privacy. Found. Trends® Theoret. Comput. Sci. **9**(3–4), 211–407 (2014)
5. Harmelink, R., Joosten, R., Topan, E., Adriaanse, A., van Hillegersberg, J.: Data: to share or not to share? A semi-systematic literature review in (rational) data sharing in inter-organizational systems. Discover Data **2**(1), 13 (2024)
6. Hummel, P., Braun, M., Tretter, M., Dabrock, P.: Data sovereignty: a review. Big Data Soc. **8**(1), 2053951720982012 (2021)
7. Kosa, T.A., El-Khatib, K., Marsh, S.: Measuring privacy. J. Internet Serv. Inf. Secur. **1**(4), 60–73 (2011)
8. Kotsev, A., Minghini, M., Tomas, R., Cetl, V., Lutz, M.: From spatial data infrastructures to data spaces–a technological perspective on the evolution of European SDIs. ISPRS Int. J. Geo Inf. **9**(3), 176 (2020)

9. Nissenbaum, H.: Privacy as contextual integrity. Wash. L. Rev. **79**, 119 (2004)
10. Pettenpohl, H., Spiekermann, M., Both, J.R.: International data spaces in a nutshell (2022)
11. Rose, S., Borchert, O., Mitchell, S., Connelly, S.: Zero Trust Architecture. NIST Special Publication 800(207), 1–52 (2020)
12. Telkamp, L., Hulsebos, M.: Towards contextual sensitive data detection (2025). arXiv:2512.04120 arXiv preprint
13. Teyssier, J., Saenko, S.V., Van Der Marel, D., Milinkovitch, M.C.: Photonic crystals cause active colour change in chameleons. Nat. Commun. **6**(1), 6368 (2015)
14. Vukusic, P., Sambles, J.R.: Photonic structures in biology. Nature **424**(6950), 852–855 (2003)

Towards an IoT Plan for Smart Buildings: Lessons from a Real-World Construction Project

Lasse Harjumaa(✉), Petri Jyrkkä, Ilkka Kivelä, and Ismo Hakala

Kokkola Univesity Consortium Chydenius, University of Jyväskylä, Jyväskylä, Finland
{lasse.m.harjumaa,petri.ao.jyrkka,ilkka.v.kivela, ismo.hakala}@jyu.fi

Abstract. The integration of Internet of Things (IoT) technologies into smart building projects introduces a range of technical, organizational, and project management-related challenges that must be addressed to ensure successful implementation. We have encountered several issues during the deployment of IoT solutions on-site in a real-life construction project, such as limited experience, absence of dedicated hardware fittings, and the lack of standardized methods for representing IoT components in blueprints. Furthermore, effective communication among stakeholders requires precision. Procurement and cost estimation for IoT-related work remains uncertain without detailed planning, and managing subcontractors is difficult in the absence of clearly defined use cases. These factors highlight the need for a structured way to document and communicate IoT-related assets and user needs throughout building projects. In this paper, we present guidelines for smart building construction projects and highlight the need for a dedicated IoT plan - a document designed to synchronize construction and IoT implementation processes and to improve coordination among stakeholders, contributing to the best practices for integrating IoT into building projects.

Keywords: Internet of Things · IoT Development · IoT Plan · Smart Buildings

1 Introduction

The Internet of Things (IoT) can play a pivotal role in achieving sustainability and enhanced safety in smart buildings, in addition to improving the overall comfort for occupants. Still, the deployment of IoT systems presents unique challenges that are not adequately addressed by traditional construction planning and design practices.

Despite the rapid expansion of IoT solutions, there remains a substantial gap in understanding the impacts of the technology on the smart building construction projects [20]. Institutional frameworks to support the adoption of technologies are often lacking, and architects' conceptualizations of smart buildings are still insufficiently defined [1], among other things. Smart building projects differ from traditional construction initiatives, requiring coordination practices that address technical and design complexities. Deployment of IoT technologies can introduce significant challenges from sensor selection to the availability of skilled technical personnel.

A. Bernasconi et al. (Eds.): RCIS 2026, LNBIP 597, pp. 16–28, 2027.
https://doi.org/10.1007/978-3-032-35043-5_2

Limited research seems to exist on formalized planning approaches that anticipate both technical and user-centered requirements, highlighting a need for dedicated IoT plans. Findings from our smart building project underscore the importance of a dedicated IoT plan as a formal project artifact. Such a plan could unify varying terminologies, guide procurement, account for software and system integration, and facilitate interdisciplinary collaboration in construction projects, thus reducing implementation risks. In this paper, we report the observations from a smart building construction project and present guidelines for improving coordination of such projects. We also suggest that a dedicated IoT plan would represent a crucial step toward efficient management of smart building projects and realization of the full potential of smart buildings.

The rest of the paper is organized as follows. Section 2 summarizes the related work. Our ongoing research project is briefly depicted Sect. 3. Section 4 lists the observations we have made during the execution of the project. Guidelines derived from these observations for smart building projects are listed in Sect. 5. Section 6 concludes the article.

2 Related Work

Research on smart building projects has identified a range of considerations that influence their successful planning and implementation. Despite the rapid spread of smart home systems and smart city projects, there remains a considerable lack of understanding regarding the potential impacts of smart buildings on quality of life and environmental sustainability. [7] Furthermore, there is a lack of institutional structures to promote and facilitate the adoption of smart materials and technologies in construction. Furthermore, architects' perceptions and understanding of the concept of smart building remain insufficiently defined. [11].

Project management in smart building initiatives differs from that in traditional construction projects, requiring a clearly defined project scope and in-depth knowledge of the available technologies and their capabilities. [17] Challenges include misalignment with existing construction techniques and procedures, the need for design modifications, unclear or rigid contractual specifications, and the extended time required for organizational approval of new technologies [9]. The increased complexity of smart building projects constitutes one of the primary obstacles in construction initiatives [9, 22]. Project stakeholders may also be concerned with cost management and the heightened financial, regulatory, security, and design risks associated with the introduction of new technologies [16]. Thus, there is a need for new project management practices [8].

In addition to increasing the occupant comfort, there are other goals in introducing smart and IoT technologies in buildings, such as enabling preventive maintenance and remote control of building facilities. Energy efficiency is another important design goal. IoT technologies can help in optimizing the use of energy while keeping a high level of service at the same time [10]. The selection of appropriate sensors, the determination of measurement accuracy suitable for the intended application, and the assessment of the energy efficiency of proposed solutions present significant challenges [3].

Typically, smart buildings incorporate a large number of interconnected systems. These systems span diverse functions, including energy management, security, lighting, and climate control, involving multiple areas of expertise [15]. Thus, smart building technologies are inherently complex [18], and the associated terminology and algorithms are often difficult to comprehend [4].

IoT is the core component of a smart building, but it has to be accompanied with meaningful analytics and applications to interpret the data [10]. The planning and management of the extensive data produced by smart buildings is a demanding task [13, 15]. Additionally, managing heterogeneous data sources demands careful planning and coordination [3]. Beyond collection, the data must be managed and processed in a sustainable way. Data processing required by smart technologies can vary greatly depending on the complexity of the system. Energy usage will increase significantly with complex analytics [14]. Cloud services can also produce significant amount of emissions [21], which are difficult to estimate when setting project goals.

Even the installation of sensors and the provision of adequate computing resources, including network capacity and on-site servers, may present difficulties [3]. Successful adoption of IoT and automation systems depends on skilled technical personnel, yet the presence of sufficiently qualified workers cannot be assumed [8, 9].

The absence of a universally accepted framework, contractual standards, and common vocabulary may hinder the widespread deployment of IoT solutions. Currently, IoT tools and platforms adhere to diverse codes and protocols, resulting in fragmentation and reduced interoperability [2]. On the other hand, ensuring interoperability by relying on technologies from a single provider can result in vendor lock-in. Furthermore, solutions provided by a single vendor may not result in optimal functionality.

Smart buildings inherently involve a high degree of multidisciplinary, particularly across interrelated technical domains. This complexity requires expertise to oversee multiple disciplines and ensure effective interoperability among systems [15]. Achieving interoperability across heterogeneous device types supplied by multiple manufacturers constitutes one of the most critical technical challenges in smart building projects [4].

Concerning he early adoption and planning of smart or IoT-specific features within building projects, Jia et al. [10] propose a set of questions that system users and designers should rigorously address prior to the development of any integration prototype, in order to guide design decisions and ensure the effective transfer of knowledge to industry. These questions encompass both technical and non-technical aspects of system development, emphasizing the clear definition of user requirements and responsibilities throughout the life cycle of the project, as well as the specification of data formats and workflows.

In a larger context, Chang et al. [5] emphasize the importance of developing IoT deployment plans that account for the needs of future users as well as the concerns of authorities and domain experts regarding infrastructure resilience. Their research concerns smart cities, and represents an urban planning tool, that is aimed to assist in integrating IoT into city infrastructure. The tool provides algorithms to generate an IoT deployment plan to fulfill the requests of community residents.

The tool support for depicting IoT capabilities has been addressed in [6], for example. They present an architecture that provides information about energy consumption, occupation and conditions of a building by integrating IoT devices and Building Information Modeling (BIM) documents. However, the work focuses more on utilizing the data than planning for the construction. Li et al. have also studied integrating IoT into BIM [12]. A more comprehensive study on integration of IoT and BIM tools is presented by Tang et al. [19]. Their survey revealed that even though several applications exist to combine the IoT characteristics with BIM information, there is still work to do in achieving universal methodology to integrate these aspects. For example, there are no standards to connect hardware, BIM data, semantic information, communication protocols, software and the actual sensor data.

3 Methodology and Project Description

The project combines the construction of a residential home with its digital twin, highlighting practical smart building applications. As a new build, it integrates IoT from the ground up. The project is a collaboration between several organizations. One main task of the project focuses on the design, implementation, and piloting of sensor systems for monitoring water and electricity consumption, structural conditions (e.g., subfloor and wall assemblies), and indoor and outdoor environments.

Researchers participated in project tasks and meetings that included both internal stakeholders - such as designers, engineers, and construction managers - and external stakeholders, such as service providers. Observations from the official meetings were complemented by informal discussions with project workers on the construction site. These interactions provided insights into practical challenges encountered during planning and installation tasks, as well as perceptions of new responsibilities introduced by IoT technologies.

During the project consultations have been held with specialized service providers, including IoT vendors and system integrators having the hands-on knowledge on integrating IoT technologies in smart buildings. These discussions helped to contextualize the project-specific challenges within broader industry practices. Additionally, project documentation, including use case scenarios and other technical requirements, architectural drawings, procurement materials, and installation guidelines, was examined, enabling comparison between formal planning artifacts and observed practices.

To manage the extensive measurement data generated, the project will establish an IoT data platform for gathering and managing the measurement data from varying sources. After the project, the measurement data will serve as an open data repository. The overall architecture of the platform is depicted in Fig. 1. The building will incorporate multiple sensor networks, including temperature, humidity, and air quality sensors connected via a wireless network to a local server running the Home Assistant framework. The lighting system is connected through a KNX bus and integrated with the local computing infrastructure. Sensor data are transmitted to a Message Queuing Telemetry Transport (MQTT) broker, which publishes the data to a cloud-based middleware component. In parallel, an additional data flow utilizes a Long-Range Wide Area Network (LoRaWAN), transmitting data directly from sensors to a LoRaWAN gateway that forwards the data to the same MQTT broker. The middleware enriches time-series data with metadata. Finally, the platform provides standardized interfaces to enable secure and efficient data access for diverse stakeholders through GraphQL queries.

The project is now approaching its final stages. Key technical systems, including heating, ventilation, and air conditioning, have been fully implemented, ensuring the

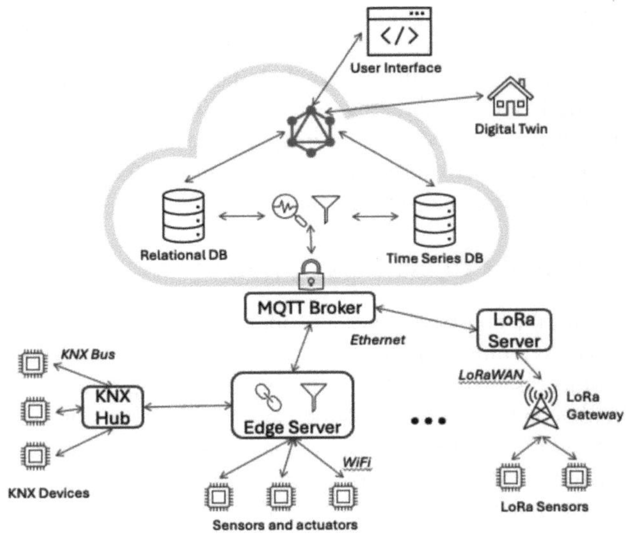

Fig. 1. Simplified IoT architecture deployed in the project.

operational readiness of core building services. In addition, most IoT-related components have been installed and integrated into the building infrastructure. These installations provide the foundation for subsequent commissioning activities, data collection, and the development of digital twin of the building.

4 Lessons Learned

4.1 Separate Design Documents

There are no standardized methods for representing IoT components within architectural blueprints and digital design tools. Current design practices are primarily optimized for describing conventional building elements such as structural components, electrical wiring, and Heating, ventilation, and air conditioning (HVAC) systems. Consequently, there is no universally accepted approach to specifying IoT devices that are either embedded within structural layers or need to be otherwise taken into account, during structural design. This complicates coordination between design and engineering teams and can lead to omission of essential IoT equipment during construction.

A related challenge is the absence of a widely adopted standard for documenting the interfaces and connections between HVAC systems and IoT equipment. While HVAC systems increasingly rely on smart controls, the documentation practices remain fragmented. Sometimes only minimal information on the interfaces is provided, which hinders planning for interoperability and creates risks of integration errors.

4.2 Insufficiently Developed Practices and Inexperience

Electrical contractors responsible for wiring the building generally lack specialized knowledge and practical experience in the installation and energy management of IoT-enabled devices. In some cases, deployment of the devices require configuration, which may vary significantly in each case. The training of electrical engineers still focuses on conventional electrical systems, even though smart appliances constantly gain more popularity, and the need to deploy IoT systems increases rapidly.

Similarly, professionals engaged in the architectural and mechanical design of buildings may have limited experience with the IoT integration and its infrastructural implications. Traditionally structural integrity, aesthetic considerations and thermal performance have been given the main priority in design, while the connectivity requirements essential for seamless IoT deployment have not been paid attention.

4.3 Device Selection and Rationale

Device selection involves evaluating the suitability of technologies, their environmental impact, and aligning of system functionality with design goals. For instance, predictive maintenance systems can greatly reduce operational costs, contributing to more sustainable living environments.

Comparative assessment of available IoT technologies in terms of functionality, reliability, and cost-effectiveness is a critical aspect of the project planning phase. Factors such as interoperability with existing building systems, compliance with safety standards, and adaptability to future upgrades need to be considered. Ensuring that these technologies meet the comfort and security expectations of future inhabitants is equally important. Availability of the devices needs to be taken into account, too. The acquisition of IoT devices can significantly affect project timelines, as long delivery lead times or supply chain constraints may delay installation and commissioning. In addition, variability in component availability can complicate procurement planning.

4.4 Physical Deployment

The absence of standardized electrical fittings and accessories specifically designed for IoT devices can make the installation difficult. Conventional socket chassis and electrical enclosures are optimized for traditional wiring configurations, leaving insufficient space for the additional hardware and connectivity modules required by IoT systems.

Installation engineers report that IoT deployment requires high precision due to tight spatial constraints. Cable routing, sensor alignment, and mounting must be carefully executed to avoid damage. Predefined wiring channels limit flexibility, often forcing ad hoc modifications that can affect aesthetics and reliability—for example, sensors in water piping may create sealing and access issues. Accurate placement and identification of devices are essential, but as device numbers grow, so do risks of misplacement. Manual labeling and spreadsheet tracking increase labor and the potential for human error.

4.5 Visibility of IoT Requirements

The desired functionalities of IoT systems should be explicitly incorporated into architectural and engineering plans to guide both design and implementation activities. Visibility enables traceability and may guide device selection, thus improving the overall system performance.

IoT infrastructure is frequently treated as an aftermarket addition rather than a core component of building design. This reactive approach may result in compromises in system performance. For example, positioning of occupancy sensors should be planned alongside built-in furniture to avoid caveat areas. Including IoT considerations in design workflows helps to achieve smooth integration and operational reliability.

4.6 Communication

The project revealed, quite unsurprisingly, that effective communication among stakeholders is a critical success factor. Coordination requires not only continuous interaction but also precise terminology to avoid misunderstandings and technical inconsistencies.

Stakeholders bring diverse professional backgrounds -such as architecture, electrical and HVAC engineering, and IoT development – which results in significant variation in vocabulary and conceptual frameworks. Semantic gaps complicate collaboration and decision-making, underscoring the need for standardized communication practices, shared vocabularies, and digital collaboration tools to bridge disciplinary boundaries and ensure coherent project execution.

4.7 Project Management

IoT devices and systems originate from numerous manufacturers, and suppliers often specialize in specific brands or ecosystems. This specialization makes it difficult to compare solutions on equal terms. As a result, competitive bidding processes may become cumbersome. Developing procurement framework that emphasizes interoperability standards and vendor-neutral specifications is necessary to mitigate these challenges.

Accurate estimation of labor and installation costs for IoT system components can be difficult. Furthermore, installation and programming tasks may be overlooked in project scheduling, if the IoT-related work is not made clearly visible in plans. Without a detailed implementation plan cost and schedule projections remain unreliable.

Managing subcontractors in IoT-enabled construction requires that use cases are clearly defined. Effective communication must include detailed documentation of device roles, integration points, and performance expectations. Structured communication protocols and digital collaboration platforms could potentially improve coordination among multidisciplinary teams.

4.8 Contextual Understanding

Interpreting the data from individual sensors requires understanding their physical context, including their exact location and surrounding materials. When this mapping is absent, visualization and data analysis are compromised, reducing the value of IoT data for decision-making. Linking sensor data to spatial and material information within building design documentation is therefore critical for enabling meaningful analytics.

Without explicit representation of sensors in Industry Foundation Classes (IFC) model the digital twin cannot fully replicate the operational environment, limiting its usefulness. This highlights the need for extending BIM standards to include IoT-specific elements. Standardized metadata schemas and automated documentation processes are essential to ensure data integrity and interoperability.

4.9 Evolving Requirements

Due to the complex nature of IoT-enabled buildings, design plans evolve during the project. Desired functionalities should be clearly defined and agreed upon during the earliest stages of the design process. These requirements must not remain buried in technical documentation but instead be explicitly reflected in architectural and engineering plans. When changes are inadequately documented, the risk of misalignment between design intent and implementation increases.

Design alterations – whether driven by changing user requirements, the adoption of alternative materials, or changes in equipment availability – should be incorporated into plans without delay. Outdated plans can lead to miscommunication among stakeholders and inconsistencies across disciplines. Establishing clear protocols for version control and change management would aid in maintaining coherence of the documentation.

4.10 Tool Support

A significant challenge identified in the project is the inadequacy of current design and coordination tools in supporting IoT integration. Architectural design platforms, workflow management systems, and BIM software primarily focus on structural components.

The project includes dozens of Shelly-based sensors. While their positions are shown in the electrical plan (Fig. 2), this is not practical for installation: sensors are connected by 1-3-m wires that are not represented, and the large number of devices reduces readability. Although modeling tools support physical assets, they lack native features for IoT aspects such as connectivity and data flows. With no standard way to embed IoT attributes, designers rely on ad hoc notes or external documents, and coordination can suffer due to mismatched IoT and construction timelines. Current tools also lack integrated simulation for IoT functionality, such as assessing network connectivity, which requires external solutions.

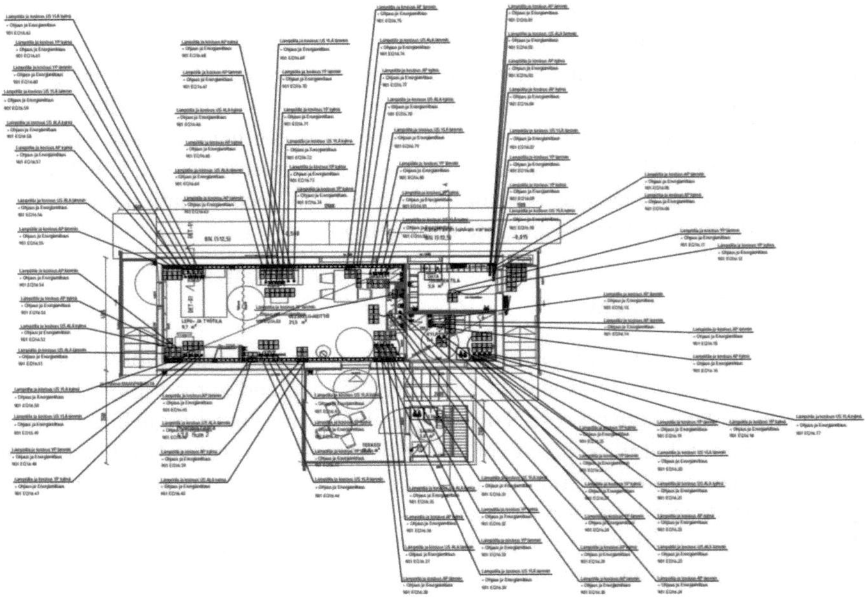

Fig. 2. An extract from the electrical plan of the building (In Finnish).

4.11 Invisible Software

Traditional design practices focus on physical structures and drawings, leaving software largely unrecognized as a core component of the built environment. As a result, software-related work remains mostly invisible in construction plans, despite its importance in modern housing. Integrating IoT data into a coherent, interoperable system requires significant software effort, yet these tasks are rarely documented since they lack physical form. Plans may specify sensor placement but often omit how data should be integrated across systems.

Tasks such as device configuration, data flow implementation, model mapping, and API integration are often identified only during commissioning and risk being excluded from project schedules. This issue is especially evident when integrating traditional building systems with digital platforms. Legacy systems rely on local control, while IoT emphasizes data-centric, remote operation, requiring programming to bridge the gap. As buildings become more data-driven, software must be recognized as an integral component.

5 Things to Address in Smart Building Projects

Table 1 outlines key topics and considerations for designing IoT-enabled smart buildings, summarizing practical recommendations on documentation, training, device selection, deployment, and management. It emphasizes treating IoT as a core building element. Establishing a formal IoT plan ensures alignment across the project lifecycle, supports procurement, and improves risk management.

Table 1. Considerations for integrating IoT into construction projects.

Topic	Considerations
Design and documentation	• Introduce dedicated IoT design documents • Standardize representation of IoT devices • Document embedded sensors explicitly • Establish a unified documentation format for HVAC–IoT interfaces • Require manufacturers to provide complete interface descriptions
Training and capacity building	• Provide specialized training on IoT connectivity, dataflow requirements • Standardize representation of IoT devices • Promote interdisciplinary planning workshops
Devices and system architecture	• Establish a transparent selection rationale for devices and services • Evaluate each solution for interoperability and scalability • Prefer devices that support open standards • Conduct pilot installations
Deployment	• Plan for precision installation • Perform spatial planning early • Supply clear mounting instructions • Use digital device maps (QR-linked or BIM-embedded)
Visibility and integration of IoT requirements	• Make the IoT plan a formal project document • Define explicit use cases for procurement • Communicate requirements and data formats early and repeatedly • Treat IoT systems as core building infrastructure • Use traceable documentation workflows • Integrate spatial requirements into structural and piping plans • Make IoT requirements explicit early
Shared understanding	• Establish unambiguous specifications for data formats and protocols • Develop shared vocabularies • Use digital collaboration tools to reduce information silos • Hold regular cross-disciplinary coordination meetings

(*continued*)

Table 1. (*continued*)

Topic	Considerations
Project management and procurement	• Use functional procurement criteria, not only device lists • Develop vendor-neutral procurement criteria • Incorporate IoT installation tasks explicitly into project schedules • Require detailed implementation plans • Ensure subcontractors receive clearly defined use cases
Contextual integration	• Extend BIM/IFC models to include representations of IoT sensors • Maintain systematic metadata management • Use enriched models to enable accurate analytics capabilities
Change management	• Implement formal version control procedures • Update documentation promptly • Notify partners about updates via systematic communication workflows
IoT-aware tools	• Extend BIM and design software to model IoT devices explicitly • Document device connectivity and data flows in design tools • Adopt standards for IoT-specific metadata • Ensure tool support for rapid iteration
Recognition of software as infrastructure	• Treat software as a first-class component of the built environment • Plan and budget for software integration tasks • Document data flows between systems • Adopt software-oriented planning artefacts • Plan for aligning legacy building systems with IoT architectures

6 Conclusion

Through a participatory research approach, this study examined emerging issues in IoT deployment. The findings highlight the importance of making IoT requirements visible and aligned with project objectives to reduce risks. Effective integration goes beyond adding devices, requiring IoT-related information to be incorporated into design documentation.

A dedicated IoT plan could support procurement, reduce deployment risks, and establish a shared vocabulary among stakeholders, enhancing collaboration. This work contributes to smart building research by highlighting practical challenges and the need for a holistic approach integrating technical, organizational, and procedural aspects throughout the building lifecycle. While the study benefits from multidisciplinary engagement, it is limited to a single case. Future work will focus on developing a standardized IoT plan template to guide smart building implementation.

Acknowledgments. This work has been partially funded by the European Union under the Horizon programme.

Disclosure of Interests. The authors have no competing interests to declare that are relevant to the content of this article.

References

1. Affonso, E.O.T., et al.: The main barriers limiting the development of smart buildings. Buildings **14**(6), 1726 (2024)
2. Ahmed, E., Yaqoob, I., Gani, A., Imran, M., Guizani, M.: Internet-of-things-based smart environments: state of the art, taxonomy, and open research challenges. IEEE Wirel. Commun. **23**(5), 10–16 (2016)
3. Alsafery, W., Rana, O., Perera, C.: Sensing within smart buildings: a survey. ACM Comput. Surv. **55**(13s), 297 (2023)
4. Benavente-Peces, C.: On the energy efficiency in the next generation of smart buildings—supporting technologies and techniques. Energies **12**(22), 4399 (2019)
5. Chang, T.C., Bouloukakis, G., Hsieh, C.Y., Hsu, C.H., Venkatasubramanian, N.: SmartParcels: cross-layer IoT planning for smart communities. In: Proceedings of the International Conference on Internet-of-Things Design and Implementation, pp. 195–207 (2021)
6. Dave, B., Buda, A., Nurminen, A., Främling, K.: A framework for integrating BIM and IoT through open standards. Autom. Constr. **95**, 35–45 (2018)
7. Froufe, M.M., Chinelli, C.K., Guedes, A.L.A., Haddad, A.N., Hammad, A.W.A., Soares, C.A.P.: Smart buildings: systems and drivers. Buildings **10**(9), 153 (2020)
8. Ghaffarianhoseini, A., et al.: What is an intelligent building? Analysis of recent interpretations from an international perspective. Archit. Sci. Rev. **59**(5), 338–357 (2016)
9. Ghansah, F.A., Owusu-Manu, D.-G., Ayarkwa, J., Edwards, D.J., Hosseini, M.R.: Exploration of latent barriers inhibiting project management processes in adopting smart building technologies (SBTs) in developing countries. Constr. Innov. **21**(4), 685–707 (2021)
10. Jia, M., Komeily, A., Wang, Y., Srinivasan, R.S.: Adopting internet of things for the development of smart buildings: a review of enabling technologies and applications. Autom. Constr. **101**, 111–126 (2019)
11. Kamal, M.M.H.S., Faggal, A.E.D.: Smart and intelligent buildings achieving architectural concepts. Eng. Res. J. (Shoubra) **50**(1), 155–163 (2021)
12. Li, X., et al.: Blockchain-enabled IoT-BIM platform for supply chain management in modular construction. J. Constr. Eng. Manag. **148**(2), 04021195 (2022)
13. Marinakis, V.: Big data for energy management and energy-efficient buildings. Energies **13**(7), 1555 (2020)
14. Ostadijafari, M., Dubey, A., Liu, Y., Shi, J., Yu, N.: Smart building energy management using nonlinear economic model predictive control. In: 2019 IEEE Power & Energy Society General Meeting (PESGM), pp. 1–5 (2019)
15. Pašek, J., Sojková, V.: Facility management of smart buildings. Int. Rev. Appl. Sci. Eng. **9**, 181–187 (2018)

16. Radziejowska, A., Sobotka, B.: Analysis of the social aspect of smart cities development for the example of smart sustainable buildings. Energies **14**(14), 4330 (2021)
17. Rodrigues, A.M., Oladimeji, O., Guedes, A.L.A., Chinelli, C.K., Haddad, A.N., Soares, C.A.P.: The project manager's core competencies in smart building project management. Buildings **13**(8), 1981 (2023)
18. Saad, S., Alaloul, W.S., Ammad, S., Altaf, M., Qureshi, A.H.: Identification of critical success factors for the adoption of Industrialized Building System (IBS) in Malaysian construction industry. Ain Shams Eng. J. **13**(2), 101547 (2022)
19. Tang, S., Shelden, D.R., Eastman, C.M., Pishdad-Bozorgi, P., Gao, X.: A review of building information modeling (BIM) and the Internet of Things (IoT) devices integration: pre-sent status and future trends. Autom. Constr. **101**, 127–139 (2019)
20. Viana Vargas, R., Soares, A., Rodrigues, A., Chinelli, C., Qualharini, E., Haddad, A.: Barriers and challenges of smart building projects in the context of construction 4.0. (2022)
21. Wang, J., Gupta, U., Sriraman, A.: Peeling back the carbon curtain: carbon optimization challenges in cloud computing. In: Proceedings of the 2nd Workshop on Sustainable Computer Systems (HotCarbon 2023), Article 8, pp. 1–7 (2023)
22. Yang, A., Han, M., Zeng, Q., Sun, Y.: Adopting building information modeling (BIM) for the development of smart buildings: a review of enabling applications and challenges. Adv. Civ. Eng., 26 (2021)

When Fake News Sounds Right: Assessing Message Credibility in AI-Generated Misinformation

Paulina Wojciechowska-Dzięcielak[1,2](✉), Pierre-Emmanuel Arduin[1], and Myriam Merad[2]

[1] Université Paris-Dauphine, PSL, DRM UMR CNRS 7088, 75775 Paris, France
{p.wojciechowska-dziecielak, pierre-emmanuel.arduin}@dauphine.psl.eu

[2] Université Paris-Dauphine, PSL, LAMSADE UMR CNRS 7243, 75775 Paris, France
myriam.merad@lamsade.dauphine.fr

Abstract. Generative Artificial Intelligence (GenAI) enables the large-scale production of fluent and coherent content that mimics legitimate news, raising critical challenges for how credibility is assessed in digital information systems. While "fake news" dominates public discourse, this study adopts the broader and analytically more precise concept of "misinformation" to describe false or misleading content regardless of intent. Although empirical research on GenAI's impact on credibility judgments is growing, findings remain fragmented across disciplines and system contexts.

This paper provides a preliminary critical synthesis of empirical and conceptual studies on AI-generated misinformation, focusing on two outcomes central to information systems research: perceived credibility and detection accuracy.

To enable meaningful comparison, findings are synthesized descriptively along two informational dimensions (source and structure) which are the only cues users can realistically access when evaluating AI-generated content. Results show that AI-generated misinformation tends to be perceived as more credible and is less accurately detected than comparable human-written content. Disclosure labels indicating AI authorship have negligible effects on credibility judgments, underscoring the limited efficacy of transparency-based interventions.

Theoretically, the paper contributes to Information Systems research by clarifying how GenAI shifts credibility heuristics away from source-based evaluation toward structurally optimized content. Practically, it highlights the limits of content-level countermeasures and points to the need for system-level, user-centered design interventions that restore epistemic salience and foster reflective trust.

Overall, the study reframes AI-generated misinformation not simply as a new type of falsehood, but as a structural transformation of the credibility landscape. It calls for a reconceptualization of credibility as a dynamic process shaped by interaction design, interface cues, and evolving epistemic norms in AI-mediated environments.

Keywords: Generative Artificial Intelligence · Fake News · Misinformation · Credibility Judgements · Detection Accuracy · Critical Analysis · Information Systems

A. Bernasconi et al. (Eds.): RCIS 2026, LNBIP 597, pp. 29–40, 2027.
https://doi.org/10.1007/978-3-032-35043-5_3

1 Introduction

"For flat-Earth followers, the perfectly straight horizon is the ultimate proof that scientists have been lying to us all along!" (Conversation with Microsoft Copilot, January 16, 2026). This sentence generated by Artificial Intelligence (AI) shows the easiness to produce misinformation, which is defined as "false and misleading information" by Altay *et al.* [1]. According to Broda & Strömbäck [3], research on misinformation and fake news has exploded since 2016, with strong disciplinary fragmentation.

AI-generated misinformation undermines long-standing user heuristics for assessing credibility, because it produces false and misleading content with a level of fluency, coherence, and stylistic authenticity that closely mimics legitimate information. The impact of misinformation depends heavily on users' cognitive shortcuts for evaluating truthfulness. However, research shows that people often rely on superficial cues [11], such as familiarity, source appearance, information structure, or presentation quality, to estimate credibility, and these cues become unreliable when generative models can effortlessly create high-quality but inaccurate content.

In this paper, we propose to clarify how Generative AI (GenAI) reshape the cues users rely on when assessing information credibility in the context of misinformation through the research question: How does the narrative structure shapes users' perception of the credibility in AI-generated misinformation compared to human-produced content?

Our contribution lies in (1) offering a critical systematic review of the literature on AI-generated misinformation, (2) integrating a dual perspective that jointly examines perceived message credibility and detection accuracy, and (3) outlining key design and governance implications for information systems.

In this paper, we first outline the conceptual foundations of misinformation by examining how both source characteristics and message structure shape credibility cues. We then synthesize findings from prior studies with a dual focus on perceived message credibility and users' detection accuracy. Building on this review, we offer a critical discussion of convergences, divergences, and moderating factors across the literature, before outlining key implications for the design and governance of information systems. The paper concludes with avenues for future research.

2 Conceptual Foundations: The Impact of Source and Structure

This review conceptualizes AI-generated misinformation primarily as a problem of information properties rather than technological infrastructure. The proposed framework distinguishes two core informational dimensions: source and structure, while treating interactional context as a secondary moderator shaping how users evaluate messages. Credibility judgments in AI-mediated systems seem to be driven mainly by how information is linguistically produced and presented.

The scope of the framework is intentionally limited to **source** and **structure** because these are the only information-related features that users can realistically observe and evaluate when interacting with AI-generated content. Source refers to the substance of the information provided, including such qualities as relevance, accuracy, and degree of aggregation. Structure, by contrast, denotes the way in which that information is arranged and presented to the user [1]. Technical components such as hardware, software architecture, or data repositories operate below the level of user perception and therefore cannot meaningfully inform human credibility judgments. Users directly encounter the AI system as an apparent source of information and evaluate credibility primarily through linguistic and structural properties of the message.

2.1 Misinformation as a Content-Level Phenomenon

Misinformation is defined as false or misleading information disseminated without intent to deceive [10]. This definition is appropriate for generative AI contexts, as Large Language Models (LLMs) lack communicative intentions and cannot strategically manipulate audiences. *"AIs are only as smart as their training, are subject to manipulation, can produce misinformation, and reproduce epistemic and ethical values encoded in data and systems"* [4, p. 2]. Instead, inaccuracies arise from probabilistic language generation and incomplete grounding. Erroneous outputs are therefore best understood as misinformation rather than disinformation.

Fake news refers more narrowly to fabricated or misleading content in journalistic form, often associated with political manipulation. However, focusing only on fake news overlooks misleading conversational explanations, summaries, and advisory responses generated by AI systems. Moreover, definitional choices affect estimates of prevalence and impact [1]. This paper therefore adopts misinformation as a broader analytical category. Table 1 presents some misinformation definitions underlying the role of source and structure. The definitions were selected based following a critical review of the relevant literature.

A closer analysis of these definitions reveals important theoretical divergences in how misinformation is conceptualized, particularly in relation to source of intent and message structure. While some accounts emphasize verifiability and factuality [8, 13], others stress that misinformation often operates through heuristic and affective pathways where the *form* of the message, its fluency, coherence, or emotional resonance, shapes user perception more than its *truth content* [7, 12, 14]. Some definitions, like that of Altay *et al.* [2], go further by decentering factual falsity altogether, embedding misinformation within broader low-quality or hyperpartisan ecosystems.

In GenAI contexts, this becomes especially relevant: LLMs are epistemically opaque sources [10] that systematically generate fluent, coherent, and stylistically consistent content. This convergence of opaque sourcing and high structural plausibility produces a credibility illusion, where users infer truthfulness not from source reliability but from narrative smoothness. Thus, the Table 1 underscores a central proposition of this paper: in AI-mediated misinformation, credibility judgments increasingly emerge from the interplay of persuasive structure and reduced source intelligibility, two dimensions that traditional credibility models tend to analyze separately. Bridging this gap is crucial for understanding how GenAI shifts the informational landscape from verifiability to plausibility as the dominant heuristic for truth.

Table 1. Some misinformation definitions underlying the role of source and structure

Reference	Misinformation conceptualization	Role of source	Role of structure
Altay *et al.* [1]	A part of a broad class of unreliable news, including false, deceptive, low-quality, hyperpartisan and junk content; scale depends on operational definition	Source intent is not central; misinformation may arise without deliberate manipulation and is embedded in broader information ecosystems	Structural and qualitative features (e.g., hyperpartisan framing, low-quality presentation) are included as part of what counts as misinformation, not only factual falsity
Koohikamali & Sidorova [8]	Not explicitly defined; misleading content is examined through perceived information quality when users cannot verify actual correctness	Source matters indirectly via platform and social signals when users assess trustworthiness	Accuracy, completeness, coherence, and format shape perceived quality and re-sharing behavior
Matheson & Zanna [12]	Persuasive input accepted under heuristic rather than systematic processing	Source credibility functions as a heuristic cue under low elaboration conditions	Message fluency and presentation increase acceptance without analytical validation
Lu *et al.* [10]	Inaccurate or misleading output produced without intent due to probabilistic text generation	AI acts as an epistemically opaque source, limiting users' calibration of reliability	Fluency and coherence create a factuality illusion that increases perceived credibility
Moravec *et al.* [13]	Operationalized as claims users cannot verify through direct epistemic access, requiring plausibility judgments	Source cues substitute for factual verification when knowledge is unavailable	Internal consistency and narrative coherence increase perceived accuracy
Rijo & Waldzus [14]	Evaluated under heuristic credibility judgments, especially under cognitive or emotional constraints	Source credibility dominates when elaboration is low	Narrative framing and emotional salience affect belief independent of truth
Jost *et al.* [7]	Its acceptance is shaped by motivated cognition and ideological congruence rather than factual accuracy alone	Trusted in-group sources amplify acceptance of inaccurate claims	Argument framing and coherence interact with ideological priors

2.2 Source as Informational Producers

Within this framework, source refers to the epistemic status of large language models as information producers, not to hardware or software infrastructure. Large Language Models do not retrieve facts from verified databases but generate outputs by predicting plausible linguistic continuations. Consequently, correctness is not guaranteed even when responses appear confident and coherent.

Users may interpret AI-generated outputs as neutral or technically authoritative while lacking accurate mental models of how such systems operate [10]. Research also indicates that credibility is not consistently reduced when content is labeled as AI-generated, particularly when linguistic quality is high [13]. Source cues therefore provide limited guidance, shifting evaluative focus toward message structure.

2.3 Structure as Primary Credibility Signal

Structural properties of messages are the central mechanism through which generative AI affects credibility judgments. Structure includes content organization, logical flow, and formatting features such as tone and narrative framing. Structural properties of messages are the main way through which generative AI influences credibility judgments [9]. These features affect how easy information is to understand and how coherent it appears, and therefore shape perceived information quality when users cannot easily verify factual accuracy [8].

Because well-structured messages are easier to process and seem more reasonable, they are often judged as more plausible even when they are incorrect. Internal coherence does not guarantee truth, as messages can be persuasive while remaining inaccurate [6]. Generative AI systems consistently produce fluent and well-organized text, which shifts credibility judgments toward how information is presented rather than whether it is factually grounded [9, 10].

When motivation or verification capacity is limited, individuals rely on surface cues such as fluency and coherence to infer reliability [12]. Generative AI systematically amplifies these cues, producing polished and contextually appropriate responses. Users therefore often perceive fluent and internally consistent messages as more credible regardless of objective accuracy [13], particularly when messages align with prior beliefs [7].

3 Synthesis of Findings: The Importance of Perceived Credibility and Detection Accuracy

This preliminary study adopts a critical, theory-driven selection of studies, consistent with common practices in Information Systems research when evidence is fragmented and technologies are rapidly evolving.

Search queries (via Web of Science, Scopus, and Google Scholar) combined terms related to generative AI and misinformation with credibility-related constructs, including "AI-generated misinformation", "synthetic news", "large language models", "machine-generated text", and "credibility judgements", "detection accuracy", and "fake news".

The search focused on 256 peer-reviewed journal papers and full conference papers published between 2016 and 2026. After removing duplicates, titles and abstracts were screened to identify studies examining user evaluation of misleading or news-like content in digital systems.

Given the limited number of GenAI-native empirical studies, a complementary critical selection phase was applied to include theoretically influential work on credibility assessment under low verifiability.

Studies focusing solely on automated detection or content generation without user evaluation were excluded. For each study, key characteristics were extracted, including outcome variables, source cues, message structure, and study design. Due to heterogeneity in methods and measures, quantitative synthesis was not feasible. Instead, a descriptive, concept-driven synthesis was conducted, structured around two informational dimensions (source and structure) and two outcome variables (perceived credibility and detection accuracy), central to Information Systems research on user interaction with AI-generated content.

3.1 Overview of the Considered Studies

The reviewed literature integrates perspectives from Information Systems, Psychology, Health Communication, and Human-Computer Interaction. The set comprises a large-scale meta-analysis of psychological inoculation against misinformation [10], a controlled information systems experiment examining believability judgments under interface-level prompts [13], an information systems study framing re-sharing behavior through perceived information quality in social networking environments [8], a conceptual synthesis of heuristic credibility processes [14], and two GenAI-oriented conceptual reviews addressing fluency, detectability, and trust in AI-generated content [4, 5].

Lu *et al.* [10] provide the empirical basis, aggregating 29 papers and 42 independent studies with a combined sample of 42,530 participants. Moravec *et al.* [13] contribute mechanistic evidence through a laboratory experiment in which participants evaluated 42 headlines formatted as social media posts. Koohikamali & Sidorova [8] situate credibility evaluation within perceived information quality under low verifiability. Rijo & Waldzus [14] synthesize cognitive explanations for susceptibility to misinformation. Claros & Dalkir [4] and Eissa [5] extend this literature to GenAI contexts by analyzing how linguistic fluency and reduced detectability affect credibility assessment. Perceived credibility.

According to the reviewed literature, perceived credibility is consistently treated as a cue-driven judgment rather than a direct function of factual verification. Rijo & Waldzus [14] emphasize that when individuals lack prior knowledge or motivation, they engage in peripheral processing and rely on heuristic cues, with source credibility being one of the most influential. Koohikamali & Sidorova [8] similarly argue that when users cannot ascertain actual information quality, perceptions of quality-shaped by coherence, completeness, and presentation-guide trust and re-sharing behavior.

Credibility judgements occur within sociotechnical information systems where interface design and interaction flows influence how users encounter structured information. When objective quality is difficult to verify, perceived information quality becomes the main determinant of trust and system use [8]. While design interventions can moderate reliance on structural cues, message construction remains the primary driver of credibility judgments.

GenAI-specific conceptual work indicates that generative systems may intensify cue-driven credibility. Claros & Dalkir [4] argue that linguistic fluency can mislead users about system intelligence and reliability. Eissa [5] further proposes that high-quality, stylistically appropriate AI content is increasingly difficult to distinguish from human-produced content, which may increase initial persuasive influence and delay skepticism. While these studies do not provide direct comparative experiments between AI- and human-written misinformation, they identify linguistic fluency and detectability as mechanisms likely to inflate perceived credibility.

Taken together, these findings reveal a critical shift in the epistemology of digital trust: when fluency and coherence consistently outweigh verifiability, credibility becomes a function of *how* information feels rather than *what* it says. This raises significant concerns in GenAI contexts, where linguistic performance is decoupled from factual grounding, and surface cues systematically reinforce plausibility. If credibility judgments remain highly sensitive to presentation rather than accuracy, then even well-intentioned users may become more vulnerable to structurally persuasive but misleading content.

3.2 Detection Accuracy

Detection accuracy, defined as the ability to distinguish false from accurate information [1], is also framed as contingent on cognitive effort and epistemic access. Lu *et al.* [10] report that inoculation improves credibility discernment, but they also highlight substantial heterogeneity and sensitivity to publication bias, suggesting that improvements in detection are context-dependent and potentially overstated in the literature.

Following Luo *et al.* [11], detection accuracy is defined as the degree to which users' credibility judgments correspond to the verified ground truth of a message. It is operationalized as the percentage of correct classifications of news headlines as real or fake, based on fact-checked veracity, and must be analyzed separately for true and false items due to systematic judgment biases affecting each category differently. It may be understood as users' ability to distinguish misleading from accurate information when interacting with AI-generated messages. It is shaped by message structure, cognitive heuristics, and contextual cues, rather than by exposure alone. Engagement with misinformation does not necessarily imply belief, and diffusion metrics cannot be equated with cognitive acceptance [1].

In GenAI contexts, detection becomes more complex because users face two overlapping tasks: identifying whether content is synthetic and evaluating its factual accuracy. Eissa [5] argues that decreasing detectability of AI-generated content shifts evaluation toward message-level cues and community norms. Claros & Dalkir [4] note that warnings about AI fallibility may be insufficient if users lack time, motivation, or expertise to verify outputs. Together, these works suggest that structural plausibility combined with opaque source cues may reduce effective detection accuracy in practice, even if users remain generally skeptical of misinformation.

Overall, the literature on detection accuracy reveals a deep mismatch between the complexity of AI-mediated misinformation and the simplicity of current evaluative metrics. Most studies still treat detection as a binary classification task, assuming users can and should discern truth from falsehood on the basis of isolated cues. However, GenAI complicates this logic: when the origin, authorship, and epistemic status of content are uncertain, the very notion of "detection" becomes unstable. Users are not just verifying facts; they are navigating plausibility under cognitive constraints, platform cues, and shifting community norms. As such, improving detection accuracy may require moving beyond content-level interventions toward system-level strategies that reinforce epistemic transparency and encourage reflexive engagement with the production context of information.

4 Critical Discussion

A key conceptual limitation of the reviewed studies is the frequent conflation of perceived credibility with detection accuracy. Several papers operationalize susceptibility mainly through credibility judgments, perceived information quality, or willingness to share, implicitly treating these outcomes as proxies for the ability to correctly identify misinformation [8, 10, 13].

However, reduced credibility ratings may reflect generalized skepticism rather than improved discrimination between true and false content, and only few studies directly model classification accuracy or separate sensitivity from response bias [10, 13]. In addition, source cues and message structure are often examined in isolation, even though theoretical accounts emphasize that credibility judgments emerge from their joint influence [14].

Fluent and well-structured misinformation may therefore appear credible despite weak or ambiguous source cues, while trusted sources can compensate for low-quality or vague content. Because most experimental designs manipulate either source or content features, it remains unclear whether observed effects reflect content-based detection, reliance on source heuristics, or shifts in general trust thresholds [8, 14]. This distinction becomes especially critical in GenAI contexts, where linguistic fluency and stylistic coherence are systematically elevated while provenance cues are often opaque, potentially further decoupling perceived credibility from actual detection accuracy [4, 5].

This suggests that credibility and detection, though empirically intertwined, must be theoretically re-differentiated in GenAI environments. Rather than viewing credibility as a proxy for accuracy, future models should treat it as a heuristic interface between human cognition and opaque algorithmic production. In this sense, credibility becomes less a judgment about truth and more a plausibility calibration process, highly sensitive to interface, style, and context. Recognizing this shift is critical for building sociotechnical systems that do not equate fluency with factuality.

4.1 Convergence and Divergence Across Studies

Across studies, there is strong convergence on three points. First, credibility judgments are largely heuristic and shaped by surface-level cues when verification is costly. Second, credibility and detection accuracy are malleable through cognitive and interface-level interventions. Third, GenAI systems amplify structural cues while reducing detectability of non-human origin, which plausibly increases early persuasive impact.

Divergence arises in methodological focus. Lu *et al.* [10] emphasize aggregate intervention effects but also document strong heterogeneity and publication bias risks. Moravec *et al.* [13] focus on mechanism and design, showing how small interface changes can alter cognitive processing. GenAI-oriented papers provide plausible mechanisms but lack comparative experiments quantifying AI versus human misinformation effects.

Overall, the synthesis indicates that AI-generated misinformation should be understood less as a novel category of falsehood and more as a structural escalation of existing credibility dynamics within sociotechnical information systems.

4.2 Discussing Credibility as Structural Performance

The synthesis indicates that GenAI fundamentally reshapes credibility judgments by altering the relationship between message structure and information source. Traditional credibility heuristics - such as grammatical correctness, logical coherence, and stylistic fluency - have historically functioned as indirect signals of expertise and editorial oversight. In AI-generated content, however, these structural cues are systematically optimized independently of both factual accuracy and source credibility. As a result, well-structured misinformation can appear credible even in the absence of a reliable or identifiable source.

This shift has important implications for cue-based models of credibility, which typically distinguish between message-based cues (*e.g.*, structure, clarity) and source-based cues (*e.g.*, expertise, authority). In AI-mediated environments, structural cues increasingly dominate credibility judgments, while source cues become less salient or harder to interpret. When generative systems produce content that mimics the linguistic and rhetorical style of expert sources, users may rely on structural plausibility as a substitute for source verification. Credibility thus emerges from the interaction between source ambiguity and optimized message structure, rather than from either dimension alone.

4.3 Limitations and Future Research

Methodologically, the existing literature offers limited insight into these dynamics due to its reliance on laboratory-based studies with small samples and simplified tasks. Many experiments isolate either message features or source labels, despite real-world information systems presenting users with both simultaneously and often ambiguously. Moreover, short-term judgment tasks fail to capture how repeated exposure to structurally fluent content may gradually erode attention to source credibility over time. These limitations suggest that current evidence may underestimate the extent to which generative AI reshapes the balance between source-based and structure-based credibility cues.

Future structured research (following PRISMA guidelines) should therefore examine credibility judgments as system-level processes shaped jointly by source signals, message structure, and platform design. Research should explore how Information Systems can reinforce expert-linked sources and contextualize structural fluency, for example through provenance indicators, expert-centered information pathways, and interface designs that make epistemic authority visible. Strategically, future research must also explore how credibility judgments evolve under conditions of prolonged exposure to AI content across platforms and contexts. This includes tracing how repeated engagement with fluent but unverifiable content reshapes not only individual beliefs but collective epistemic norms. What counts as "credible" may itself be shifting in response to the aesthetic and interactive norms of GenAI, demanding longitudinal studies that treat credibility not as a fixed judgment but as a moving target in dynamic information environments. Longitudinal and field-based studies are needed to understand how these mechanisms influence credibility judgments over time and support epistemic resilience in AI-mediated information environments.

5 Conclusion

This paper examined how GenAI disrupts the informational architecture of credibility by amplifying structural fluency while obscuring source transparency. Through a dual analytical lens (perceived credibility and detection accuracy) we show that traditional user heuristics, once grounded in the alignment of surface quality and reliable sourcing, are increasingly destabilized in GenAI environments.

Our review reveals a critical epistemic shift: credibility judgments are not merely vulnerable to deception, but structurally predisposed to overtrust fluent, internally coherent, yet unverifiable content. Studies such as Lu *et al.* [10] and Moravec *et al.* [13] demonstrate the malleability of these judgments through cognitive and interface-level interventions, while Claros & Dalkir [4] highlight how linguistic fluency systematically inflates trust in AI outputs. Yet few studies jointly model the interaction of source and structure, despite this interplay being essential in GenAI contexts where source cues are often opaque.

This paper makes three core contributions. First, it provides the first structured synthesis of empirical and conceptual work across disciplines focused explicitly on user-facing credibility in AI-generated misinformation. Second, it clarifies a key theoretical distinction often conflated in the literature: credibility is not detection. Rather, credibility functions as a heuristic interface shaped by structural performance, interface cues, and evolving user expectations. Third, it introduces a dual-dimensional framework that integrates source ambiguity and structural fluency as co-determinants of user judgments, offering a conceptual lens to reframe detection challenges beyond binary classification.

Practically, this reframing has implications for how we design and govern Information Systems. Efforts to counter misinformation must move beyond factual corrections and automated filters to address the semiotic performance of trust itself. Interfaces should make epistemic authority more visible and intelligible, for example, by foregrounding provenance indicators, strengthening expert-linked pathways, and counterbalancing stylistic fluency with source accountability.

Looking ahead, we call for a new research agenda that recognizes credibility as dynamic and system dependent. Future work should investigate how repeated exposure to AI-generated content shapes credibility heuristics over time, through longitudinal field studies, real-world interface experiments, and cross-cultural analyses of trust in multilingual environments. Methodologically, combining behavioral data, qualitative inquiry, and system instrumentation will be key to capturing how structural fluency, source ambiguity, and trust interact under real-world conditions.

In short, GenAI does not merely produce more misinformation; it reshapes the conditions under which information is believed. The central challenge is no longer whether content is true or false, but whether our systems are designed to make that distinction meaningful, visible, and actionable for users. Addressing this requires not only technical solutions, but a paradigm shift in how we conceptualize, study, and support credibility in AI-mediated information ecosystems.

Disclosure of Interests. The authors have no competing interests to declare that are relevant to the content of this paper.

References

1. Ahituv, N.: A systematic approach towards assessing the value of an information system. MIS Q. **4**(4), 61–75 (1980). https://doi.org/10.2307/248961
2. Altay, S., Berriche, M., Heuer, H., Farkas, J., Rathje, S.: A survey of expert views on misinformation: definitions, determinants, solutions, and future of the field. Harvard Kennedy School Misinf. Rev. **4**(4), 1–34 (2023)
3. Broda, E., Strömbäck, J.: Misinformation, disinformation, and fake news: lessons from an interdisciplinary, systematic literature review. Ann. Int. Commun. Assoc. **48**(2), 139–166 (2024)
4. Claros, C., Dalkir, K.: The role of ai in the misinformation ecosystem: virtuous or vicious? In: Proceedings of the 3rd Workshop on Adverse Impacts and Collateral Effects of Artificial Intelligence Technologies (AiOfAi'23), Macao, China, pp. 36–40 (2023)
5. Eissa, M.E.: The influence of AI-generated content on trust and credibility within specialized online communities: a brief review on proposed conceptual framework. J. Artif. Intell. **2**(2), 1–14 (2025). https://doi.org/10.29121/shodhai.v2.i2.2025.40
6. Godfrey-Smith, P.: Misinformation. Can. J. Philos. **19**(4), 533–550 (1989). https://doi.org/10.1080/00455091.1989.10716781
7. Jost, P., Pünder, J., Schulze Lohoff, I.: Fake news – does perception matter more than the truth? J. Behav. Exp. Econ. **85**, 101513 (2020). https://doi.org/10.1016/j.socec.2020.101513
8. Koohikamali, M., Sidorova, A.: Information re-sharing on social network sites in the age of fake news. Inform. Sci. Int. J. Emerg. Transdiscip. **20**, 215–235 (2017). https://doi.org/10.28945/3871
9. Kujur, A.: A comparative analysis of AI-generated and human-written text: linguistic patterns, detection accuracy, and implications for modern communication. SSRN (2025). https://doi.org/10.2139/ssrn.5833302
10. Lu, C., Hu, B., Li, Q., Bi, C., Ju, X.D.: Psychological inoculation for credibility assessment, sharing intention, and discernment of misinformation: systematic review and meta-analysis. J. Med. Internet Res. **25** (2023). https://doi.org/10.2196/49255
11. Luo, M., Hancock, J.T., Markowitz, D.M.: Credibility perceptions and detection accuracy of fake news headlines on social media: effects of truth-bias and endorsement cues. Commun. Res. **49**(2), 171–195 (2020)

12. Matheson, K., Zanna, M.P.: The impact of computer-mediated communication on self-awareness. Comput. Hum. Behav. **4**(3), 221–233 (1988). https://doi.org/10.1016/0747-5632(88)90015-5
13. Moravec, P., Kim, A., Dennis, A., Minas, R.: Do you really know if it's true? how asking users to rate stories affects belief in fake news on social media. SSRN Electron. J. (2022). https://doi.org/10.2139/ssrn.3271057
14. Rijo, A., Waldzus, S.: That's interesting! the role of epistemic emotions and perceived credibility in the relation between prior beliefs and susceptibility to fake news. Comput. Hum. Behav. **141**, 107619 (2023). https://doi.org/10.1016/j.chb.2022.107619

From Data Governance Policies to Policy-as-Code: An LLM-Based Framework

Matteo Falconi[1(✉)], Valeria Maria Fortina[1], Marco Tonnarelli[2,3,4], Indika Kumara[2,3], and Pierluigi Plebani[1]

[1] Politecnico di Milano, Piazza Leonardo da Vinci 32, 20133 Milan, Italy
{matteo.falconi,valeriamaria.fortina,pierluigi.plebani}@polimi.it
[2] Eindhoven University of Technology, Het Eeuwsel 53, 5612 AZ Eindhoven, Netherlands
{m.tonnarelli@tue.nl, i.p.k.weerasingha.dewage}@tue.nl
[3] Tilburg University, Warandelaan 2, 5037 AB Tilburg, Netherlands
{m.tonnarelli, i.p.k.weerasingha.dewage}@tilburguniversity.edu
[4] NXP Semiconductors N.V., High Tech Campus 60, 5656 AG Eindhoven, Netherlands

Abstract. This paper explores the possibility to reduce the gap between data governance policies, typically expressed in natural language, and the technical representations of these policies required for their enforcement. Focusing on data sharing policies, the objectives of this paper are: (i) a framework supporting the translation from high-level to low-level policy description; (ii) an initial reference architecture to support the early steps toward the implementation of such a platform.

Keywords: Computational Data Governance · Policy as Code · LLM-based translation

1 Introduction

Within an organizational ecosystem, policies of diverse origins coexist and interact dynamically. Internal policies reflect organizational governance needs, covering areas such as data management, operational procedures, security protocols, and staff conduct guidelines. Simultaneously, the organization must adhere to broader regulatory frameworks, including national and international legislation that in the last years are more and more increasing [11]. This is particularly important when it comes to defining the ways in which data can be shared. Data sharing policies are based on internal decisions as well as external regulations: e.g., GDPR or Data Act defined by the EU or the HIPAA in the USA.

With the increase in the amount of data and the considerable increase in the possible users of such data, especially to feed the AI models, it becomes

M. Falconi, V. M. Fortina and M. Tonnarelli: These authors contributed equally to this work.

A. Bernasconi et al. (Eds.): RCIS 2026, LNBIP 597, pp. 41–53, 2027.
https://doi.org/10.1007/978-3-032-35043-5_4

fundamental to automate policy compliance mechanisms as much as possible. One factor preventing automation is the difference between the business-level policies, usually expressed in natural language and often based on *legal jargon*, and the technical-level policies that use computing-oriented formalisms and are dictated by specific technologies and platforms in use.

Several attempts have been proven to be effective [7], and a trend in this field is making a tradeoff between precision (typically higher in rule-based approaches) and generalizability (typically higher in AI-based approaches). This tension causes these systems to be applied only under specific constraints, which they depend on the application context [12]. However, a end-to-end solution for automating the translation of high-level policies (the natural language source) to low-level ones (the technical representation of a policy statement), general enough but yet precise, to the best of our knowledge, cannot be found in the literature.

On this basis, with specific emphasis on the data sharing policies, the contribution of this paper is twofold. Firstly, inspired by the approaches in the state of the art that tackle only portions of the whole problem, the paper proposes a unified framework for translating high-level policies, defined at business level, into Policy-as-Code [8,15] that can be enforced at platform level. In addition to LLMs, the proposed framework relies on Controlled Natural Languages (CNL) to increase the precision in the translation. Secondly, this paper provides an architecture supporting the proposed framework and a validated implementation of its basic components.

The rest of the paper is organised as follows. Section 2 presents the related work and describes the research gap that is addressed by the framework as discussed in Sect. 3. Sections 4 and 5 describe, namely, the current implementation of the framework and the evaluation. Finally, Sect. 6 concludes the work by discussing the threats to the validity.

2 Related Work

Fragmentation in Data Governance Frameworks. Current literature indicates that there is no shared or standardized approach to defining and structuring data governance policies. Existing studies report a fragmented landscape in which governance frameworks are developed for domain-specific applications, such as cloud computing, big data, and interorganizational networks. Each framework adopts its own conceptual and policy components. Systematic reviews emphasize the lack of a consolidated framework for data governance across domains and the absence of a common reference point [2]. Several studies surveying data and big data governance frameworks highlight that there are substantial differences in the scope and components, confirming that no shared structural reference has emerged across them [1]. This fragmentation is even more evident in inter-organizational settings, due to limited formalization and an absence of transferable governance structures across organizational boundaries [13].

Formalization Through CNLs. The lack of a standard policy structure makes it difficult to translate natural language policies into formal representations. This challenge has led to research on policy harmonization and translation using CNLs and LLMs. Policies and regulations are typically expressed in natural language, which is inherently ambiguous. To enable comparison, they need to be translated into a more formal and interpretable representation, a task commonly addressed through the use of CNLs and LLMs. CNLs have been designed for diverse purposes and domains [16], with studies addressing the conversion of policies using CNL, proposing new languages or comparing existing solutions [19–21]. In particular, CNL4DSA [17] is a simple, readable language for converting data sharing agreements that is sufficiently expressive and translatable into formal languages for automatic analysis. Its applicability has already been demonstrated in policy translation and in compliance studies [6,22].

Compliance Automation and the Role of LLMs. Several works have addressed the issue of compliance with specific regulations, such as the GDPR [3,4,10]. Recent literature has also begun to investigate the use of LLMs to support compliance checking with privacy regulations [5,8,9]. These approaches leverage the capabilities of LLMs to extract obligations and permissions from legal texts.

3 Policy Translation Framework

While existing work addresses policy representation through CNLs and compliance checking with regulations like GDPR, using both rules-based and LLM-based approaches, these aspects are frequently treated in isolation. To the best of our knowledge, a unified framework that ensures data governance compliance by reasoning across organizational policies, data regulations, and technical requirements is currently missing from the literature.

With specific reference to data sharing policies, the objective of such a framework is to simplify the compliance of data governance policies that deal with this aspect, to allow the data owner to expose the data that can be shared without violating the terms of regulations and company policies.

To address this gap, we propose the framework reported in Fig. 1. The goal of simplifying compliance leverages on two main aspects. Firstly, the use of LLMs that have proven to be effective for understanding and managing texts written in natural range. Secondly, the interplay between the policy and the technical aspects as the same policy can be enforced in different ways if two organizations rely on different underlying platforms.

The framework (represented using the C4 notation) is designed as a modular pipeline that transforms high-level organizational policies into executable technical rules while maintaining a structured representation of the policy landscape. The framework is based on two main inputs: (i) the policies expressed in natural language, and (ii) the description of the digital infrastructure responsible to enforce the rules of the policies that are expressed using technical documents, e.g., Infrastructure as Code (IaC) recipes.

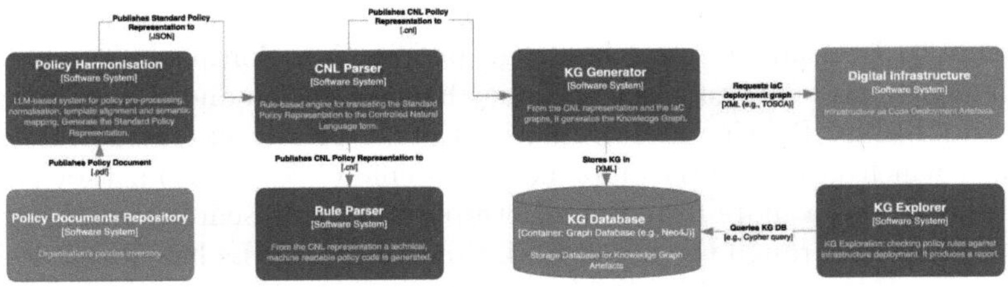

Fig. 1. Framework Architecture Overview, C4 model.

The proposed framework is defined in terms of a software architecture composed of loosely coupled software systems, each responsible for a distinct transformation step. At the core of the architecture is the *Policy Harmonisation* system, an LLM-based component in charge of transforming heterogeneous organisational policy documents into a unified semantic representation. It interacts with the *Policy Documents Repository*, which, we assume, stores policies in their original document (e.g., PDF file) form within an organisation. The harmonisation process produces a *Standard Policy Representation* that can be expressed in readable formats like JSON, YAML, or XML.

The *Standard Policy Representation* is consumed by the *CNL Parser*, a rule-based software system that translates structured policy data into a Controlled Natural Language (CNL). The CNL serves as a rigorous intermediate step that is both human-readable and computationally parsable. Two complementary systems operate on the CNL representation. The *Rule Parser* generates a fully technical, machine-readable policy specification suitable for automated reasoning or enforcement mechanisms (e.g., Rego). In parallel, the *Knowledge Graph Generator* integrates the CNL policies with infrastructure deployment information obtained from the *Digital Infrastructure system*. This infrastructure system exposes Infrastructure-as-Code deployment artefacts, expressed as structured graphs using domain-specific notations like TOSCA[1], capturing resources and their relationships.

The *Knowledge Graph Generator* produces a unified *Knowledge Graph* that encodes both policy constraints and infrastructure topology. This graph is persisted in the *Knowledge Graph Database*, implemented as a graph database system that supports expressive querying. The *Knowledge Graph Database* serves as the central persistence and reasoning substrate for compliance analysis. The *KG Explorer* interacts with the Knowledge Graph Database to perform policy compliance checks: it analyses whether deployed infrastructures satisfy the constraints specified in organisational policies and produces reports.

The framework proposed in this paper includes both rule-based and LLM-based components, an end-to-end pipeline for translating high-level policies into technical rules. We argue that combining the LLM's expressiveness with the rule-based engine precision is crucial for a reliable and industry-grade automated

[1] https://docs.oasis-open.org/tosca/TOSCA/v2.0/csd06/TOSCA-v2.0-csd06.html.

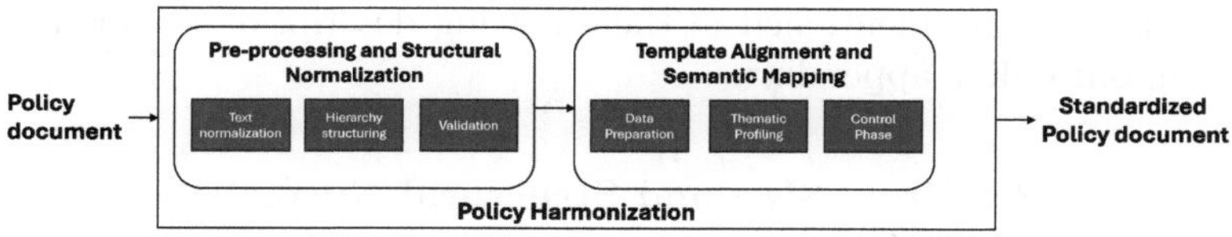

Fig. 2. Policy Harmonization Overview.

tool. Our pipeline, at each stage, generated intermediate representations of the original policy text by extracting individual policy statements and producing formulations that shift from legal language to a technical one, employing a knowledge graph and CNL (a subset of natural language) to aid the translation. Although the structure of this framework leads to the many research questions, this paper focuses on the following one: **RQ1**: *How accurately are LLMs able to understand what data-related statements are present in policy documents?* LLMs are prone to allucinations, bias, and limited reasoning. Thus, it is required to assess if LLMs are capable of identifying what sections of a policy are relevant for data-related activities.

4 Policy Harmonization

One of the central objectives of the framework is to ensure compliance across disparate layers, including organisational governance rules, or regulatory requirements, and the constraints imposed by IT infrastructures. Achieving such alignment requires a systematic harmonisation effort of the organisational policies: extracting a common structure from heterogeneous organizational policies to support subsequent compliance activities. The proposed harmonisation pipeline consists of two main phases:

- *Pre-processing and Structural Normalization.* Organizational policies are often stored in unstructured formats lacking clear semantic divisions, hindering automatic analysis. This step cleanses and restructures the content to transform raw text into a coherent, hierarchical representation ready for downstream tasks.
- *Template Alignment and Semantic Mapping.* Once cleaned and structured, the policies are benchmarked against a foundational Data Governance Policy Template (DGPT) [23]. This reference provides a comprehensive framework for data management, codifying guidelines and responsibilities across domains such as integrity, security, and regulatory compliance. The goal of this phase is to evaluate the coverage of each organizational policy relative to the template's requirements. The framework normalizes diverse inputs into a unified, harmonized schema.

An overview of the different steps performed during these two phases is represented in Fig. 2. All the material, including LLM prompts, JSON schemas, results and analysis, that are mentioned in the following description of this process is available in our online appendix[2].

4.1 Phase 1: Pre-Processing and Structural Normalization

First, the policies are extracted from their original location and converted into a common format, such as JSON. Among the various organizational policies, those relevant to data governance are selected. The text then undergoes a normalization process, which includes decoding escape sequences and special characters (e.g., from `\u201c` to standard quotation marks) to ensure encoding consistency.

Furthermore, the policies are decomposed into their constituent topics to create a hierarchical structure according to the structure described in Fig. 3. This structure supports recursive nesting, where sections may contain multiple levels of incrementally numbered subsections. Beyond the body text, the pipeline extracts critical metadata, including policy titles, source URLs, categories, revision histories, and governing bodies. Given the inherent structural heterogeneity of these documents, we employ the *GPT-4o-mini* model to automate the restructuring process. Utilizing a prompt the model cleans and maps the raw text onto a predefined schema, resulting in a consistent and machine-readable output.

The output of this process is then checked to mitigate common errors in LLMs. A verifier checks that the expected structure is respected, for example, by checking for the presence of the required fields in each section and subsection. A human-in-the-loop audit is performed to verify semantic completeness, ensuring the absence of omissions or hallucinations. This rigorous pipeline ensures that the processed policies are functionally ready for the subsequent alignment phase.

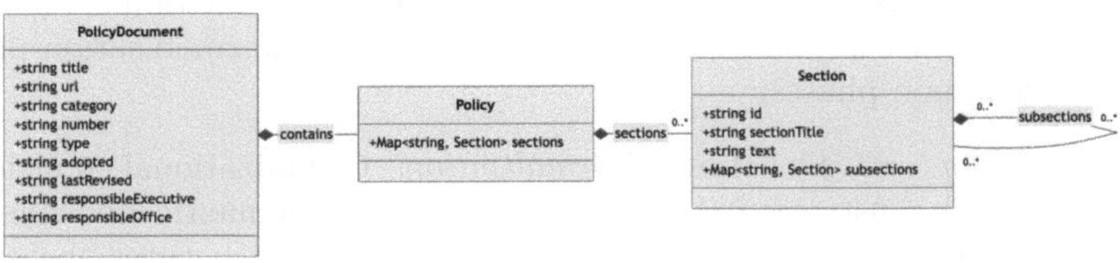

Fig. 3. Structure Of The Cleaned Policy.

4.2 Phase 2: Template Alignment and Semantic Mapping

The *Template Alignment and Semantic Mapping Phase* is designed to evaluate the structural and thematic alignment of an organizational policy with the DGPT. To facilitate this assessment, we defined a specialized JSON schema (compliant with the structure in Fig. 3) that quantifies the coverage of each policy relative to the template's requirements. The schema consists of the title of the

[2] https://figshare.com/s/07e78135b35921879370.

```
{"title": "Data Governance Policy",
  "sections": [ {
     "id": "1",
     "title": "Introduction",
     "confidence": 90,
     "evidence": "Outlines purpose and strategy but lacks
         implementation timelines.",
     "covered_subsections": ["1.1"],
     "missed_subsections": ["1.2"],
     "subsection_summary": { "total_subsections": 2 },
     "raw_subsection_scores": { "1.1": 90, "1.2": 40 }
   },"..." ]}
```

Listing 1.1. JSON schema for automated policy alignment and scoring.

policy and an array called *sections*, containing a dictionary for each section of the template. Each dictionary includes the section number, its name, a score from 1 to 100 for each subsection (see Table 1), a self-reported confidence level associated with those scores, an explanatory textual comment, and a control variable. This process is executed through three functional stages: (i) Data Preparation, (ii) Thematic Profiling, and (iii) Quality Control. This pipeline consumes the structured policies generated in Phase 1 and outputs a populated JSON schema representing the policy's alignment with the DGPT.

Data Preparation Phase. The preparation phase involves extracting the full text of the policies from their structure. As the evaluation requires analysis of the entire clean policy, it is necessary to reconstruct the text in its entirety by recursively processing the generated clean structure.

Thematic Profiling Phase. The clean text of the policy, extracted in the preparation phase, is provided to the model as a variable within a specially defined prompt, obtained after a three-cycle refinement process, described in Sect. 5.1. This prompt contains: (i) the task to be performed; (ii) the structure of the JSON output; (iii) the scoring scale used; and (iv) a summary of the template structure. Points (i) and (ii) have already been described in previous paragraphs. Regarding point (iii), the 1100 rating scale is described in Table 1. This approach requires the model to evaluate each subsection based solely on textual evidence, while providing an explicit rubric. These factors allow for consistent and interpretable profiling. Regarding the template summary (point iv), we implemented a context-optimization strategy to remain within the model's token limits. Rather than providing the full DGPT, we supply a condensed version that retains the hierarchical nomenclature of all sections and subsections. This summary includes brief functional descriptions for subsections that are not self-explanatory, effectively mirroring the semantic benchmark of the original template without exhausting the context window.

Table 1. Policy Alignment Scoring Rubric

Score Range	Compliance Level	Evaluation Criteria
1–20	Absent / Minimal	The topic is not mentioned or appears only as an incidental phrase providing no guidance.
21–40	Minimal Mention	A brief mention exists but the content lacks procedures, defined roles, or clear standards.
41–60	Partial Coverage	The topic is present with some specifics or examples, but significant procedural details are missing.
61–80	Substantial	Coverage is substantial with measurable guidance, though minor elements remain unaddressed.
81–100	Full Coverage	Content is comprehensive and actionable, perfectly matching the template subsection requirements.

Control Phase For each policy, the Thematic Profiling Phase's output is a JSON file that adheres to the structure of Fig. 3. This file contains the evaluations generated by the model. The control phase removes any unwanted characters generated by the model and ensures correct formatting.

5 Evaluation

This section describes the method used to evaluate the consistency and reliability of policy coverage scoring performed by three human reviewers and an LLM-based reviewer. The evaluation examines both continuous scores (1–100 scale) and categorical labels derived from those scores (*missing*, *partial*, *covered*). We assess: (i) inter-rater reliability among human reviewers, (ii) agreement between LLM-based reviewer and human reviewers, (iii) agreement patterns across policies and sections. The goal is to determine whether the LLM-based assessor can consistently reproduce human-like scoring behavior.

5.1 Settings

Dataset. The dataset used consists of 18 policies extracted from a set of 68 policies stored in three policy repositories. The selection has been performed considering the relevance of the policy with respect to data-related topics: data governance, data security/privacy, data quality, data lifecycle management, roles and responsibilities. Each document was normalized and cleaned to extract text in a structured hierarchical format, as described in Sect. 4.

Scoring Framework. Three trained human annotators (R1, R2, R3) independently scored all policies using identical spreadsheets, following the scoring system defined in Table 1. The LLM-based annotator (GPT), which relies on GPT-4o-mini, was prompted with: (i) the full policy text, (ii) the scoring template, (iii) strict instructions describing how the scoring should be performed. LLM-based annotator outputs were parsed into the same structure as human scores.

Metrics. To evaluate the consistency and robustness of the scoring process, we adopt three complementary reliability metrics, each chosen to capture a specific aspect of agreement across reviewers. First, we use the Intraclass Correlation Coefficient (ICC(2,1)) [14] to assess agreement on the continuous numeric subsection scores. ICC(2,1) is appropriate because each subsection receives a score on a 1–100 scale, and the model assumes that both raters and items are random effects. This formulation measures whether reviewers use the scoring scale in a comparable way and penalizes absolute differences in magnitude. Second, we compute Fleiss' Kappa [18], which evaluates agreement on categorical section labels (*missing*, *partial*, *covered*) after aggregating subsection scores. Fleiss' k measures the degree of consensus beyond chance among more than two raters, making it suitable for the three-human and GPT setting. Finally, we include descriptive reliability statistics (raw exact agreement, per-policy agreement, and confusion matrices) to provide insight into scoring patterns, reveal systematic biases (e.g., tendencies toward more conservative or optimistic scoring), and highlight which sections or policies generate the most divergence.

Experimental Conditions. We conducted our evaluation by iteratively refining the prompt across three successive optimization cycles to progressively improve output quality. In each cycle, we generated model responses and had them assessed them in comparison with the results from three human reviewers in order to identify strengths and weaknesses in the outputs. The feedback and comparative judgments from these reviewers and GPT were then used to update and sharpen the prompt before proceeding to the next iteration. After obtaining a satisfactory prompt, we performed an additional validation phase in which a fourth, independent human reviewer, together with GPT, evaluated the final scoring and its associated outputs to confirm the accuracy of the LLM. Across all stages of the process, both iterative refinement and final validation, we used the gpt-4o-mini model with a fixed temperature setting of 0.1.

5.2 Results

We report the scoring results across all three refinement cycles, assessed by both human reviewers and GPT. Metrics' scores are summarised in Table 4, while the voting class distributions are depicted in Fig. 5

Score	3 Reviewers	3 Reviewers and GPT	Final Validation
ICC	0.581	0.523	0.424
Fleiss' Kappa	0.466	0.436	N/A
Raw Exact Agreement	0.521	0.418	0.591

Fig. 4. Evaluation Scores.

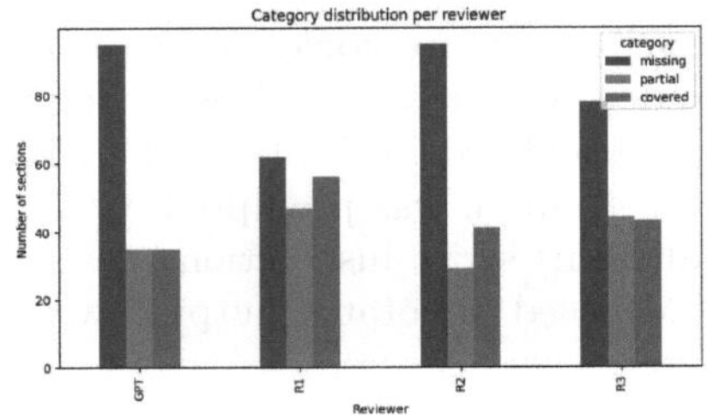

Fig. 5. Voting Class Distribution.

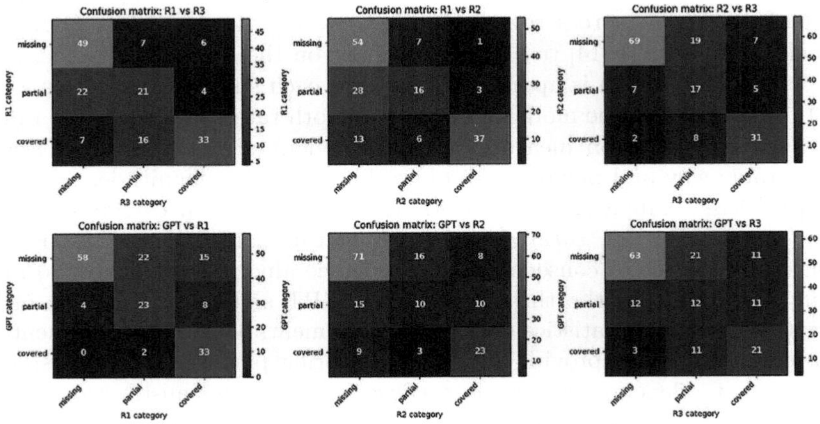

Fig. 6. Confusion Matrices Among Reviewers' Scoring Classes.

Test Results. Policies were scored at the subsection level by multiple human reviewers and by the LLM, and the resulting scores were aggregated and compared across raters. The subsection-level scores show a moderate degree of alignment, with an ICC(2,1) of 0.523. This value indicates that, while raters often converged on similar numeric assessments, there remains substantial variability in how specific subsections are interpreted and scored. This variability becomes more pronounced once subsection scores are aggregated into section-level categories: the resulting Fleiss' Kappa of 0.436 reflects only partial categorical agreement, suggesting that small differences at the subsection level frequently propagate into different final section labels.

The confusion matrices at the section level reinforce this pattern. They show that disagreements are not uniformly distributed across categories but are concentrated among conceptually adjacent sections of the template. In particular, sections that share overlapping governance concepts are more likely to be confused with one another, whereas sections representing more clearly delimited policy dimensions show higher consistency. This indicates that disagreement is driven less by random noise and more by structural ambiguity in how certain sections are operationalized through scores.

Validation Results. In the final validation view, focusing on the agreement between the LLM scores and the human reference (R4), subsection-level consistency decreases slightly, with an ICC(2,1) of 0.424. This reduction suggests that the LLM captures the overall scoring tendencies of human reviewers but diverges more frequently on fine-grained numeric judgments at the subsection level. Despite this, when subsection scores are aggregated, the exact agreement on section categories reaches 0.591, indicating that the LLM often assigns the same final section labels as the human reference, even when numeric subsection scores differ.

The confusion matrices (Fig. 6) in this final evaluation show a clearer structure than in the initial analysis. Most disagreements are concentrated in a limited set of off-diagonal cells, again corresponding to neighboring or semantically similar sections. Importantly, there is little evidence of systematic confusion between distant or conceptually unrelated sections, implying that the LLMs errors are largely local rather than global. Taken together, the score distributions and confusion patterns indicate that performance limitations are primarily associated with boundary cases between similar sections, rather than with a failure to recognize the correct governance dimension altogether.

6 Concluding Remarks

This paper aims to contribute to the improvement of data sharing by providing a framework capable of reducing the gap between those who define policies and those who are responsible for configuring platforms so that such policies become enforceable. The growing adoption of LLM-based tools offers new opportunities, which the proposed framework seeks to systematize and make effective within the considered context.

At present, the implementation of the framework has focused on a subset of its aspects, leaving its completion to future work. Moreover, the set of policies considered for the evaluation is low due to the limited availability of such documents. Therefore, it is necessary to expand the policy dataset for further validation and comparison with other approaches that may emerge. Finally, although several measures were adopted to reduce bias in the evaluation of the translations, the analysis was conducted internally within the research group, thus additional persons are therefore desirable in the near future.

To conclude, it is important to emphasize that the proposed framework can also be seen, to some extent, as a roadmap for the scientific community interested in automated data governance.

References

1. Al-Badi, A., Tarhini, A., Khan, A.I.: Exploring Big Data Governance Frameworks. Procedia Computer Science **141**, 271–277 (2018). 10.1016/j.procs.2018.10.181, EUSPN/ICTH 2018
2. Al-Ruithe, M., Benkhelifa, E., Hameed, K.: A systematic literature review of data governance and cloud data governance. Pers. and Ubiquitous Comput. **23**(5), 839–859 (2019). https://doi.org/10.1007/s00779-017-1104-3
3. Amaral, O., Abualhaija, S., Sabetzadeh, M., Briand, L.: A Model-based Conceptualization of Requirements for Compliance Checking of Data Processing against GDPR. In: 2021 IEEE 29th International Req. Eng. Conference Workshops
4. Cejas, O.A., Azeem, M.I., Abualhaija, S., Briand, L.C.: NLP-Based Automated Compliance Checking of Data Processing Agreements Against GDPR. IEEE Trans. on Software Eng. **49**(9), 4282–4303 (2023)
5. Cory, T.: Word-level Annotation of GDPR Transparency Compliance in Privacy Policies using Large Language Models. CoRR. 10/q64m
6. Costantino, G., Martinelli, F., Matteucci, I., Petrocchi, M.: Efficient Detection of Conflicts in Data Sharing Agreements. ICISSP. pp, 148–172 (2017) Springer
7. Das, S., Deb, N., Chaki, N., Cortesi, A.: A Multi-Agent RAG Framework for Regulatory Compliance Checking of Software Requirements. ACM Trans. Softw. Eng. Methodol. (2025). https://doi.org/10.1145/3785472
8. Elluri, L.: A BERT Based Approach to Measure Web Services Policies Compliance With GDPR. IEEE Access. 10/gnzrgn
9. Garza, L., Elluri, L., Piplai, A., Kotal, A., Gupta, D., Joshi, A.: PrivComp-KG: Leveraging KG and LLM for Compliance Verification. In: TPS-ISA 2024
10. Hamdani, R.E., Mustapha, M., Amariles, D.R., Troussel, A., Meeùs, S., Krasnashchok, K.: A combined rule-based and machine learning approach for automated gdpr compliance checking. In: ICAIL 2021, pp. 40–49
11. Harmeling, T.: Over 150 Data Privacy Statistics Companies Need to Know About in 2025 (2025), https://usercentrics.com/guides/data-privacy/data-privacy-statistics/
12. Hunsicker, T., König, C.J., Langer, M.: Investigating choices regarding the accuracy-transparency trade-off of ai-based systems across contexts. CHBAH **6**, 100216 (2025). https://doi.org/10.1016/j.chbah.2025.100216
13. Jagals, M., Karger, E.: Inter-Organizational Data Governance: A Literature Review. In: ECIS 2021 Research Papers, (2021)
14. Koo, T.K., Li, M.Y.: A Guideline of Selecting and Reporting Intraclass Correlation Coefficients for Reliability Research. J. Chiropr. Med, **15**(2) (2016). https://doi.org/10.1016/j.jcm.2016.02.012
15. Kruiper, R., Kumar, B., Watson, R., Sadeghineko, F., Gray, A., Konstas, I.: A platform-based Natural Language processing-driven strategy for digitalising regulatory compliance processes for the built environment. Adv. Eng. Informatics **62** (2024). https://doi.org/10.1016/j.aei.2024.102653
16. Kuhn, T.: A Survey and Classification of Controlled Natural Languages. Comput. Linguist. **40**(1), 121–170 (2014). https://doi.org/10.1162/COLI_a_00168
17. Matteucci, I., Petrocchi, M., Sbodio, M.L.: CNL4DSA: a controlled natural language for data sharing agreements. In: Proceedings of the 2010 ACM SAC
18. McHugh, M.L.: Interrater reliability: the kappa statistic. Biochemia Medica **22**(3), 276–282 (2012). https://doi.org/10.11613/BM.2012.031

19. Shi, L., Chadwick, D.W.: A controlled natural language interface for authoring access control policies. In: Proceedings of the 2011 ACM SAC
20. Taniguchi, S., Nakajima, S., Michikata, T., Seike, H., Koshizuka, N.: DSPOL: A High-Level Language for Defining Data Policies in Data Spaces. IEEE Big Data
21. Tanoli, I.K., Pais, S.: Modeling Natural Language Policies into Controlled Natural Language: A Twitter Case Study. In: ICIMP 2020
22. Tanoli, I.K., Petrocchi, M., De Nicola, R.: Towards automatic translation of social network policies into controlled natural language, p. RCIS. (2018)
23. Velikii, D.: Template of data governance policy (07 2024). https://doi.org/10.13140/RG.2.2.16959.52649

MAPS-AI a Model-driven AI-Assisted Tool for IT Project Plan and Scope: Case Study and Evaluation

Megija Krista Milune(✉) and Oksana Nikiforova

Institute of Information Technologies, Riga Technical University, Riga, Latvia
{megija-krista.milune,oksana.nikiforova}@rtu.lv

Abstract. The Advanced IT project development practices are often incorrectly associated with the absence of formal documentation, despite the fact that clients typically require a complete IT project management plan with clearly specified requirements. The use of Generative Artificial Intelligence for automating project documentation introduces new opportunities, but also raises concerns regarding the quality, consistency, and completeness of generated results. This paper presents the first practical evaluation of MAPS-AI, which is a tool for generating IT project artefacts, including project backlog, user stories, roadmap, and project scope information, which are subsequently combined into a draft IT Project Management Plan aligned with ISO, PMBOK, and SWEBOK standards. The evaluation is conducted using a Case Study Research Methodology within an educational project context. The quality of the generated documentation is assessed through expert evaluation and by comparison against manually created project document. The results demonstrate the feasibility of the developed tool and highlight its potential to support standards-compliant IT project documentation.

Keywords: IT Project Management · IT Project Plan · Evaluation of GenAI · Model Transformations · Large Language Model

1 Introduction

When working on IT projects, project teams often mistakenly associate Agile methodology with the absence of documentation [1]. Most clients require project processes to be documented, requirements to be clearly specified, and a complete IT project plan to be prepared. During the documentation process, project teams frequently encounter challenges, such as incorrect or incomplete information in documented requirements, overlooking important sections, and creating misunderstandings within the team. In addition, poorly defined requirements, project budgets, deadlines, or risks can later lead to serious issues and may even result in project failure [2].

Currently generative AI is being adopted across many domains and project phases. Researchers are actively exploring its use for different tasks. The authors address the problem described above by investigating how to automate the generation of IT project documentation. The first results of this research were presented in [3]. As a result,

A. Bernasconi et al. (Eds.): RCIS 2026, LNBIP 597, pp. 54–65, 2027.
https://doi.org/10.1007/978-3-032-35043-5_5

MAPS-AI was developed as a model-driven, AI-assisted tool that generates project artefacts (project backlog, user stories, roadmap, and project scope information) [4].

These artefacts are combined into an artefact pool, which provides the context for producing a draft IT project management plan. The resulting draft is a structured document that describes project processes and is aligned with recognized project management and software engineering standards, such as ISO [5], PMBOK [6], and SWEBOK [7]. The Generative AI component supports the project manager by helping generate descriptive requirements and textual content, while reducing the risk that important information is missing during documentation.

This paper presents the first practical evaluation of the MAPS-AI tool using a case study research methodology. The quality of outputs produced with GenAI remains one of the main challenges in this area. Many researchers have proposed several approaches for evaluating AI-generated content. In this study, two approaches are applied: (1) expert evaluation of the generated output and (2) comparison against a manually created artefact. MAPS-AI is evaluated using both approaches.

The paper is structured as follows: Sect. 2 describes the background of the MAPS-AI tool's development and reviews related work. Section 3 presents the research design and the case study context. Section 4 discusses the evaluation of MAPS-AI. Section 5 presents discussion on MAPS-AI evaluation. And finally, Sect. 6 provides conclusions and outlines directions for future work.

2 Background and Related Work

Project documentation plays a crucial role in IT projects [8]. Clearly defined requirements, user stories, acceptance criteria, and deadlines help project teams work efficiently during development and deliver a product that meets stakeholder expectations [9]. However, producing precise and detailed documentation requires significant time and effort from the project manager, stakeholders, clients, and whole project team [1].

To address this issue, teams use GenAI tools (Copilot [10], ChatGPT [11], Gemini [12]) to support documentation by saving time and generating text. In practice GenAI tools are mainly used for small fragments (user stories). Tools struggle with large documents (50–100 pages) and may produce hallucinated or inconsistent content [13].

A possible solution is to constrain GenAI through defined rules and structured inputs. In our previous research, we developed MAPS-AI [4], a model-driven, AI-assisted tool for generating IT project plan and scope documentation. The tool starts with model-driven transformations. From a project business process model provided by the user, MAPS-AI generates project backlog and user stories with descriptions and acceptance criteria. It also produces a project roadmap by prioritizing user stories and estimating their start and end dates, and it generates project scope information. These artefacts are first generated by the tool and then reviewed by the project manager, who can edit and approve the results. After the artefacts are finalized, MAPS-AI generates an IT project management plan. The user provides a file that defines the plan structure in as much detail as possible, and the GenAI component generates the textual content for the required sections using the approved artefacts as input.

In this way, the GenAI component and prompt is constrained by rules and detailed context [14]. However, the process still requires human review at each step. This means that IT project documentation generation cannot yet be fully automated while also ensuring a trustworthy outcome. MAPS-AI currently functions as an assistant to the project manager.

However, there is extensive discussion between researchers about how to evaluate and validate GenAI-produced content [15, 16]. Prior studies suggest that it is difficult to measure the accuracy and precision of generated text using quantitative metrics, and that human judgment is often required [17]. As discussed in [18], a common approach is to involve domain experts in the evaluation process. Experts can review the generated content, assess whether it aligns with the project context, verify stated facts where possible, and determine whether the output is usable. They can also assess whether using GenAI for project documentation genuinely saves time for the project team or instead requires additional effort for correction and content refinement.

3 Research Design and Case Study Context

MAPS-AI applies Generative AI to support the generation of project artefacts and, later, the IT Project Scope and Plan [4]. This solution is implemented through five engines that handle the main processing tasks described in Sect. 4.2.

MAPS-AI, the object under evaluation, is developed using the Design Science Research Methodology (DSR) [19], including Relevance, Design, and Rigor phases. In accordance with DSR the evaluation of the developed artefacts should be done through practical application and empirical evidence. In the paper such methodology requirement is addressed through a case study, enabling the evaluation of the tool in a realistic context.

The problem domain used for the case study is IT project management plan documentation for a Carbon Emission Reporting Solution (CERS) for Small and Medium Enterprises (SMEs). The documentation used as the project context is created within a university course titled IT Project Management Planning, which requires students to develop an IT project plan for an industry-provided problem domain that includes the project budget, timeline, and other project plan components defined by the course.

In this study, the tool is expected to replicate the main student activity of creating an IT project plan document by using the same problem domain process model as input for the MAPS-AI tool and generating a comparable project plan automatically.

3.1 Case Study Research Methodology

Following a case study research methodology, this study evaluates MAPS-AI by applying it to a project planning context, while analyzing the documents and artefacts the tool generates. The main goal is to assess whether MAPS-AI can produce an IT Project Management Plan draft that is comparable to a manually created plan and whether the draft is usable as a starting point for a project manager.

The unit of analysis is the IT Project Management Plan draft generated by MAPS-AI for the selected problem domain. Analysis considers the main inputs and generated artefacts that contribute to the final plan, because these outputs form the contextual artefact pool used for content generation.

The case is defined by a single problem domain, CERS for SMEs, and a single required plan structure defined by the course template. The same problem-domain process model is used as the tool input, and the same template defines the target plan structure, enabling a direct comparison between the tool-generated plan and a manually created document produced in the same context.

Multiple data sources support the evaluation. These include the process model used as input, the artefacts generated by MAPS-AI, and the final plan draft. A manually created project plan for the same case is used as a reference for comparison. In addition, expert feedback is collected to assess usability and quality before the plan can be accepted as a draft for stakeholders.

The procedure follows a clear sequence. First, the problem-domain business process model and the plan structure are prepared. Second, MAPS-AI is executed to generate project artefacts such as a project backlog, user stories, acceptance criteria, a roadmap, and project scope information, which then supports the generation of the final plan draft. Third, the output is reviewed against the required structure. Finally, the plan is evaluated in two ways by comparing it with the manually created plan and by collecting expert assessments of document quality.

The analysis examines whether the generated plan follows the required structure, covers key plan sections, and reflects the case context. To support quality, identical input process model and the same template are used during tool execution, and the evaluation procedure is documented. The main limitation is that the evaluation is based on a single case in an educational setting.

3.2 Case Study Design and Data Sources

IT project plan structure can be defined using standards, organizational practices, or academic guidelines. MAPS-AI is designed in a way that allows any project plan structure to be used for the generation of project documentation. The structural elements of a plan can be provided to the tool, and the quality of the generated result does not depend on the chosen structure. The key requirement for better generated results is that the plan structure is defined in detail so that the rules and constraints are better set for the GenAI and the output can be in better quality.

MAPS-AI uses knowledge about the problem domain. This domain-specific content forms the dynamic part of the prompts and directly affects the quality of the AI-generated result. The content is obtained through model transformations, based on conceptual metamodeling of artefacts used in the initial phases of IT projects.

The study program, used for the case study, has established IT Project plan and it serves as the input data and structure for the experiments conducted with the MAPS-AI tool and for the generation of the IT project plan content. The plan used for content generation consists of nine first-level sections, and each section is structured into up to three hierarchical subsection levels [20].

Related work on the validation of AI-generated outputs indicates that current validation approaches are still limited. The literature shows that a common method relies on expert involvement [21]. Due to the lack of standardized automated validation for

GenAI-generated artefacts, human-centered evaluation remains the dominant approach. In this paper domain experts are provided with AI-generated text and asked to evaluate its quality, correctness and whether the generated plan complies with the required structure and rules.

3.3 Use Case Definition and Context

The problem domain is a system that enables the company to report its carbon emissions based on invoice data. It supports the accounting, accumulation, and storage of sustainability data. CERS system also collects and summarizes this information to prepare the required reports and statements on the company's greenhouse gas emission consumption for a defined reporting period.

As illustrated in Problem domain's business process model [20] the use of the CERS system begins with user authentication through the existing online banking system. After accessing CERS, three main options are available.

First, the company accountant can select the invoice processing function. This allows the accountant to upload invoice documents, after which the system extracts and displays the scanned invoice data. The accountant can review the data, make corrections if necessary, and submit it for confirmation. The company manager then verifies and confirms the accuracy of the data. Once confirmed, business users can view the system-calculated emission data and save the results.

Second, the accountant chooses the manual data entry option, which enables emission-related data to be entered manually. Data is reviewed and confirmed by the company manager, after business users can view and save the calculated emission data.

Third, the accountant can select the emission report menu, where previously recorded data is displayed. The accountant can define the reporting period, select the data content, choose visualization options, and generate a preview of the emission report. Finally, the company manager can download the finalized emission data report.

4 MAPS-AI Evaluation

To evaluate the IT project plan and scope generated by MAPS-AI, the student-created project plan documentation is used as a baseline and compared with the MAPS-AI-generated plan. The expert group consisted of five participants with both industry and academic backgrounds. Three experts are IT professionals from industry, including two IT project managers with approximately 2 and 5 years of experience, and one senior IT project manager/technical lead with over 10 years of experience. Two of the experts are academic professionals - university professors specializing in the software development life cycle and IT project management. The inclusion of experts with varying levels of experience ensures that the evaluation combines practical industry expertise with an academic and methodological perspective.

A second evaluation approach is a comparative method. The artefact being under evaluation is produced in two ways independently: the one by a human expert and by a Generative AI tool. Analysis is conducted by an IT project management professional, who acted as an independent evaluator. The two results are then compared to identify differences in the text, while comparing the quality and contextual alignment.

Across all these evaluation methods, an open question remains regarding the trustworthiness and quality of GenAI-produced outputs. It is still unclear to what extent AI-generated results can be trusted and whether the generated results are qualitative. The evaluation of GenAI products remains an open research area, with researchers continuing to explore approaches that can ensure reliable and trustful outputs.

Metamodel-based evaluation is another approach for validating GenAI-produced outputs. If a metamodel exists for the evaluated artefact, the generated results can be checked against this metamodel. During the development of the MAPS-AI tool, a conceptual model of artefacts used in the early phases of an IT project was defined [22]. This model was applied in all later model transformations and prompt designs. It was based on established IT engineering methods and standards. Based on this, it can be assumed that if a GenAI solution is built on metamodeling principles, the artefacts it produces will follow the same metamodel. However, in this paper, this assumption is only a hypothesis. The authors plan to examine it further in future research.

4.1 Evaluation Dimensions

The evaluation of MAPS-AI is performed using two methods. First, experts assess the quality and usability of the GenAI-generated IT Project Management Plan draft by evaluating the nine first-level plan sections defined in the project plan template [20]. Second, the generated plan is compared with a student-generated plan created for the same problem domain and based on the same template.

To support both evaluation methods, a set of evaluation dimensions is defined. These dimensions represent key quality aspects expected from an IT project plan and address typical risks of GenAI usage, such as missing information, overly generic content, hallucinations or incorrect and out-of-context statements.

The evaluation dimensions focus on whether the generated content complies with the required plan structure, whether it is logically structured and readable, and whether it fits the specific problem-domain context and project goals. In addition, the evaluation considers whether the generated plan can be used in practice as a starting draft for a project manager. In the expert-based evaluation, each of the nine sections is assessed using a decision, where experts indicate whether the section complies or does not comply with the expected plan content and typical project planning practices.

4.2 MAPS-AI Transformation Chain and Generated Document

The MAPS-AI transformation chain is an automated workflow that starts with a business process model. First, the model is transformed automatically into a UML use case diagram using model transformations. The use cases are then transformed into project backlog. The Project Artefacts Creation Engine produces user stories using a predefined template and GenAI enriches them with context and acceptance criteria. The Project Artefacts Refinement Engine prioritizes user stories, assigns start and end dates and generates a project roadmap. Then project scope information is generated, including purpose, background, deliverables, resources, operations and support, safety, security, risks, and stakeholders.

After that, the Prompt Generation Engine creates structured prompts for each project plan section based on the artefact pool and section rules aligned with standards (ISO, PMBOK, and SWEBOK). Meta-prompting adds a control layer to keep the generated content consistent with these standards while also supporting Agile practices. The prompts include relevant project artefacts and section-specific instructions to improve output precision. A meta-level LLM generates section prompts that encode the document structure, compliance rules, and references to standards, while integrating project semantics: background, user story constraints, and roadmap details.

The prompts guide the IT Project Document Generation Engine to generate section texts a structured project management plan draft. Finally, the Document Formatting Engine applies formatting rules to produce a clear and readable IT Project Management Plan with sections, subsections, headings and descriptions for each section.

4.3 Expert Evaluation

The first evaluation method applies expert-based assessment to determine whether the generated IT Project Plan complies with the required plan structure and problem domain context, and whether the produced content is useful and suitable for further use by a project manager when completing and improving the plan. Five experts evaluate each of the nine plan sections based on the previously defined project plan structure. The evaluation results are presented in Table 1.

Table 1. Expert evaluation on GenAI created IT Project plan.

Sec No	Expert Nr.1	Expert Nr.2	Expert Nr.3	Expert Nr.4	Expert Nr.5
1	Complies	Complies	Complies	Complies	Complies
2	Does not comply	Does not comply	Complies	Complies	Complies
3	Complies	Does not comply	Does not comply	Complies	Complies
4	Complies	Complies	Complies	Complies	Complies
5	Complies	Does not comply	Complies	Does not comply	Complies
6	Does not comply	Does not comply	Complies	Does not comply	Does not comply
7	Complies	Complies	Complies	Complies	Complies
8	Complies	Does not comply	Does not comply	Complies	Does not comply
9	Complies	Complies	Complies	Complies	Complies

Expert evaluation shows that generated content complies with structure and meets expectations for Sect. 1: Project Overview, Sect. 4: Compliance with Task Requirements, Sect. 7: Project Implementation Plan, Sect. 9: Information about the Project Team Members. For these sections, experts agreed that the generated text is suitable and that they would be willing to use MAPS-AI to support content creation.

For Sect. 2: Proposed Solution, Sect. 3: Technical Solution, Sect. 5: Project Schedule and Workload Estimation, the results are mostly positive but less consistent. In each of these sections, three expert ratings indicate compliance, two ratings indicate non-compliance, suggesting that the generated content is generally acceptable but may require additional refinement to match the expected level of detail and context.

The weakest results are in Sect. 6: Financial Proposal and Sect. 8: Additional Proposal. For Sect. 8: Additional Proposal, three experts indicate non-compliance, showing mixed agreement. Section 6: Financial Proposal receives the lowest evaluation, with four experts stating that the content does not comply, indicating that this section needs the most improvement to align with the required structure.

The second evaluation method is a comparative analysis between the GenAI-generated IT project management plan and the student-generated plan for the same problem domain illustrated in Table 2. The evaluator compared both documents. The comparison was performed at the 9 section levels. For each section, analyzing the content in terms of structure, level of detail, and alignment with the problem domain. Based on this assessment, the evaluator identified the main differences.

The comparison in Table 2 shows that MAPS-AI performs best in sections where the output is mainly descriptive. In Sect. 1: Project Overview, Sect. 4: Compliance with Task Requirements, and Sect. 7: Project Implementation Plan, the GenAI-generated text is comparable to the student-generated plan in terms of coverage and structure. These sections rely on explaining goals, requirements, and planned activities, and the tool can generate content that matches the expectations.

Differences appear in sections that require visual artefacts or specific values. In Sect. 2: Proposed Solution and Sect. 3: Technical Solution, the student plan includes prototypes and required diagrams, while the GenAI plan does not generate visual items. However, the GenAI output describes the key elements that would be shown in the diagrams, such as user roles, flows, and main system components. This means the text can support the project team when preparing diagrams manually, even though the diagrams are missing from the tool generated project documentation.

Section 5: Project Schedule and Workload Estimation show similar results. The student plan provides a detailed schedule with effort estimates, duration in workdays, timeframes, and a clear plan defined by phases, while the GenAI plan provides workload only as percentages and does not include the visualization of roadmap that was generated earlier in the process of artefact generation.

The largest gap is observed in cost-related sections. In Sect. 6: Financial Proposal, the GenAI output defines cost categories and items, but it does not provide cost values. This is consistent with the prompt constraint not to invent new information. As a result, the GenAI plan functions more as a cost framework, where the project manager adds the values. The same applies in Sect. 8: Additional Proposal. The GenAI plan defines

Table 2. Comparative analysis for AI-Generated project plan.

Sec No	GenAI-Generated Plan	Student-Generated Plan
1	Provides a detailed project overview, including the problem statement and project objectives	Provides a detailed project overview, including the problem statement and project objectives
2	Describes system users and use cases. No visual models are included, however, the text explains the key elements captured in such diagrams. The listed use cases are consistent with artefacts generated earlier by the tool	Includes system prototypes and required visual models, such as user interface flow diagrams, activity diagrams, and a context diagram
3	Describes the intended technical architecture in text form, without an architecture diagram. The section outlines the main system components at a high level	Presents a defined technical architecture and specifies concrete solution components. Alternative components are discussed and compared to justify the selected approach
4	Lists requirements and describes how they will be addressed. Includes an execution approach aligned with the plan structure	Lists requirements and describes how they will be addressed. Includes an execution approach aligned with the plan structure
5	Provides workload information as percentages. The previously generated roadmap is not included. Team composition and deliverables are listed	Provides a detailed schedule including estimated effort, duration in workdays, and timeframe. Activities are described per project phase, deliverables are specified, and the project team is presented visually
6	Identifies cost categories, items and detailed descriptions but does not provide cost values	Provides cost values and shares of the total budget. Includes a detailed cost estimate with defined cost items
7	Presents a detailed implementation plan covering the steps required by the plan template	Presents a detailed implementation plan covering the steps required by the plan template
8	Describes additional cost considerations but does not provide a specific additional budget amount	Proposes an additional budget; however, the cost breakdown is not fully transparent across all items
9	Provides detailed descriptions of project team members, including experience and expected involvement in project activities	Lists project team members and their prior experience but provides limited detail on their roles and involvement during project execution. Lists project team members and their prior experience but provides limited detail on their roles and involvement during project execution

what should be considered in additional costs, while the student plan proposes a budget amount, even though its breakdown is not fully transparent.

Finally, Sect. 9: Information about the Project Team Members shows a difference in level of detail. The GenAI plan provides more explicit descriptions of team members' experience and involvement, while the student plan mainly lists team members and their prior experience with limited explanation of roles in execution.

5 Discussion

Students develop project documentation over an entire semester. In industry, project managers also build this knowledge over a long period of practice. If GenAI can provide similar structured information, project managers can use it as practical support during planning. Therefore, MAPS-AI should be seen as an assistant that helps the project manager create and structure documentation, while the generated content still requires human review before it can be trusted and approved.

The expert evaluation results in Table 1 shows that the sections "does not comply" are mainly those where the expected output includes visualized diagrams. Although in the project plan generation section MAPS-AI does not generate diagrams, the GenAI component can still provide structured textual descriptions of the diagram elements, such as actors, flows, and key entities. The generated information can be used by the project team as input when creating the required diagrams manually.

The comparison in Table 2 shows that MAPS-AI is effective for generating well-structured and descriptive content. It can produce sections comparable to student work when the content is template-driven and based on available context. The main limitations appear in sections that require visual diagrams or specific values such as costs and detailed schedules.

6 Conclusion and Future Work

This paper presented the first practical evaluation of MAPS-AI, a model-driven, AI-assisted tool for generating IT project planning artefacts and producing a draft IT Project Management Plan and Scope. The evaluation was conducted as a single-case study in the CERS for SMEs problem domain. Two methods for evaluation were applied: expert-based assessment of nine plan sections and a comparative analysis against a student-generated plan created for the same template and context.

The results indicate that MAPS-AI is most effective in sections where the required output is mainly descriptive and can be derived from the provided problem-domain context and generated artefacts. Both the expert evaluation and the comparison with the student plan show that the tool can produce text that follows the required structure for sections (Project Overview, Compliance with Task Requirements, Project Implementation Plan). Findings support the view that MAPS-AI can be used as a practical assistant for project managers by accelerating documentation work.

At the same time, the evaluation highlights clear limitations. Sections that require visual artefacts (prototypes, diagrams), were assessed less positively because MAPS-AI currently generates textual descriptions rather than visual models. In addition, sections that require specific values, especially cost amount and detailed schedule information, remain incomplete when the tool is constrained not to invent missing data. In these cases, MAPS-AI still provides value by structuring the content and identifying what should be included, but human input is required to complete and validate the plan.

Future work will focus on extending the evaluation across more problem domains and plan templates. Improvements will be made to the integration of generated artefacts into the final project plan, by ensuring that roadmap and scheduling outputs are included in the document. Further research will address the systematic handling of quantitative inputs by enabling user-provided parameters for cost and schedule estimation. Moreover, future work will include the evaluation of multiple GenAI models to identify which produce the highest-quality outputs. Finally, the prompt engineering process will be refined to reduce generic content and minimize the risk of hallucinations.

Acknowledgments. The research leading to these results has been funded by Riga Technical University 2025/2026 Project for Strengthening Scientific Personnel Capacity Nr. ZM-2026/1" Model Transformation-Based Artificial Intelligence-Assisted Management of the Initial Phase of IT Projects for Achieving Scientific Objectives and Results.

Disclosure of Interests. The authors have no competing interests to declare that are relevant to the content of this article. Authors have received student-developed IT Project Management Plan materials for this research from the university course IT Project Management Planning, led by lecturer L. Jokste.

References

1. Stettina, C.J., Heijstek, W.: Necessary and neglected? An empirical study of internal documentation in agile software development teams. In: Proceedings of the 29th ACM International Conference on Design of Communication (SIGDOC 2011), pp. 159–166. ACM, New York (2011). https://doi.org/10.1145/2038476.2038509
2. Akhavan Tabassi, A., Bryde, D., Mustafa Kamal, E., Dowson, J.: Challenges for project management in the 21st century. In: Proceedings of the 7th International Conference on Built Environment in Developing Countries (ICBEDC) (2019). https://doi.org/10.15405/epms.2019.12.63
3. Nikiforova, O., Grabis, J., Pastor, Ó., Babris, K., Miļūne, M.K., Bobkovs, R.: Model-based methodology for development of IT project management plan and scope using artificial intelligence: project in progress. In: Research Challenges in Information Science (RCIS 2025). CEUR Workshop Proceedings, vol. 3987, pp. 1–7 (2025). https://ceur-ws.org/Vol-3987/paper7.pdf
4. Nikiforova, O., Bobkovs, R., Miļūne, M.K., Babris, K., Pastor, Ó., Grabis, J.: MAPS-AI – A tool for AI-assisted model-driven generation of IT project plan and scope. In: Scanniello, G., Lenarduzzi, V., Romano, S., Vegas, S., Francese, R. (eds.) Product-Focused Software Process Improvement (PROFES 2025). LNCS, vol. 16361, pp. 1–16. Springer, Cham (2026). https://doi.org/10.1007/978-3-032-12089-2_34
5. International Organization for Standardization: Project, programme and portfolio management – guidance on project management. ISO 21502:2020. ISO, Geneva (2020)
6. Project management institute: A Guide to the Project Management Body of Knowledge (PMBOK® Guide), 7th edn. PMI, Newtown Square (2021)

7. Bourque, P., Fairley, R.E. (eds.): Guide to the software engineering body of knowledge (SWEBOK Guide), version 4.0. IEEE Computer Society (2024)
8. Jarzębowicz, A., Weichbroth, P.: A qualitative study on non-functional requirements in agile software development. IEEE Access **9**, 40458–40475 (2021). https://doi.org/10.1109/ACCESS.2021.3064424
9. Ferreira, A., Silva, A., Paiva, A.: Towards the art of writing agile requirements with user stories, acceptance criteria, and related constructs. In: Proceedings of the International Conference, pp. 477–484 (2022). https://doi.org/10.5220/0011082000003176
10. Microsoft: Copilot. https://copilot.microsoft.com/. Accessed 26 Mar 2026
11. OpenAI: ChatGPT. https://chat.openai.com/. Accessed 26 Mar 2026
12. Google: Gemini. https://gemini.google.com/. Accessed 26 Mar 2026
13. Huang, L., et al.: A survey on hallucination in large language models: principles, taxonomy, challenges, and open questions. ACM Trans. Inf. Syst. **43**(2), 42 (2025). https://doi.org/10.1145/3703155
14. White, J., et al.: A prompt pattern catalog to enhance prompt engineering with ChatGPT. arXiv preprint arXiv:2302.11382 (2023). https://doi.org/10.48550/arXiv.2302.11382
15. Belzak, W.C., Naismith, B., Burstein, J.: Ensuring fairness of human- and AI-generated test items. In: Wang, N., Rebolledo-Mendez, G., Dimitrova, V., Matsuda, N., Santos, O.C. (eds. Artificial Intelligence in Education. Springer, Cham, pp. 701–707 (2023). https://doi.org/10.1007/978-3-031-36336-8_108
16. Barcaui, A., Monat, A.: Who is better in project planning? Generative artificial intelligence or project managers? Project Leadership Soc. **4**, 100101 (2023). https://doi.org/10.1016/j.plas.2023.100101
17. Zhu, T., Weissburg, I., Zhang, K., Wang, W.Y.: Human bias in the face of AI: the role of human judgement in AI-generated text evaluation. arXiv preprint arXiv:2410.03723 (2024). https://doi.org/10.48550/arXiv.2410.03723
18. Ahmed, A., Kerr, E., O'Malley, A.: Quality assurance and validity of AI-generated single best answer questions. BMC Med. Educ. **25**, 300 (2025). https://doi.org/10.1186/s12909-025-06881-w
19. Bichler, M.: Design science in information systems research. Wirtschaftsinformatik **48**(2), 133–135 (2006). https://doi.org/10.1007/s11576-006-0028-8
20. Miļūne, M.K., Nikiforova, O.: Data package on GenAI-supported IT project management artefacts. Google Drive (2026). https://drive.google.com/drive/folders/1U1-3bjpIc2dLoCqj0ERcGGQBEByjZyJP
21. Treviño, M., Arias-Carrión, O.: Artificial intelligence in systematic reviews: overcoming reproducibility, bias and validation challenges. Preprints 2025061895 (2025). https://doi.org/10.20944/preprints202506.1895.v1
22. Nikiforova, O., Babris, K., Miļūne, M.K., Tanguturi, N., Pastor, Ó.: Key artefacts in the initial phases of IT project management: systematic mapping study. In: Proceedings of the 20th International Conference on Evaluation of Novel Approaches to Software Engineering (ENASE), pp. 773–781. SciTePress (2025). https://doi.org/10.5220/0013471000003928

Low-Code Platforms Adoption: A Motivational Adoption Process Perspective

Elisa De Coster[1(✉)], Anthony Simonofski[1], and Monique Snoeck[2]

[1] Université de Namur – NaDI, Rue Grandgagnage 21, 5000 Namur, Belgium
elisa.decoster@unamur.be

[2] Katholieke Universiteit Leuven – LIRIS, Oude Markt 13, 3000 Leuven, Belgium

Abstract. The growing demand for digital transformation in organizations, coupled with a shortage of skilled-IT workers, presents a challenge that can be addressed through the utilization of low-code platforms (LCPs). While existing literature examines organizational drivers of LCPs adoption, our human-centered study shifts the focus toward the individu, investigating motivations of non-technical users adopting LCPs on their own initiative for professional needs. Through qualitative interviews and by refinement of Keller's ARCS Motivational Model, an adoption framework emerged in the form of a five-stage motivational adoption process (interest, need, ability, learning, reinforcement). By effectively articulating key motivational factors in the adoption of LCPs by citizen developers in this cohesive framework, it can help organizations better foster low-code initiatives.

Keywords: Citizen Developers · Low-code · Adoption · Motivation

1 Introduction

The rapid pace of digital transformation has created a gap, where the demand for software exceeds the supply of professional developers. Low-code platforms (LCPs) have emerged as a solution, specifically designed to bridge the gap between technical complexity and usability by prioritizing understandability. These platforms based on visual interfaces make complex system logic approachable and transparent to non-IT professionals, hence empowering them to build functional software solutions without the traditional need of coding [24]. Those non-expert users building software solutions with LCPs are referred to as "Citizen Developers" (CDs) in the literature [16]. Nowadays, CDs could consider using LLM-based platforms instead, but it brings along significant risks [26] where LCPs are safer given their deterministic approach to code generation.

While high understandability and visual interfaces of such platforms effectively reduce the technical barrier to entry, it does not inherently guarantee adoption. A system's design dictates how easily a human can comprehend how it works, but it is the user's intrinsic motivation that dictates his willingness to explore it. Existing literature has mainly focused on organizational drivers and required skills, leaving the motivational landscape largely unexplored [2]. Neglecting employees' motivations to become

A. Bernasconi et al. (Eds.): RCIS 2026, LNBIP 597, pp. 66–77, 2027.
https://doi.org/10.1007/978-3-032-35043-5_6

CDs constitutes a theoretical and practical blind spot. It prevents organizations from understanding why some employees take on this time-consuming role, and therefore potentially causes failed attempts of fostering low-code initiatives.

We conducted nine in-depth semi-structured interviews with CDs from different organizations. The relevance of these interviewees lies in the profile of the respondents. Despite having little to no prior experience in software development, these participants working in very diverse fields (e.g., pharmaceuticals) learned to use LCPs to develop a solution that is still in use today in their organizations, demonstrating its quality. Those interviews hence constitute a valuable case study for analysing the practical adoption of such tools in real professional settings, gaining insights into the motivational factors fostering the learning journeys of non-experts using LCPs. Using Keller's ARCS Motivational Model (AMM) as a lens [14], we explore how factors like Attention, Relevance, Confidence and Satisfaction foster adoption. This paper contributes to a novel motivational adoption process structured across five stages (Interest, Need, Ability, Learning, and Reinforcement) offering a comprehensive look at the journeys of CDs.

2 Background

2.1 Organizational Factors

The organizational culture influences the emergence of CDs within a company. Research by [10] highlights that factors such as active Top Management support and centralized IT Governance decision structures can contribute to the adoption of LCPs. The adoption can be further facilitated by a corporate culture with strong change management. Best practices for organizations to benefit from low-code initiatives shed light on the importance of social interaction during the adoption process, as it enables CDs to obtain guidance, while fostering a feeling of belonging to the community [7]. [22] articulated 21 organizational capabilities into a coherent framework to help organizations successfully implement low-code solutions. For a more comprehensive exploration of the organizational perspective, one may refer to [12] which distinguished 13 drivers and 19 inhibitors of the adoption of LCPs in organizations.

2.2 Individual Factors

When trying to understand potential motivations of CDs to take on this time-consuming role on top of their core responsibilities, [21] found out that perceived usefulness positively influences the intention to use. They also showed that the level of perceived usefulness was partially explained by job relevance and by the status or influence provided by the use of LCPs. It appeared that basic programming knowledge seems needed to correctly wield those tools and that the learning curve is steep, even with the decrease of the hand-coding component [1]. As for the character traits of those CDs, curiosity, tech enthusiast, willingness to improve and to learn but also critical thinking skills are qualities often cited [2]. [18] showed that benefiting from feedback increased motivation to learn LCPs. Both autonomy in the process and the perceived added value for their future job research were viewed as incentives.

2.3 Research Gaps

Recent studies take place in environment where the decision to use LCPs is not made on a voluntary basis, for example focusing on how students interact with those tools in a dedicated course and the impact on their metacognitive skills [19, 25]. Those studies, however, fail to consider real-life settings where becoming a CD is a voluntary decision, concept referred to as "community-based citizen development" [3].

Literature highlights the need for intrinsic qualities such as adaptability and proactivity [9]. [20] mention both individual skills and organizational factors as influencing the effectiveness of CDs using LCPs. These studies directly dive into the skills needed, failing to address what motivates CDs. Exploring their learning journey and motivations is decisive to discover how this decision can be replicated and/or fostered.

The few studies that mention motivational factors are either business reports or focus on a few intrinsic qualities a CD should preferably display. Both thus lack the necessary granularity to analyze what factors triggered and fostered those individual learning journeys and led to a sustained adoption [2]. Furthermore, only [13] and [21] provide a theory-based approach to the topic, but the former does not solely research motivational factors of CDs and the latter demonstrates statistically only the impact of perceived usefulness on the intention to use. [1] stands out as the sole study combining motivational factors with a rigorous methodology grounded in a multiple regression of survey data. While this constitutes a significant initial stride, the quantitative nature of the survey limits the discovery of novel motivations. The part of the study focusing on CDs principally demonstrates that the adoption of LCPs correlates with programming skills and knowledge about those tools, yet it falls short in delving into the underlying reasons and nuances behind this observation. Furthermore, attempts to establish correlations between the adoption of LCPs and other variables, such as performance, code quality, etc., proved inconclusive. Given the inconclusiveness of most identified motivational factors and the impossibility of exploring new ones given the quantitative methodology, there arises a need for a qualitative investigation to understand and uncover these motivations. As the lack of case-based theoretical study focused on the personal motivations of CDs constitutes our identified gap, the research question is the following: *What are the motivations of citizen developers adopting low-code platforms?*

3 Methodology

3.1 Theoretical Framing of the Study

To identify new motivational factors of CDs and to study those already identified, nine exploratory in-depth interviews were conducted with a panel of CDs [4]. To ensure the rigor and theoretical foundation of the study, those interviews were based on Keller's AMM [14]. It hinges on four main categories: attention, relevance, confidence and satisfaction that depict what motivates learners to start and sustain their study of a new subject [15]. While this model has never been used on technology adoption, it has been proven insightful in prior research in IT-related domains [6, 11], suggesting it holds promises for uncovering novel pathways within the field of IT adoption.

The AMM has been favored for several reasons. First, it is designed to investigate the learner's motivation in situations where learning is non-compulsory. Second, its objective is to explore and understand how motivation is fostered and sustained. And third, some of the subcategories of the model (such as the importance of goal orientation) have already been mentioned in the literature [13, 18] although not identified as such. This suggests an alignment of the model with previous findings, making it encouraging for further exploration of CDs' motivations.

3.2 Data Collection and Analysis

The selection criteria for respondents were straightforward: they must have no formal professional experience in software development and must have successfully deployed at least one functional LCP solution within a professional context. Both criteria were intended to retain individuals who were not merely curious about these platforms but had acquired sufficient experience in their learning process to overcome inherent obstacles and apply this acquired knowledge. The demographic and educational background of our nine interviewees can be found in Table 1.

The interview guide for the semi-structured in-depth interviews consisted in three main points: the respondents' background both regarding their prior knowledge of coding and their project(s) deployed thanks to LCPs, the subcategories of the AMM, and supplementary motivations that were already identified in the literature. The respondents were selected employing theoretical sampling approach [5].

For the data analysis, the transcripts of those interviews were analysed according to the thematic content analysis method using open coding where relevant parts of transcripts were associated to self-defined codes, hence retaining the possibility of encountering motivational factors beyond those of the theoretical framework. To assist this analysis and ensure a rigorous approach, the Taguette software was utilized, notably facilitating the tracking, and revisiting of code definitions. This first phase was iterative and left us with extracts being categorized into themes. In the second phase, deductive axial coding was performed [27] to uncover patterns between the different themes and classify those into the existing categories of the AMM. In the final phase, the structure of the proposed five-steps model was empirically derived by extracting causal markers and precedence relationships from the interviews' transcripts, which were then synthesized into a graph and clustered.

4 Results

4.1 Identified Motivational Factors

The qualitative analysis of the interviews led to the identification of 34 motivational factors (see Table 2; letters refer to the respondents identified in Table 1), hence providing qualitative insights along with illustrations extracted from the interviews. Among those factors, only two could not be mapped to an existing subcategory of the AMM, namely initial curiosity and initial confidence, but only to a meta-category (respectively Attention and Confidence).

Table 1. Profile of the respondents.

	Gender	Age	Country	Highest diploma	Job	LCPs
A	M	[40;49]	BE	Master's in economics	Prison director	Glide, Bubble
AC	F	[30;39]	FR	PhD in pharmacy	Regulatory affairs manager	AirTable
B	M	[20;29]	BE	Master's in Business Analytics	Business process analyst	Jira, Confluence
C	M	[30;39]	AUT	Master's in Business Engineering	Process wizard focused on SBPM	Automation Anywhere, Power Automate, UIPath
H	M	[20;29]	US	Bachelor's in Business Administration, Management & Operation	Solution specialist	Microsoft Power Platform
J	F	[20;29]	FR	Master's in Business Engineering	President of a student club	WebFlow
MA	M	[40;49]	BE	Teacher certification in Economics & postmaster in CS	CTO, teacher	AirTable, Odoo Studio, WebFlow, WordPress
MU	F	[50;59]	FR	Professional certificate in web programming	Trainer & self-employed low-code developer	Bubble
S	M	[30;39]	BE	Master's in Business Engineering	CTO, consultant in BI & ERP	KNIME

Table 2. Summary of the identified motivation factors, classified according to the meta categories of Keller's ARCS Motivational Model: Attention, Relevance, Confidence, Satisfaction.

	Category	Instantiation to LCPs	Concrete Example
Attention	**Perceptual Arousal**	Concreteness	"there's a final project, something concrete, this what I find most enjoyable in the learning process" (J)
		Wow Factor	"You quickly obtain a result, a little wow" (AC)
	Inquiry Arousal	Proactive	"that's when I did some digging of my own" (AC)
	Variability	Resources	"the documentation is very extensive" (B)
	Initial Curiosity		"I ask myself what is it, how could I use it? Because else, you hear about a lot of things everyday but without curiosity, you never dig deeper" (C)
Relevance	**Goal Orientation**	Gain Status	"when you gain the knowledge, you are seen in the company with this it, so that you can get the new position" (C)
		Create Opportunities	"it may bring me other opportunities" (MU)
		Financial Incentive	"I felt taken advantage of, I was getting underpaid" (H)
		Consultancy/ Advising	"I want to be able to help people and to feel specialized in a field, so I can give advice" (J)
		Mixing IT & Business	"it's very satisfying to combine IT and business, whereas if I were purely IT, I'd probably have been more into code or infrastructure implementation, and I wouldn't have had this vision of people's business needs" (B)
		Increase quality	"my goal is to get the process established (*thanks to LCPs*) so that they (*employees*) can be much more autonomous/ faster, and concentrate on what really adds value" (AC)
		Used by others	"for me, the most important thing is that the (*developed*) tool is used by others" (A)
	Motive Matching	IT could not deliver	"the ICT service is supposed to develop the solutions we need but is not able to do so" (A)
		Time saving	"the implementation of process was much faster" (C)
		Money saving	"I asked him (*the boss*) for 1/10 of the budget planned by the developers" (AC)
		Integration	"It connects a lot of software/applications together, so you can push information, connect a lot of tools together" (AC)
		Easier requirements gathering	"they (*IT*) don't understand the words I'm using, and they have their technical layer that I don't understand. So I think that doing it myself is a big gain" (AC)
		Dynamic	"I hang out in start-ups, and these are fast moving environments so you can't start thinking for 2 months" (MA)
		Visualization	"We love using KNIME because after each node you can see a preview of your table" (S)
		Perceived usefulness	"If you're not using such tools for getting rid of mundane tasks, you will lose at the end of the day" (C)
	Familiarity	Quick & Easy	"easy to use, super intuitive" (J)
		Experience with coding	"When you know a programming language, it's very easy to learn a new one" (S)

(*continued*)

Table 2. (*continued*)

		Related Background	"I did an additional year in IT" (MA)
Confidence	**Learning Requirements**	Practical Learning	"I think they (*tutorials*) are really essential" (A)
	Success Opportunities	Self-assessment	'I learned so much because of my errors" (H)
		Accomplishment	"We have this feeling that we really did it, it really was my website" (J)
		Practical cases	"the thing that kept me really motivated was those practical exercises" (C)
	Personal Control	Autonomy	"I am a self-taught learner" (A)
	Initial Confidence		"I told myself that I already had the programming logic, so it might be less complicated for me to learn" (MU)
Satisfaction	**Intrinsic Reinforcement**	Willingness to learn	"the main driving force is this taste for intellectual challenge" (A)
		Enjoyable	"I've never understood those who did it differently, because it's really great" (S)
	Extrinsic Reinforcement	Positive Feedback	"the appreciation you receive for the work you do" (B)
		Rewards	"I thought that after automating it, I would get more than a pat on the back, you know" (H)
	Equity	Worth it	"the knowledge I gained, it really was worth spending these 4-5 months with intense learning" (C)

4.2 Interconnected Motivations

The motivational factors underpinning LCPs adoption are interconnected, both within and across categories. Initial curiosity often serves as a catalyst for one's willingness to learn, while interviewees who harbour financial or status-oriented goals tend to place a greater emphasis on tangible rewards. Accomplishment is frequently intertwined with the wow factor as the respondent is impressed and proud of the result, but also with positive comments from others as it affirms the legitimacy of one's accomplishment.

Moreover, initial confidence is frequently strengthened by either a moderate to high level of familiarity, an abundance of available resources or a perceived ease of use. However, this initial confidence did not consistently translate into a seamless learning experience. The easiness with which one uses LCPs indeed emerged as an area of disagreement between the interviewees[1]. While some individuals reported a relatively straightforward learning journey, others described the process as difficult but worth it in the end. Some interviewees with a moderate to high level of familiarity with coding encountered unexpected challenges and declared that their learning process has been more intricate than anticipated. They attributed those challenges to differences in logic between conventional programming languages and low-code, and to the level of complexity of the chosen platform. Interviewees emphasized that coding skills may be beneficial, but what really matters is the programming logic and problem-solving abilities.

[1] https://zenodo.org/records/11612709.

The link between autonomy and proactivity is straightforward as the full autonomy of the learner forces them to be proactive to master those platforms. This autonomy empowers CDs to take charge of their learning journey and engage in self-assessment through practical cases. A successful self-assessment is a significant milestone that, in turn, gives rise to the accomplishment. While this autonomy permeates a greater involvement of the learners, it is nuanced with a desire for mentorship. One participant articulated the importance of belonging to a community, while others acknowledged the value of community primarily in terms of potential assistance in times of need.

4.3 Towards a Motivational Adoption Process

We were able to summarize the findings in a motivational adoption process that unveils five pivotal stages (interest, need, ability, learning, reinforcement), depicted in Fig. 1.

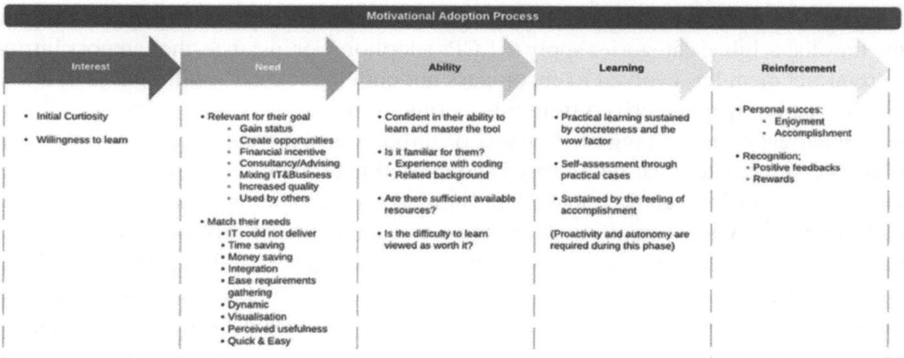

Fig. 1. Five stages of the motivational adoption process of LCPs by CDs.

Need. After the cultivation of curiosity and willingness to learn, individuals are propelled by a discernible rationale for engaging with these tools—a motivation grounded in their long-term objectives, or the fulfilment of immediate business needs.

Ability. Paramount to the adoption is the initial confidence, intricately linked with the availability of resources and perceived ease of use as depicted in Sect. 4.2. While familiarity is not an absolute prerequisite for initial confidence, the abundance of resources emerges as a crucial determinant as insufficient resources may instil doubt regarding the tool's reliability and the learner's ability to navigate its intricacies. Note that a minimal level of initial confidence is required for the individual to begin his learning journey, but the rest can be substituted by the "worth it" motivator, i.e. the idea that even if it appears difficult, it will be worth it, hence the importance of the rationale.

Learning. Learners commence their engagement with the chosen tool, propelled by inherent proactivity and autonomy, intrinsic to the learning process itself rather than constituting standalone motivational factors. Progress is sustained by tangible learning experiences (practical learning + concreteness), characterized by the swift manifestation of impressive outcomes (the wow factor) and punctuated by self-assessment through practical cases, fostering a sense of accomplishment and pride.

Reinforcement. Upon the completion of the learning process, remaining motivational factors such as enjoyment and positive feedback dictate the user's decision to continue engaging with those tools and sustain his learning. While the importance of rewards varies among participants, those valuing it the most exhibit a propensity to change environment to obtain them rather than discontinuing platform usage altogether.

5 Discussion

5.1 Theoretical Contributions

This study facilitated a nuanced comprehension of the overarching categories delineated within the AMM, applied to the context of LCPs. It entailed an in-depth comprehension and refinement of these categories in more granular subdivisions, leading to new insights into how the original categories apply to LCPs adoption. For instance, the category intrinsic reinforcement has evolved to encompass not only the learner's willingness to learn, but also the enjoyment derived from this learning process, whereas extrinsic reinforcement manifest either in tangible rewards or through positive comments. Additionally, it has been elucidated that familiarity is composed not only of past coding experience, but also of domain-specific knowledge. Similarly, perceptual arousal associated with the learning of LCPs is manifested through the concreteness of this process, potentially coupled with the learner's awe at the attainment of a result early in the process. Moreover, the success opportunities mentioned in the model are composed of self-assessment opportunities embedded within practical applications, fostering a sense of achievement and pride upon successful completion. Finally, a plethora of concrete examples pertaining to goal orientation and motive matching were revealed by the analysis. The interconnectedness of these categories enhances our understanding of this phenomenon and has primarily contributed to the development of a five-step motivational adoption process. This framework is corroborated by the antecedent relationships defined in the Technology Acceptance Model [17]. In this model, *perceived ease of use* and *perceived usefulness* influence the *attitude toward using*, constituting together the user motivation part of this model. Where it resembles our *ability* and *need* phases leading to the *learning* phase, our framework provides a granular and context-specific approach. Tailored specifically to the nuances of LCPs, our Motivational Adoption Process offers a comprehensive and theoretical framework for understanding the motivational factors leading to their adoption by CDs.

This contribution, originally grounded in Keller's AMM can be used to enhance the results of previous studies. Our findings are consistent with those of [1], which highlight the positive effect of prior programming skills and background on the adoption of those tools. However, our study reveals that the underlying reason for this effect lies in the initial confidence instilled in the learner, rather than simply the ease of use of the platform. Additionally, the correlation uncovered by [21] between perceived usefulness and the intention to use can be directly explained by the importance of having a clear goal in the adoption process. Moreover, their study partially unveiled the influence of goal orientation, as they established that the potential for higher status positively impacts the perceived usefulness, consequently influencing the intention to use. The lack of correlation with the perceived ease of use in [21] research can similarly be understood, as our study reveals that ease of use is not a mandatory factor; it can be offset by the learner's trust in his ability to succeed and his perception that the difficulty is worthwhile. [18] underscored the importance of students' beliefs in their own propensity to succeed, which is influenced by feedback and the availability of resources.

While certain motivational factors were already mentioned in the literature, our contribution lies in identifying them as such, in improving their comprehension, and in discovering new ones. Consistence with previous studies is a positive indicator, confirming the relevance of our perspective focused on individual motivations, and making our motivational adoption process a useful framework for articulating previously and newly found motivational factors influencing the adoption process of LCPs by CDs.

5.2 Practical Implications

The proposed motivational adoption process offers valuable insights for organizations. By gaining a thorough understanding of how CDs adopt LCPs, organizations can concentrate their efforts on decisive motivational factors and thus effectively support their employees through their learning journey. Organizational measures can now focus on what will really motivate employees to learn and adopt LCPs (e.g., cultivating their curiosity). Moreover, the organizational focus is often how to foster those LCPs initiatives, but their sustained use is just as important. Hopefully, our model will raise awareness on the importance of giving positive feedback and rewards, as otherwise employees may change jobs. More generally, our process could be the starting point of a framework for assessing the progress of LCPs initiatives in organizations. By identifying in which phase employees are, the organizations know what key motivational factors to consider and can therefore adjust their measures in a consistent and informed manner.

5.3 Limitations

About our interviews, the impact of social desirability bias must be considered, as respondents may (un)consciously adjust their responses to conform to socially desirable norms. The small sample size must also be considered, despite our efforts to diversify it both in terms of demographics and LCP used. Although some could argue that nine respondents are not enough to generalize, it is important to remember that the objective of qualitative research is rather exploratory. Finally, while a platform-specific study is beyond the scope of this paper, the major differences in LCPs (in terms of complexity, abstraction level, usability, etc.) used by respondents might have influenced the results.

5.4 Future Research

Addressing generalization, the next step will be to empirically validate the presented adoption process, potentially through adapting the ARCS's Instructional Material Motivational Survey [15]. Future research could evaluate if our results can be extrapolated to

LLM-based software: while those have their own specificities, the motivational factors at play may be similar. Afterwards, drawing from [23] who integrated personality, attitude, motivation, cognitive and social factors with organizational ones into a cohesive Psychological Technology Adoption Framework, there is a need to study how already identified organizational factors of LCPs adoption can augment our motivational adoption process. Finally, exploring methods to stimulate CDs' motivations for each of the five steps of the motivational adoption process would provide invaluable insights.

6 Conclusion

Our findings contribute to research by introducing a motivational adoption process that articulates previously and newly found motivational factors influencing the adoption process of LCPs by CDs. By extending Keller's ARCS Motivational Model, our five-stages framework (interest, need, ability, learning, reinforcement) allows a human-centered vision of this subject. This motivational perspective on LCPs adoption can be leveraged by organizations to better foster LCPs initiatives, considering what will motivate employees to become CDs rather than solely focusing on organizational factors.

References

1. Alsaadi, H., Radain, D., Alzahrani, M.M., Alshammari, W.F., Alahmadi, D.H., Fakieh, B.: Factors that affect the utilization of low-code development platforms: survey study. Revista Română de Informatică Şi Automatică (2021)
2. Binzer, B., Winkler, T.J.: Democratizing software development: a systematic multivocal literature review and research agenda on citizen development. In: A. and W. X. and S. V. Carroll Noel and Nguyen-Duc (Ed.), Software Business, pp. 244–259. Springer International Publishing (2022)
3. Binzer, B., Fuerstenau, D., Winkler, T.: Bridging Business and IT Through Low-Code/No-Code: Insights into Business-IT Collaboration in Enterprise Citizen Developer Programs (2025)
4. Boyce, C., Neale, P.R.: Conducting in-depth interviews: a guide for designing and conducting in-depth interviews for evaluation input (2006)
5. Breckenridge, J., Jones, D.: Demystifying theoretical sampling in grounded theory research. Grounded Theory Rev. **8**, 113–126 (2009)
6. Burgos, A., Branch-Bedoya, D., The, J.W., Amores-Valencia, A., Burgos, D., Branch-Bedoya, J.W.: The influence of augmented reality (AR) on the motivation of high school students. Electronics **12**(22), 4715 (2023)
7. Carroll, N., Maher, M.: How shell fueled digital transformation by establishing DIY software development. MIS Q. Exec. **22**, 99–127 (2023)
8. Daniel, P., Ademar, A., Vasco, A.: What about the usability in low-code platforms? A systematic literature review. J. Comput. Lang. **74** (2023)
9. Elshan, E., Binzer, B., Winkler, T.: From software users to software creators: an exploration of the core characteristics of the citizen developer role and the related re- and upskilling programs. Bus. Inf. Syst. Eng. (2024)
10. Hoogsteen, D., Borgman, H.: Empower the Workforce, Empower the Company? Citizen Development Adoption (2022)

11. Huang, D.-W., Diefes-Dux, H.A., Imbrie, P.K., Daku, B.L.F., Kallimani, J.G.: Learning motivation evaluation for a computer-based instructional tutorial using ARCS model of motivational design. In: 34th Annual Frontiers in Education, 2004 (2004)
12. Käss, S., Strahringer, S., Westner, M.: Drivers and inhibitors of low code development platform adoption. In: 2022 IEEE 24th Conference on Business Informatics (2022)
13. Käss, S., Strahringer, S., Westner, M.: Practitioners' perceptions on the adoption of low code development platforms. IEEE Access **11**, 29009–29034 (2023)
14. Keller, J.M.: ARCS Model of Motivation. In: Seel, N.M. (ed.) Encyclopedia of the Sciences of Learning 304–305. Springer, US (2012)
15. Keller, J.M.: Motivation, Learning, and Technology: Applying the ARCS-V Motivation Model (2016)
16. Lebens, M., Finnegan, R.J., Sorsen, S.C., Shah, J.: Rise of the citizen developer. Muma Bus. Rev. **5**(12), 101–111 (2021)
17. Marangunić, N., Granić, A.: Technology acceptance model: a literature review from 1986 to 2013. Univ. Access Inf. Soc. **14**, 81–95 (2015)
18. Matook, S., Wang, Y.M., Koeppel, N., Guerin, S.: Metacognitive skills in low-code app development: work-integrated learning in information systems development. J. Inf. Technol. (2023)
19. Matook, S., Wang, Y.M., Axelsen, M.: Experiential learning for citizen developers. Bus. Inf. Syst. Eng. **67**, 7–30 (2025)
20. Muhammad, S., Prybutok, V., Sinha, V.: Unlocking citizen developer potential: a systematic review and model for digital transformation. Encyclopedia (2025)
21. Ploder, C., Bernsteiner, R., Schlögl, S., Gschliesser, C.: The Future Use of LowCode/NoCode Platforms by Knowledge Workers – An Acceptance Study. In: Uden, L., Ting, I.-H., Corchado, J.M. (eds.) Knowledge Management in Organizations 445–454. Springer International Publishing (2019)
22. Prinz, N., Huber, M., Leonhardt, J., Riedinger, C.: Unleash the Power of Citizen Development: Leveraging Organizational Capabilities for Successful Low-Code Development Platform Adoption. HICCS (2024)
23. Roberts, R., Flin, R., Millar, D., Corradi, L.: Psychological factors influencing technology adoption: a case study from the oil and gas industry. Technovation **102** (2021)
24. Rokis, K., Kirikova, M.: Challenges of low-code/no-code software development: a literature review. In: Nazaruka, Ē., Sandkuhl, K., Seigerroth, U. (eds.) Perspectives in Business Informatics Research. BIR 2022. Lecture Notes in Business Information Processing, vol. 462. Springer, Cham (2022)
25. Sońta, M., Przegalinska, A.: Say 'yes' to 'no-code' solutions: how to teach low-code and no-code competencies to non-IT students (2024)
26. Terragni, V., Vella, A., Roop, P., Blincoe, K.: The future of AI-driven software engineering. ACM Trans. Softw. Eng. Methodol. **34**(5) (2025)
27. Williams, M., Moser, T.: The art of coding and thematic exploration in qualitative research. Int. Manag. Rev. **15**, 45 (2019)

Quantifying ICT Third-Party Concentration Risk Using DORA Reporting Data

Beāte Krauze[1,2] and Laura Kruglova[2(✉)]

[1] Institute of Information Technology, Riga Technical University, Riga, Latvia
beate.krauze@rtu.lv
[2] Latvijas Banka, Riga, Latvia

Abstract. Digital transformation in the financial sector has increased dependence on information and communication technology (ICT) third-party service providers, creating concentration risk and resilience challenges. This paper proposes a quantitative method for measuring concentration risk using Register of Information (RoI) data required under the Digital Operational Resilience Act (DORA). As standardised reporting does not provide calculation approaches for quantifying concentration risk, this paper addresses the gap by developing a method that measures: (1) provider dependence, (2) service-type concentration and (3) subcontracting exposure. The method structures dependencies between providers, ICT service types and subcontracting levels and links them to critical or important functions. It also applies risk weights based on substitutability and impact of discontinuation. The results show that RoI data can be used not only as a compliance artefact, but also as a computable dependency model that supports residual risk assessment and mitigation prioritisation.

Keywords: Enterprise management · risk management · resilience · third-party ICT risk · concentration risk · DORA

1 Introduction

The Digital Business Ecosystem (DBE) in the financial sector includes banks, payment institutions, investment firms, asset managers, financial market infrastructures, *fintechs*, regulators, payment and transaction networks and a wide range of ICT third-party service providers (TPPs). Although this structure enables rapid innovation, growing dependence on ICT TPPs also creates operational risk and resilience challenges [10].

In the context of concentration risk, established risk assessment approaches include qualitative and semi-quantitative methods such as risk matrices, RAG ratings, dependency mapping, key risk indicators, threshold based assessments and scenario analysis [15]. These methods help identify whether critical or important functions depend on the same ICT provider, cloud platform, subcontractor,

A. Bernasconi et al. (Eds.): RCIS 2026, LNBIP 597, pp. 78–89, 2027.
https://doi.org/10.1007/978-3-032-35043-5_7

technology stack or geographic location [12]. However, these methods have limitations when applied to material concentration risks. Risk matrices and RAG ratings support communication, but they may rely heavily on expert judgement and may not capture the full scale of exposure or systemic dependencies. For critical services, they should be complemented by quantitative or scenario-based techniques, such as dependency ratio, financial impact estimates, expected loss modelling, recovery cost analysis and stress testing. This enables a more evidence based assessment of the concentration risk [12,16].

Regulation (EU) 2022/2554, the Digital Operational Resilience Act (DORA), increases the relevance of this challenge by requiring organisations to map third-party ICT dependencies and demonstrate effective ICT risk management and operational resilience. Prior research highlights that DORA implementation requires financial entities to align regulatory requirements with operational resilience practices and information systems governance [4]. The Register of Information (RoI) provides a standardised dataset for this purpose by capturing providers, contracts, services, functions and risk attributes. However, the RoI alone does not ensure consistent concentration risk measurement. This paper addresses this gap by converting RoI fields into comparable concentration indicators and risk-weighted scores.

Related research shows that enterprise architecture (EA) can be quantified through metrics on components, heterogeneity, density and relationships across EA layers, but that limited consensus on methods and tools makes concentration risk difficult to measure consistently [6]. These developments emphasise the need for a consistent and comparable method for quantifying third-party concentration risk within the ICT risk management framework under DORA.

In order to address this gap, the following research questions are defined.

RQ1. How can ICT third-party concentration risk be quantified using data from the Register of Information (RoI)?
RQ2. How can concentration indicators be risk-weighted using RoI attributes, such as substitutability and impact of discontinuation, to prioritise dependencies with the greatest operational impact?

The contribution of this paper is to transform RoI reporting data from a compliance artefact into a computable dependency model that supports repeatable measurement and prioritisation of concentration risk.

The paper is structured as follows. Section 2 reviews related work and research gaps. Section 3 presents the proposed quantification method, including provider, service-type and subcontracting concentration, risk weighting and the composite score. Section 4 discusses practical application, residual risk governance and limitations. Section 5 concludes the paper and outlines future work.

2 Related Work and Research Gaps

Concentration risk arises when critical or important functions depend on a limited number of ICT service providers, cloud platforms, subcontractors or locations, creating potential single points of failure and reducing substitutability [4].

From a regulatory perspective, the assessment should identify excessive dependence in ICT arrangements, evaluate substitutability and potential systemic impact, and complement qualitative risk assessment with indicators, limits, scenarios, exit strategies or stress testing where dependencies are material [12,16].

Related work shows that reliance on ICT TPPs creates not only traditional outsourcing risk, but also concentration risk, vendor lock-in and limited supply-chain transparency [5,10,14]. These risks are particularly relevant for cloud services, where critical services may be provided by a small number of dominant providers and exit may be difficult because dependencies are deeply integrated into business processes [3,8]. Supervisory guidance therefore emphasises the need to assess concentration risk at institution level, maintain up-to-date registers and dependency mappings, and monitor supply-chain concentration, lock-in and substitutability [7,8].

Prior studies have proposed governance frameworks for third-party and cloud outsourcing risk, covering due diligence, contractual safeguards, monitoring, audits and exit planning [2,14]. However, dependency risk becomes visible only when relationships between services, providers, subcontractors and business functions are explicitly modelled. Enterprise architecture and information systems research therefore highlights the need for structured dependency analysis to reduce hidden coupling and architectural complexity [1,2,6]. Multi-layered subcontracting further increases this complexity, since incidents at upstream providers may affect services indirectly and remain difficult to detect or control [10,14].

At the financial sector level, common reliance on a small number of global ICT TPPs, including hyperscalers, may transform operational incidents into wider disruption, especially where substitutability is low and subcontracting chains are opaque [3,5,9]. A joint ESA exercise identified approximately 15,000 ICT TPPs directly serving around 1,600 EU organisations in the analysed sample, illustrating the scale of interconnectedness in the EU financial sector [8]. In a further step under DORA, the ESAs published the first EU-level list of designated critical ICT TPPs on 18 November 2025, based on data submitted through the RoI [11]. Under DORA, the RoI has therefore become a central supervisory artefact for identifying, monitoring and overseeing ICT third-party dependencies [11–13].

The literature and current practice reveal three gaps. First, most approaches remain qualitative or procedural and do not provide a consistent quantitative method for measuring concentration risk. Second, concentration is often discussed at a high level in relation to cloud market structure, vendor lock-in or systemic dependency, without an operational method for translating regulatory reporting data into repeatable indicators. Third, existing inventories and registers identify providers and services, but usually do not convert dependency structures into comparable metrics that combine provider dependence, service-type concentration and subcontracting exposure, nor risk-weight these metrics using operational attributes such as criticality, impact or substitutability [7].

This paper addresses these gaps by operationalising the RoI data model as a quantitative assessment model for concentration risk. In contrast to existing work that focuses mainly on governance expectations or qualitative assessment, the proposed approach transforms structured RoI fields into computable indicators for: (1) provider dependence, (2) service-type concentration and (3) subcontracting exposure. It further extends baseline concentration measures by applying risk weights derived from RoI attributes on substitutability and impact. This enables concentration risk to be assessed using reporting artefacts that are already required under DORA and supports repeatable measurement, comparison and mitigation prioritisation.

3 The Concentration Risk Quantification Method

A central transparency instrument under DORA is the mandatory Register of Information (RoI) on contractual arrangements with ICT TPPs. This requirement is supported by the data model defined in the implementing technical standards set out in Commission Implementing Regulation (EU) 2024/2956 [13].

This research develops a method for quantifying concentration risk from RoI data by mapping contractual arrangements, ICT TPPs, service types, subcontracting structures and risk attributes into a computable dependency model. The method then calculates provider, service-type and subcontracting concentration indicators, applies risk weights based on substitutability and impact, and evaluates the results using an illustrative RoI case.

Data Source. The key RoI elements used for quantification are listed in Table 1 [13].

3.1 Dataset

Unit of Observation. The unit of observation is the contractual arrangement, indexed by c.

RoI Variables. For each contractual arrangement c, the method uses the following variables:

- Annual expense x_c: annual expense or estimated annual cost of contractual arrangement c. This is the monetary exposure used for the concentration measurement.
- Direct provider $p(c)$: the direct ICT TPP associated with contractual arrangement c.
- Service type $t(c)$: the ICT service type associated with contractual arrangement c.
- Supply-chain rank $Sub(c)$: an indicator used to determine whether subcontractors are present for contractual arrangement c.
- Substitutability indicator $S(c)$: the RoI category describing how easily the ICT service supporting critical or important functions can be substituted.

Table 1. RoI elements

Data	Type	Description
Annual cost of the arrangement	Numeric	Annual expense or estimated annual cost for the contractual arrangement with an ICT TPP.
Reference of the arrangement	Alphanumeric	Unique reference used to identify the contractual arrangement.
ICT TPP	Categorical	Information identifying the ICT TPP.
Ultimate parent	Categorical	Information identifying the ultimate parent undertaking of the ICT TPP.
Type of ICT service	Categorical	Type or nature of the ICT service provided.
Use of subcontractors	Categorical	Information indicating whether subcontractors are used for the provision of the ICT service.
Rank in the ICT service supply chain	Ordinal	Position of the ICT TPP in the supply chain, distinguishing direct service providers and wider networks.
Link to critical or important function	Relational	Information classifying whether the ICT service supports a function classified as critical or important.
Substitutability of the ICT service	Categorical	Assessment of the substitutability of the ICT service.
Impact of discontinuation of the ICT service	Categorical	Assessment of the impact of discontinuing the ICT service supporting a critical or important function.

- Impact indicator $I(c)$: the RoI category describing the impact of discontinuing the ICT service.

The illustrative RoI used to evaluate the method contains 40 contractual arrangements, total annual ICT expense of EUR 5,468,882 and 29 distinct direct ICT TPPs. These values instantiate the contract set $\mathcal{C}$, the annual expenses x_c and the provider mapping $p(c)$.

3.2 Baseline Concentration

Provider Exposure. Provider exposure measures the annual expense attributed to each direct ICT TPP:

$$E_i = \sum_{c:p(c)=i} x_c \tag{1}$$

where E_i is the total annual expense attributed to provider i, x_c is the annual expense of contractual arrangement c, and $p(c)$ maps contractual arrangement c to its direct ICT TPP.

Provider Share. Provider exposure is converted into a share of total ICT spend:

$$s_i = \frac{E_i}{\sum_j E_j} \tag{2}$$

where s_i is the fraction of total annual expense attributable to provider i.

ICT Risk Concentration Ratio. The concentration ratio summarises the share of total ICT spend accounted for by the largest ICT TPPs. The top-k concentration ratio is:

$$CR_k = \sum_{i \in \mathrm{Top}k(s)} s_i \tag{3}$$

where $\mathrm{Top}k(s)$ denotes the set of the k largest provider shares. CR_1 captures dependence on the largest provider, while CR_3 captures reliance on the three largest providers.

For the illustrative RoI, the baseline concentration ratios are:

$$CR_1 = 0.2194, \quad CR_3 = 0.5464, \quad CR_5 = 0.7386, \quad CR_{10} = 0.9178. \tag{4}$$

This means that the three largest ICT TPPs account for 54.64% of the measured annual ICT expense.

3.3 Analysis of Largest Service Types

Service-Type Concentration. Service-type concentration measures whether ICT expenditure is concentrated in one dominant service category, such as cloud infrastructure, core banking SaaS or payment processing. This is especially relevant where the service supports critical or important functions and is difficult to replace.

For each ICT service type t, exposure is defined as:

$$E_t = \sum_{c:t(c)=t} x_c \tag{5}$$

where E_t is the total annual expense attributed to service type t.

Largest Service Type Share. The largest service type share is defined as:

$$MS = \max_t \frac{E_t}{\sum_u E_u}. \tag{6}$$

where MS is the maximum service type share. A higher MS indicates stronger dependence on one type of ICT service.

For the illustrative RoI, the largest service type share is:

$$MS = 0.4324. \tag{7}$$

This indicates that 43.24% of the annual ICT expense is concentrated in the single largest ICT service type.

3.4 Subcontracting Exposure

Subcontracting Indicator. The subcontracting indicator identifies contractual arrangements that depend on providers beyond the direct ICT TPP, capturing indirect dependencies that may increase monitoring and recovery complexity.

Using RoI supply-chain ranks, subcontracting exposure for contractual arrangement c is defined as:

$$Sub(c) = 1\left(\exists \text{ row for } c \text{ with rank} > 1\right), \tag{8}$$

where $1(\cdot)$ is an indicator function. In the RoI, rank 1 corresponds to the direct provider, while ranks greater than 1 correspond to subcontractors.

Subcontracting Spend Share. The subcontracting spend share measures the proportion of total ICT spend associated with contractual arrangements that involve subcontractors:

$$DS = \frac{\sum_c x_c \cdot Sub(c)}{\sum_c x_c} \tag{9}$$

For the illustrative RoI, subcontracting exposure is:

$$DS = 0.1368. \tag{10}$$

This means that 13.68% of annual ICT expense is associated with contractual arrangements that include subcontractors.

3.5 Assessing Risk Weight

Baseline concentration treats all spend equally. To prioritise dependencies with greater operational relevance, the method applies a minimal risk weight using substitutability $S(c)$ and impact of discontinuation $I(c)$.

Category-to-Weight Mapping. RoI categorical values are mapped as follows:

$$a(S) = \begin{cases} 1, & \text{if easily substitutable} \\ 2, & \text{if medium complexity for substitutability} \\ 3, & \text{if highly complex or impossible to substitute} \end{cases} \tag{11}$$

$$b(I) = \begin{cases} 1, & \text{if low or if assessment was not performed} \\ 2, & \text{if medium} \\ 3, & \text{if high} \end{cases} \tag{12}$$

where the RoI does not contain a substitutability or impact assessment, the contractual arrangement is assigned a default weight of 1. This avoids artificially increasing or decreasing the measured concentration risk where the assessment is not available.

Contract Risk Weight and Weighted Exposure. The contract risk weight adjusts the annual cost of each contractual arrangement by combining its substitutability and impact assessments. The contract risk weight adjusts annual expense by combining substitutability and discontinuation impact:

$$w_c = a(S(c)) \cdot b(I(c)), \qquad x_c^w = x_c \cdot w_c. \tag{13}$$

Higher weights are assigned to arrangements that are harder to substitute and have a higher discontinuation impact.

All the concentration metrics can then be recomputed by replacing x_c with x_c^w:

– Weighted provider exposure:

$$E_i^w = \sum_{c:\, p(c)=i} x_c^w \tag{14}$$

– Weighted provider share:

$$s_i^w = \frac{E_i^w}{\sum_j E_j^w} \tag{15}$$

– Weighted concentration ratios:

$$CR_k^w = \sum_{i \in \mathrm{Top}k(s^w)} s_i^w \tag{16}$$

– Weighted service concentration:

$$MS^w = \max_t \frac{\sum_{c:\, t(c)=t} x_c^w}{\sum_c x_c^w} \tag{17}$$

– Weighted subcontracting share:

$$DS^w = \frac{\sum_c x_c^w \cdot Sub(c)}{\sum_c x_c^w} \tag{18}$$

For the illustrative RoI, the resulting weighted metrics are:

$$CR_3^w = 0.5998, \qquad MS^w = 0.4150, \qquad DS^w = 0.1502. \tag{19}$$

Compared with the baseline metrics, the weighted results show whether operationally important, hard-to-substitute or high-impact dependencies increase the measured concentration risk.

3.6 Prioritisation of Dependencies Based on Concentration Risk

The weighted metrics can be combined into a normalised concentration risk score to support prioritisation of mitigation measures:

$$Score = 100 \cdot (0.6 \cdot CR_3^w + 0.3 \cdot MS^w + 0.1 \cdot DS^w) \tag{20}$$

The weights are illustrative. They place the greatest emphasis on dependence on a small number of providers, followed by service type concentration and subcontracting exposure. Alternative weightings can be applied depending on supervisory or internal governance priorities.

For the illustrative RoI, the score is calculated as follows:

$$Score = 100 \cdot (0.6 \cdot 0.5998 + 0.3 \cdot 0.4150 + 0.1 \cdot 0.1502) = 49.94. \tag{21}$$

The value 49.94 is a normalised concentration risk score on a scale from 0 to 100. It indicates that the concentration risk in the illustrative RoI is material and is primarily driven by dependence on the largest ICT TPPs, as reflected in the weighted top-three provider concentration ratio.

4 Discussion

Using the illustrative example described in Sect. 3.1, the results show that quantified structural dependencies can support residual risk governance.

The composite score of 49.94 (Eq. (21)) summarises three weighted components: dependence on a small number of providers $\left(CR_3^w = 0.5998\right)$, the concentration of a single type of ICT service $\left(MS^w = 0.4150\right)$ and exposure to subcontracting $\left(DS^w = 0.1502\right)$. From a governance and prioritisation perspective, the results of illustrative example align with the review logic of the method.

- Since CR_3^w remains high after weighting, the main drivers are likely contractual arrangements that are the most expensive and difficult to substitute and have a higher impact. These arrangements should be prioritised for exit planning, resilience testing and specific clauses in contracts that are required by DORA.
- A high MS^w indicates a concentrated reliance on one type of ICT service, suggesting that diversification options may be limited in that type. The organisation can use this as input to the vendor strategy and target-state EA decisions.

- Although DS^w has a smaller weight in the score, subcontracting exposure can create hidden single points of failure. Contracts supporting critical or important functions should be assessed for shared subcontractors and whether subcontracting increases recovery complexity.

Practical Application in the Risk Register. As a practical application, the approach supports the assessment of third-party ICT risk through RoI. Since RoI is submitted annually under DORA, the same indicators can be recalculated each year and incorporated into the risk register. This enables year-on-year comparison of risk assessments and highlights changes in the concentration, criticality and dependency exposures.

In organisations where third-party ICT risks are already quantitatively assessed through a likelihoodimpact matrix, the proposed method is intended to refine residual risk by adding a measurable dependency dimension.

Limitations. The approach has two main limitations. First, it depends on the quality and consistency of RoI reporting, including how substitutability and impact assessments are performed and maintained. Second, annual expense is used as the primary exposure measure, which may not fully reflect operational criticality.

The illustrative case study shows the value in an already quantitative risk register with matrix scoring. The contribution to residual risk governance is that the register can justify and prioritise risk mitigation actions based on quantified structural dependency: arrangements that are difficult to substitute and operationally high impact dominate exposure even when they do not dominate nominal spend. This supports a more argumentative prioritisation of residual risk actions (for example, exit planning, targeted resilience testing, diversification).

5 Conclusions and Future Work

This paper shows that RoI data can be used as a structured representation of third-party ICT dependencies and transformed into comparable concentration risk indicators. The proposed method measures dependence on providers, concentration by ICT service type and exposure to subcontracting. These indicators are then included in the residual risk assessment as a quantitative dependency dimension. The results show that the concentration risk cannot be assessed from the number of providers alone. Apparent diversification may still hide dependence on a dominant service type, a provider that is difficult to replace or an upstream subcontractor. From an EA perspective, the method supports the identification of vendor lock-in, dominant ICT service types and upstream single points of failure.

Regarding RQ1, the method quantifies the concentration risk by deriving a computable dependency model from RoI data. It links contractual arrangements to ICT TPPs, ICT service types and subcontracting structures, enabling concentration to be measured through provider exposure, concentration ratios, the largest service-type share and subcontracting spend share.

Regarding RQ2, RoI attributes are used to risk-weight these indicators based on substitutability and discontinuation impact. The weighted results prioritise dependencies that are both operationally significant and difficult to substitute, while the composite score supports mitigation prioritisation.

Future work should therefore focus on extending exposure measures beyond annual expense, for example, dependency criticality at function level, processing volumes, resilience testing results or EA debt.

References

1. K. Wehling, D. Wille, C. Seidl and I. Schaefer: decision support for reducing unnecessary IT complexity of application architectures. In: IEEE International Conference on Software Architecture Workshops (ICSAW) (2017). https://doi.org/10.1109/ICSAW.2017.47
2. Kaul, M., Storey, V.C., Woo, C.: A framework for managing complexity in information systems. J. Database Manag. (JDM) (2017). https://doi.org/10.4018/JDM.2017010103
3. Cloud Adoption in the Financial Sector and Concentration Risk, Financial Stability Board / Program on International Financial Systems (2023). https://www.fsb.org/uploads/PIFS.pdf. Accessed 08 April 2026
4. Neumannová, A., Bernroider, E.W.N., Elshuber, C.: The digital operational resilience act for financial services: a comparative gap analysis and literature review. In: Information Systems. EMCIS (2023). https://doi.org/10.1007/978-3-031-30694-5_40
5. Digital dependence of the financial sector. Risks, resilience and European autonomy. https://www.dnb.nl/media/antp5cj3/digital-dependence-of-the-financial-sector.pdf. Accessed 02 Dec 2025
6. Stoica, E., Moreira, J., Piest, J.P.S., Bukhsh, F.: Drivers and metrics for quantifying IT landscape complexity. In: Enterprise Design, Operations and Computing (EDOC) (2025). https://doi.org/10.1007/978-3-031-78338-8_12
7. Basel Committee on Banking Supervision: Principles for the sound management of third-party risk, Bank for International Settlements (2025). https://www.bis.org/bcbs/publ/d605.pdf. Accessed 08 April 2026
8. ESAs Report on the landscape of third-party ICT providers in the EU. https://www.eba.europa.eu/publications-and-media/press-releases/esas-publish-report-landscape-ict-third-party-providers-eu. Accessed 16 Dec 2025
9. Depner, S., Richter, A.: Cloud meets customer: IT service providers in the public cloud ecosystem. Bus. Horiz. (2025). https://doi.org/10.1016/j.bushor.2025.04.008
10. Seid, E., Ilter, D., Blix, F., Popov, O.: Cyber supply chain resilience: analyzing ISO, NIST and NIS2 frameworks for mitigating third-party risks. In: International Conference on Industry Sciences and Computer Science Innovation (iSCSi'24). Procedia Computer Science (2025). https://doi.org/10.1016/j.procs.2025.07.071
11. European Supervisory Authorities designate critical third-party ICT providers under the Digital Operational Resilience Act. https://www.eiopa.europa.eu/european-supervisory-authorities-designate-critical-ict-third-party-providers-under-digital-2025-11-18_en. Accessed 10Jan 2026

12. Regulation (EU) 2022/2554 of the European Parliament and of the Council of 14 December 2022 on digital operational resilience for the financial sector and amending Regulations (EC) No 1060/2009, (EU) No 648/2012, (EU) No 600/2014, (EU) No 909/2014 and (EU) 2016/1011. https://eur-lex.europa.eu/eli/reg/2022/2554/oj/eng. Accessed 21 Dec 2025
13. Commission Implementing Regulation (EU) 2024/2956 of 29 November 2024 laying down implementing technical standards for the application of Regulation (EU) 2022/2554 of the European Parliament and of the Council with regard to standard templates for the register of information. https://eur-lex.europa.eu/legal-content/EN/TXT/?uri=CELEX:32024R2956. Accessed 10 Jan 2026
14. Kun, E.: Challenges in regulating cloud service providers in EU financial regulation: from operational to systemic risks and examining challenges of the new oversight regime for critical cloud service providers under the digital operational resilience act. Comput. Law Secur. Rev. (2024). https://doi.org/10.1016/j.clsr.2023.105931
15. BaFin: Circular 05/2023 (BA) - Minimum Requirements for Risk Management (MaRisk). Federal Financial Supervisory Authority, annotated text in the version of 29 June 2023. https://www.bafin.de/SharedDocs/Downloads/EN/Rundschreiben/dl_rs_0523_marisk_ba_en.html. Accessed 29 April 2026
16. Commission Delegated Regulation (EU) 2024/1773 of 13 March 2024 supplementing Regulation (EU) 2022/2554 of the European Parliament and of the Council with regard to regulatory technical standards specifying the detailed content of the policy regarding contractual arrangements on the use of ICT services supporting critical or important functions provided by ICT TPPs. https://eur-lex.europa.eu/legal-content/EN/TXT/?uri=CELEX%3A32024R1773. Accessed 29 April 2026

Probabilistic Dempster–Shafer Fusion for Multimodal Breast Cancer Analysis

Abir Baâzaoui[1,2](✉), Malek Baâzaoui[2], Ikram Bellara[3], and Walid Barhoumi[1,4]

[1] Higher Institute of Computer Science, University of Tunis ElManar, LR16ES06 Laboratoire de recherche en Informatique, Modelisation et Traitement de l'Information et de la Connaissance (LIMTIC), Ariana, Tunisia
walid.barhoumi@enicarthage.rnu.tn

[2] Higher Institute of Computer Science and Management, University of Kairouan, Kairouan, Tunisia
abir.baazaoui@isigk.rnu.tn

[3] Liouane Medical Imaging Clinic, Kairouan, Tunisia

[4] National Engineering School of Carthage, University of Carthage, Tunis, Tunisia

Abstract. This paper proposes a hybrid system for computer-aided diagnosis of breast cancer based on deep learning and multimodal fusion. The architecture integrates mammographic images from two views (CC and MLO), analyzed using a Variational Autoencoder (VAE) to extract probabilistic latent representations, followed by supervised classification. In parallel, textual clinical reports are processed using a transformer-based language model (all-MiniLM-L6-v2) to generate semantic embeddings and binary indicators, which are fed into an ensemble classifier. The probabilities produced by these modules are aggregated using Dempster–Shafer Theory (DST), enabling uncertainty-aware decision fusion. Experiments conducted on four real-world datasets suggest the potential of the proposed multimodal approach compared to unimodal baselines, with improvements in accuracy, F1-score, and AUC. However, these results should be interpreted with caution due to dataset limitations, particularly the small size and imbalance of the private Tunisian dataset. Therefore, this work should be considered as a preliminary step toward clinically applicable multimodal diagnostic systems.

Keywords: Breast cancer · CADx · Deep Learning · Multimodal Fusion · VAE · Dempster-Shafer Theory · Mammography · Clinical reports · Natural Language Processing

1 Introduction

Breast cancer persists as one of the most significant oncological challenges globally, representing approximately 25% of all cancer diagnoses in women and constituting a leading cause of cancer-related mortality worldwide [1]. The current diagnostic tool relies heavily on mammographic screening, although it has certain

A. Bernasconi et al. (Eds.): RCIS 2026, LNBIP 597, pp. 90–101, 2027.
https://doi.org/10.1007/978-3-032-35043-5_8

shortcomings such as substantial inter-observer variability, elevated false-positive rates, and inconsistent integration of heterogeneous data sources [2]. The emergence of artificial intelligence in medical imaging has catalyzed a transformative shift in breast cancer diagnosis, with deep learning methodologies demonstrating unprecedented capabilities in detecting mammographic abnormalities and classifying lesions [3]. Convolutional Neural Networks and their architectural variants have achieved remarkable performance in identifying subtle patterns indicative of malignancy that may escape human observation However, the majority of existing AI diagnostic systems operate within isolated modalities, focusing exclusively on either image analysis or clinical data processing, thereby failing to leverage the synergistic potential of multimodal integration

Several critical challenges remain unaddressed in contemporary breast cancer diagnosis systems. First, the predominant reliance on unimodal approaches neglects the comprehensive clinical context essential for accurate diagnostic decisions [2]. Second, the mathematical and computational complexity of integrating disparate data types, particularly the fusion of imaging derived features with clinical variables presents substantial methodological hurdles [4].

To address these limitations, we propose a hybrid fusion framework that bridges the gap between deep learning-based image analysis and clinical data interpretation. This work explores three main research directions. Firstly, a dual-stream architecture, a system processes mammographic images using a Variational Autoencoder while analyzing clinical reports through a transformer-based pipeline. Secondly, we investigated then evidence-based fusion using Dempster—Shafer theory to combine heterogeneous sources of evidence while explicitly modeling uncertainty. Lastly, practical relevance in clinical settings and interpretability is ensured through transparent mass function assignments and explicit conflict resolution mechanisms, enabling clinicians to understand the underlying reasoning behind diagnostic decisions.

The remainder of this paper is organized as follows: Sect. 2 details the related work. Section 3 elaborates our proposed methodology, including the VAE architecture, clinical data processing, and DST fusion mechanism. Section 4 presents comprehensive experimental results and performance analysis. Section 5 discusses the clinical implications of our findings, acknowledges limitations, and outlines their potential impact on clinical practice. Finally, Sect. 6 concludes with a summary of our contributions and some future research directions.

2 Related Work

By analogy with radiological reasoning, modern deep learning systems are no longer limited to analyzing a single source [5], but are moving towards multimodal architectures capable of simultaneously integrating several types of data [6]. These models combine medical images, free text, previous reports, clinical metadata, and sometimes even biological signals to obtain a richer and more consistent representation of the patient's condition. A growing body of reviews and surveys has systematized multimodal fusion approaches in medical imaging.

These works classify fusion methods into feature fusion, intermediate fusion, and late fusion.

2.1 Feature Fusion in Multimodal CADs (MM-CADs)

Feature fusion is an early multimodal integration strategy that combines low-level features from multiple sources (*e.g.*, mammography, ultrasound, MRI, histopathology, and clinical data) into a shared representation for end-to-end learning. Recent studies demonstrate its effectiveness, as early fusion of mammography and ultrasound improves detection in dense breasts [8], representation-level fusion enhances malignancy prediction in BIRADS 4A lesions [9], and deep feature concatenation across mammography, ultrasound, and MRI yields high sensitivity and robustness [10]. Hybrid CNN-based feature fusion further improves benign/malignant discrimination [11], while fusing multi-view mammograms with clinical data strengthens CAD performance [12]. Despite these benefits, feature fusion remains challenging due to data heterogeneity and high dimensionality, requiring careful alignment and normalization [13].

2.2 Late Fusion in MM-CADs

Late fusion (decision-level fusion) combines the outputs of modality-specific models, making it well suited to heterogeneous, noisy, or partially missing data while improving robustness and interpretability [13]. It relies on simple aggregation rules (e.g., averaging, voting), stacking with meta-classifiers (logistic regression, LightGBM, XGBoost), or evidence-theoretic approaches. Its effectiveness is illustrated in breast cancer studies, including multimodal survival prediction integrating genomic and clinical data with weighted averaging and LightGBM stacking [14], multitask mammographic screening where task- and view-specific networks are aggregated via pooling and high-level classifiers [15], and ultrasound-based classification combining B-mode and elastography using CNN ensembles and SVM-based fusion [16]. Differently, [10] shows that late fusion, particularly stacking, achieves the highest AUC and specificity when combining mammography, ultrasound, and MRI. Owing to its reliability and capacity to manage heterogeneity, late fusion is well suited to combining mammographic images with textual clinical reports in our work.

2.3 Intermediate Fusion in MM-CADs

Intermediate (hybrid) fusion combines feature-level integration with decision-level aggregation within a unified pipeline, enabling early cross-modal interaction while preserving robustness to noise and missing data through modular ensemble mechanisms [13]. Its effectiveness is illustrated in breast cancer applications, including ultrasound micronodule classification where deep features from InceptionV3 and MobileNetV2 are concatenated and refined through classifier ensembles [11]. In multimodal CAD, this work [12] integrates multi-view mammographic features with clinical metadata via early fusion, followed by weighted decision aggregation, yielding improved sensitivity and specificity over unimodal systems. Similarly, authors in [10] demonstrate that a hybrid architecture combining deep radiomic feature fusion with stacking-based decision aggregation across mammography, ultrasound, and MRI achieves an optimal sensitivity–specificity trade-off. Beyond imaging, this paper [17] introduces BioFusionNet, which jointly integrates histopathological, genomic, and clinical representations using VAEs, cross-attention, and weighted meta-decision mechanisms, significantly improving survival risk stratification. Recent work [20] presents the relevance of hybrid fusion strategies in multimodal CAD systems, showing that combining feature-level representations with decision-level aggregation can enhance diagnostic robustness and improve generalization across heterogeneous clinical data sources.

In summary, despite considerable progress, several gaps persist in the literature. First, few studies perform systematic comparisons between intermediate fusion and uncertainty-aware decision-level fusion, particularly under conditions of conflicting modality outputs. Second, the combination of latent-regularized visual representations (e.g., VAE encoders) with evidence-theoretic fusion mechanisms remains largely unexplored. Third, many works lack comprehensive ablation studies quantifying the contribution of each module.

Our work directly addresses these gaps by proposing a hybrid multimodal architecture that integrates: (i) a VAE-based dual-view image encoder, (ii) a clinical information extraction module combining transformer embeddings with rule-based features, and (iii) a Dempster–Shafer decision fusion strategy designed to handle uncertainty and inter-modality conflict. This positions our framework at the intersection of latent regularization and uncertainty-aware fusion for breast cancer diagnosis.

3 Proposed Method

The suggested method proposes a hybrid multimodal fusion architecture for breast cancer diagnosis, integrating feature learning from mammograms with expert clinical knowledge. It is structured in three core components (Fig. 1): an image feature extraction module using VAE, a clinical knowledge processing module, and a final decision fusion module that synthesizes the evidence from both modules using Dempster-Shafer Theory (DST) to yield a consolidated and robust diagnostic prediction. In fact, the system integrates dual-view mammographic analysis and clinical text understanding. CC and MLO mammograms are processed independently through two VAE encoders to extract probabilistic latent representations, which are then combined and classified to estimate the probability of malignancy. In parallel, the clinical report is processed using the all-MiniLM-L6-v2 language model to extract key clinical information relevant to the diagnostic task. Finally, both image-based probabilities and text-based key information are fused using DST to produce a robust benign/malignant decision.

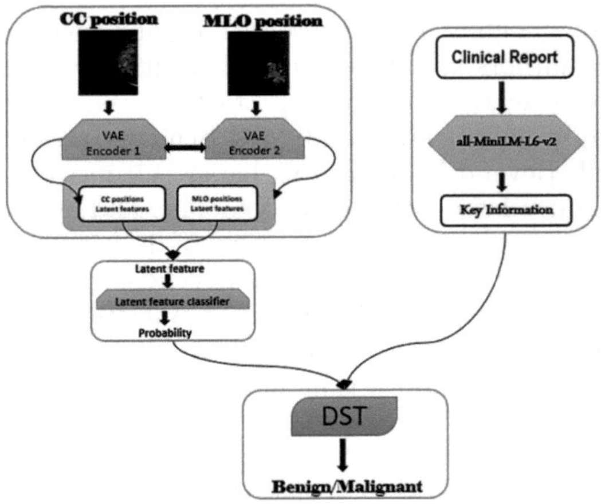

Fig. 1. Overview of the proposed multimodal breast cancer diagnosis method.

3.1 Image Classification Module

This module presents a hybrid framework combining unsupervised representation learning and supervised classification for mammographic images. A Variational Autoencoder (VAE) [19] is first trained to extract compact and robust latent representations from CC and MLO views. Unlike classical autoencoders [20], the VAE models the latent space as a probability distribution and optimizes a reconstruction loss combined with a Kullback–Leibler divergence regularization, leading to a structured latent space that improves generalization and supports data augmentation. Following training, latent vectors from both views are concatenated to form a multiview representation, which is used as input to a fully connected classifier. Several backbone architectures, including GoogLeNet, ResNet50, and Vision Transformer, were evaluated to assess the impact of feature extraction strategies (Table 1). The results indicate that a frozen GoogLeNet with a 128-dimensional latent space provides the best trade-off between performance and computational efficiency. The final classifier consists of dense layers with Batch Normalization and dropout, trained using Adam optimizer and binary cross-entropy loss. This configuration achieved 74.60% test accuracy and an F1-score of 0.843, and was therefore selected for the multimodal framework.

3.2 Clinical Report Key Information Extraction Module

An advanced pipeline for processing unstructured clinical reports was developed to extract diagnostically relevant information through a dual-stream feature extraction approach. During model selection, all-MiniLM-L6-v2 was chosen for its strong trade-off between semantic performance and computational efficiency, achieving near-equivalent MTEB scores compared to larger models while providing significantly faster inference and lower memory usage. Its lightweight archi-

Table 1. Comparative study of image classifiers.

Latent Dimension	Encoder State	Associated Architecture	Accuracy (test)	F1-Score
128	Frozen	GoogLeNet	**74.60%**	**0.843**
128	Frozen	ResNet50	71.43%	0.824
128	Fine-tuned	GoogLeNet	73.02%	0.831
256	Frozen	GoogLeNet	72.22%	0.819
256	Frozen	Vision Transformer	74.60%	0.828
128	Fine-tuned	GoogLeNet	69.84%	0.796

tecture enables efficient deployment in real-time clinical settings and demonstrates good transfer capability to French medical texts despite being originally trained on English data. The all-MiniLM-L6-v2 transformer model generates semantic embeddings from clinical narratives. This distilled BERT variant (6 layers, 384 dimensions, ~22M parameters) was trained on over one billion sentence pairs using contrastive objectives, enabling efficient semantic similarity capture. The *CLS* token embeddings are normalized and reduced to 6 principal components.

A complementary rule-based system identifies clinically significant concepts (e.g., mass or tumor descriptors, pain indicators, skin changes, nipple discharge, and symptom duration), producing a 5-dimensional binary vector encoding explicit clinical findings. The concatenation of semantic and rule-based features results in an 11-dimensional representation (6 semantic components + 5 keyword indicators), which is used to train a Random Forest classifier (100 estimators, maximum depth 5). The model achieved 100% training accuracy; however, this result should be interpreted cautiously, as potential overfitting may be related to the use of partially synthesized clinical descriptions in this study.

3.3 Dempster-Shafer Theory (DST)-Based Fusion

Decision-level fusion is implemented using Dempster-Shafer evidence theory to combine predictions from heterogeneous modalities. The framework operates through three systematic phases. The former refers to basic probability assignment (BPA). Output probabilities from image (p_{image}) and clinical (p_{clinical}) classifiers are transformed into mass functions. Each modality allocates belief masses to singleton hypotheses ({Benign}, {Malignant}) while reserving a portion of mass (0.15 for imaging, 0.20 for clinical) for uncertainty ($\theta = \{\text{Benign}, \text{Malignant}\}$), reflecting their relative reliabilities. The second phase concerns the evidence combination. In fact, Dempster's rule of combination merges BPAs from both modalities, mathematically expressed as:

$$m_{\text{combined}}(A) = \frac{1}{K} \sum_{B \cap C = A} m_{\text{image}}(B)\, m_{\text{clinical}}(C), \tag{1}$$

where, K represents the conflict between evidences:

$$K = 1 - \sum_{B \cap C = \emptyset} m_{\text{image}}(B)\, m_{\text{clinical}}(C). \tag{2}$$

Decision Making is the last phase, where final classification selects the hypothesis with maximum belief after fusion, computed as:

$$\text{Bel}(A) = \sum_{B \subseteq A} m_{\text{combined}}(B). \tag{3}$$

This approach maintains robustness against conflicting predictions and quantifies uncertainty through the conflict measure K. The implementation handles the complete set of focal elements $\{\emptyset, \{\text{Benign}\}, \{\text{Malignant}\}, \theta\}$ and includes conflict resolution mechanisms for cases where $K \approx 1$.

4 Experimental Results

This section presents the experimental evaluations of our hybrid framework for breast cancer diagnosis. Experiments were conducted on four datasets, comprising multi-view mammograms (CC and MLO) paired with corresponding clinical reports. We analyze individual module performance (image classification and clinical information extraction) before examining the impact of DST-based multimodal fusion. Finally, a comparative study validates the robustness of the proposed method across heterogeneous datasets and populations.

4.1 Dataset Description

Our method was evaluated on four multi-view mammography datasets. The Chinese Mammography Database (CMMD) comprises 3,728 mammograms from 1,775 Chinese patients with histopathologically confirmed benign or malignant lesions, acquired using GE Senographe DS between July 2012 and January 2016. The second dataset is the Chinese Breast Disease Clinical Imaging Database (CBCID), containing 176 mammographic images with 84 corresponding clinical complaints from 84 patients. The third one, Mammography-Complaints Breast Database (MCBD), includes 181 multimodal cases with mammographic images, clinical complaints, and pathological results from Harbin Medical University, comprising 117 benign and 64 malignant cases. The last one, which is Liwen Database, is a private collection from Liwen Radiology Center in Kairouan, Tunisia, containing 200 paired mammograms (CC and MLO views) from 100 patients with 95 benign and 5 malignant cases, at resolutions ranging from 2364×2964 to 4728×5928 pixels. Notably, the Liwen database exhibits different tissue texture characteristics compared to the Chinese databases due to demographic, genetic, and risk profile variations, providing a robust cross-population validation challenge.

4.2 Evaluation Metrics

The proposed method was evaluated using standard classification metrics, including accuracy, precision, recall, F1-score, and AUC. Accuracy measures overall correctness, while precision and recall assess the quality of positive predictions and the detection of actual positive cases, respectively. The F1-score provides a balance between precision and recall, and AUC evaluates the model's ability to discriminate between classes.

4.3 Qualitative Results

To demonstrate the effectiveness of the proposed method, two contrasting clinical cases are presented: a benign case and a histologically confirmed malignant case, highlighting multimodal evidence convergence.

Figure 2 shows the complete multimodal analysis pipeline for a typical benign case from the Liwen database (Patient LW-089). CC and MLO mammograms display normal breast architecture with ACR1 density. The VAE-based image classifier produces low malignancy probabilities ($p_{\mathrm{CC}} = 0.12$, $p_{\mathrm{MLO}} = 0.15$), yielding a fused image probability of $p_{\mathrm{image}} = 0.14$ (86% benign confidence). The clinical report confirms no suspicious lesions; NLP-based feature extraction assigns $p_{\mathrm{clinical}} = 0.06$ (94% benign). Dempster–Shafer fusion shows minimal conflict ($K = 0.12$), resulting in a final benign classification with 91.9% confidence and 3.9% residual uncertainty.

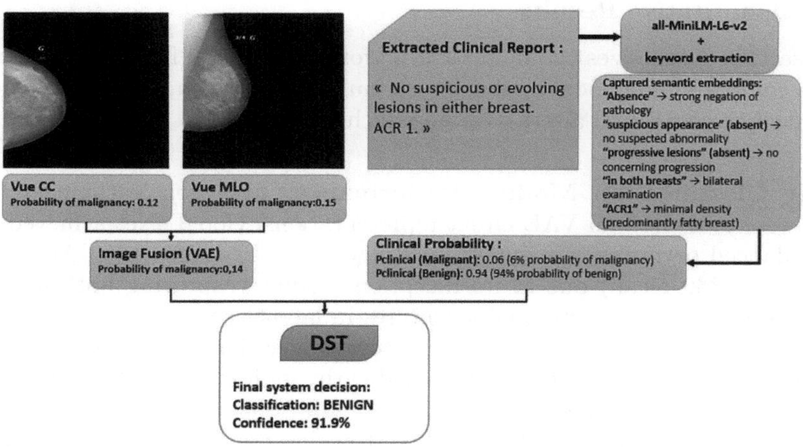

Fig. 2. Illustrative example of a complete multimodal analysis pipeline for a benign case.

Figure 3 presents a malignant case from the CMMD database (Patient D1-0124). CC and MLO views reveal a 15-mm spiculated mass with architectural distortion. The VAE-based classifier predicts high malignancy ($p_{\mathrm{CC}} = 0.79$,

$p_{\text{MLO}} = 0.85$), fused to $p_{\text{image}} = 0.82$. Clinical data identify major risk factors (age ¿50, family history, ACR3 density), producing $p_{\text{clinical}} = 0.94$. Dempster–Shafer fusion confirms strong evidence convergence ($K = 0.08$), resulting in a final malignant classification with 96% confidence and 1% residual uncertainty.

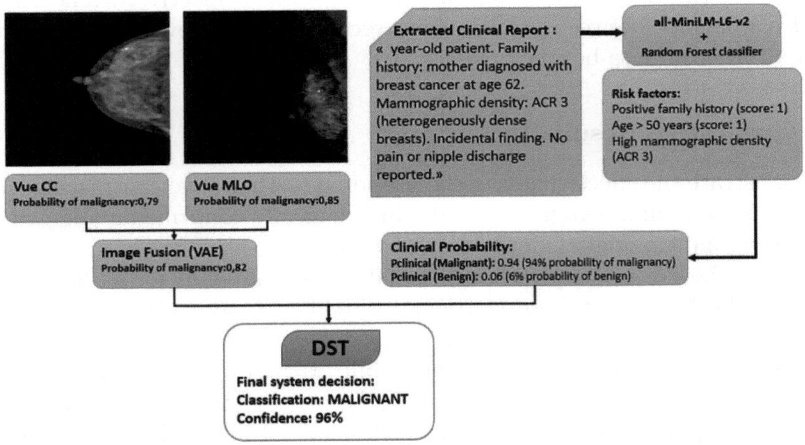

Fig. 3. Illustrative example of a complete multimodal analysis pipeline for a malignant case.

4.4 Quantitative Results

This section synthesizes the numerical performance of the framework's components, comparing unimodal modules with multimodal fusion, with a focus on the impact of the VAE and alternative approaches.

Image Classification Module Performance: five deep learning architectures were evaluated on VAE latent representations. GoogLeNet achieved optimal balanced performance (74.60% accuracy, 0.843 F1-score, 0.603 AUC, 0.827 precision, 0.860 recall) due to its Inception modules capturing both local fine patterns and broader contextual structures in latent space. Vision Transformer achieved highest precision (0.804) but showed precision-recall imbalance. Simpler models (AlexNet, VGG16) plateaued early, indicating insufficient capacity for latent space complexity modeling.

Furthermore, without VAE, GoogLeNet achieved only 68.50% accuracy (F1-score of 0.762), and Vision Transformer 70.00% accuracy (F1-score of 0.781). The VAE's probabilistic latent space regularization acts as a powerful denoising mechanism, filtering DICOM artifacts and encoding fundamental morphological lesion characteristics. The probabilistic nature—representing each image as a distribution rather than a fixed point—confers intrinsic robustness, improving generalization by learning to associate latent regions with classes rather than specific pixels.

Table 2. Unimodal vs. multimodal DST-based performance

Modality	Accuracy	Precision	Recall	F1-Score	AUC
Images only	24.0%	24.0%	100%	0.387	0.386
Clinical only	100%	100%	100%	1.000	0.566
Multimodal (DST)	**98.0%**	**92.3%**	**100%**	**0.960**	**1.000**

Clinical Information Extraction Module the Random Forest classifier (100 estimators, max depth 5) trained on all-MiniLM-L6-v2 embeddings + keyword features achieved 100% accuracy, F1-score, precision, and recall. The relatively low AUC (0.566) resulted from limited test set size and synthesized clinical reports creating near-perfect feature space separability with highly confident predictions (probabilities near 0 or 1), where AUC may not faithfully reflect discriminative quality. Validation on larger prospective clinical cohorts with natural reports is necessary.

Multimodal Fusion Performance despite substantial tissue texture differences between Tunisian (Liwen) and Chinese populations, our method demonstrated robust cross-population generalization, with DST fusion capturing diagnostic invariants beyond superficial texture variations (Table 2).

The multimodal system achieved high recall (100%) on the evaluated datasets, ensuring detection of all malignant cases within this experimental setting. The AUC reached 1.000, indicating strong class separability. However, these results should be interpreted cautiously due to the limited size and imbalance of certain datasets, particularly the private Tunisian dataset, which includes only a small number of malignant cases. This may lead to optimistic performance estimates and limits the statistical reliability of the evaluation.

Table 3. Performance comparison with reference studies.

Study	Datasets	Modalities	Fusion Method	Accuracy	F1-Score	AUC
[20]	CMMD + CBCID + MCBD	Images + Clinical Reports	AE + GPT-4 + Fusion	72.5%	0.798	0.825
[12]	Private Dataset	Multi-view Images + Clinical Metadata	Feature Concatenation + MLP	93.0%	–	0.98
Our Method	CMMD + CBCID + MCBD + LIWEN	Images + Clinical Reports	DST (Hybrid Fusion)	**98.0%**	**0.960**	**1.000**

Comparative Study Our DST framework outperforms simple fusion approaches (weighted average, majority voting) by 7–12% in F1-score, particularly managing ambiguous cases with divergent modal predictions. Systematic uncertainty management via mass assignments proved crucial for avoiding misclassifications occurring with deterministic fusion rules (Table 3).

Despite promising results, several limitations must be acknowledged. The private Tunisian dataset is highly imbalanced, with only five malignant cases, and partially synthesized clinical reports may bias performance estimates. Additionally, the lack of external clinical validation and the risk of overfitting due to limited data suggest that this framework remains a preliminary contribution, requiring validation on larger and more diverse real-world datasets.

5 Conclusion

This paper presents a hybrid multimodal fusion framework for breast cancer diagnosis, integrating mammographic images and clinical reports using Dempster–Shafer Theory. The proposed system demonstrates promising results and highlights the potential of combining imaging and clinical data within an uncertainty-aware framework.

However, due to dataset limitations and the lack of large-scale validation, these findings should be interpreted as preliminary. Future work will focus on validating the approach on larger and more representative datasets, incorporating clinician feedback, and improving the interpretability of the model to support real-world clinical deployment.

Acknowledgements. This work was carried out with the support and funding of the Tunisian Ministry of Higher Education and Scientific Research as part of the program to encourage young researchers (PEJC).

References

1. Sung, H., et al.: Global cancer statistics 2020: GLOBOCAN estimates of incidence and mortality worldwide for 36 cancers in 185 countries. CA Cancer J. Clin. **71**(3), 209–249 (2021)
2. Yala, A., Lehman, C., Schuster, T., Portnoi, T., Barzilay, R.: A deep learning mammography-based model for improved breast cancer risk prediction. Radiology **292**(1), 60–66 (2019)
3. Lotter, W., et al.: Robust breast cancer detection in mammography and digital breast tomosynthesis using an annotation-efficient deep learning approach. Nat. Med. **27**(2), 244–249 (2021)
4. Liang, G., Zheng, L., Wang, L.: A review of multimodal medical image fusion techniques. Comput. Math. Methods Med. **2021**, 9953348 (2021)
5. Abhisheka, B., Biswas, S.K., Purkayastha, B.: HBMD-net: feature fusion based breast cancer classification with class imbalance resolution. J. Imaging Inform. Med. **4** (2024)

6. Jouirou, A., Baazaoui, A., Barhoumi, W.: Multi-view information fusion in mammograms: a comprehensive overview. Inf. Fusion (2019)
7. Abderrahim, M., Baâzaoui, A., Barhoumi, W.: Survey on multi-source medical imaging fusion for classification and retrieval: current status and available datasets. SN Comput. Sci. **6** (2025)
8. Wang, Y.-M., Wang, C.-Y., Liu, K.-Y., et al.: CNN-based cross-modality fusion for enhanced breast cancer detection using mammography and ultrasound. Tomography **10** (2024)
9. Yang, Y., Zhong, Y., Li, J.: Deep learning combining mammography and ultrasound images to predict the malignancy of BI-RADS US 4A lesions in women with dense breasts. Int. J. Surg. **110** (2024)
10. Lu, G., Tian, R., Yang, W.: Deep learning radiomics based on multimodal imaging for distinguishing benign and malignant breast tumours. Front. Med. **11** (2024)
11. Alhajlah, M.: A hybrid features fusion-based framework for classification of breast micronodules using ultrasonography. BMC Med. Imaging **24** (2024)
12. Sunba, A., AlShammari, M., Almuhanna, A., Alkhnbashi, O.S.: An integrated multimodal-based CAD system for breast cancer diagnosis. Cancers **16** (2024)
13. Nakach, F.Z., Idri, A., Goceri, E.: A comprehensive investigation of multimodal deep learning fusion strategies for breast cancer classification. Artif. Intell. Rev. **57** (2024)
14. Othman, N.A., Abdel-Fattah, M.A., Ali, A.T.: A hybrid deep learning framework with decision-level fusion for breast cancer survival prediction. Big Data Cogn. Comput. **7** (2023)
15. Wimmer, M.: Multi-task fusion for improving mammography screening data classification. IEEE Trans. Med. Imaging **41** (2022)
16. Misra, S., Jeon, S., Managuli, R.: Ensemble transfer learning of elastography and b-mode breast ultrasound images. arXiv preprint arXiv:2102.08567 (2021)
17. Mondol, R.K., Millar, E.K., Sowmya, A., Meijering, E.: BioFusionNet: deep learning-based survival risk stratification in ER+ breast cancer. IEEE J. Biomed. Health Inform. **28** (2024)
18. Black, S.: Multi-view classification using hybrid fusion and mutual distillation. In: Proceedings of the IEEE/CVF Winter Conference on Applications of Computer Vision (WACV) (2024)
19. Kingma, D.P., Welling, M.: Auto-Encoding Variational Bayes. arXiv preprint arXiv:1312.6114 (2013)
20. Du, J., Zhang, Z., Tao, W.: MultiBCD: a multimodal model that simulates the human diagnostic process for automated breast cancer detection. Biomed. Sig. Process. Control **104** (2025)

Adoption of AI-Based Recruitment Systems: Benefits, Challenges, and User Trust

Hao Yue Liu and Rébecca Deneckère(✉)

Paris-1 Panthéon-Sorbonne University, Paris, France
Haoyue.Liu@etu.univ-paris1.fr, denecker@univ-paris1.fr

Abstract. This research studies the application of Artificial Intelligence (AI) in the recruitment process and examines how such technologies are perceived and adopted by users. The study combines a literature review with an empirical investigation based on quantitative surveys and qualitative interviews in order to analyze both the benefits and the limitations of AI-based recruitment tools. The results show that AI contributes to improving efficiency in candidate screening, reducing processing time, and supporting decision-making during recruitment. However, the findings also highlight significant concerns related to algorithmic bias, lack of transparency, data privacy, and trust in automated systems. Despite the potential of AI to enhance objectivity and standardization in recruitment practices, it reveals that human oversight remains essential to mitigate risks and support responsible use. This work contributes to a better understanding of the impact of AI on recruitment processes and provides insights for organizations seeking to adopt AI-driven solutions in a responsible and informed manner.

Keywords: Artificial Intelligence · Recruitment · Information Systems · Algorithmic Bias · User Perception · Decision Support

1 Introduction

The recruitment process has undergone significant transformation with the increasing adoption of Artificial Intelligence. Traditionally characterized by manual and time-consuming procedures, recruitment is progressively supported by intelligent systems designed to automate, structure, and optimize key activities. The integration of AI into recruitment practices reflects broader changes in Human Resource Management and organizational processes [1–3]. This transformation has accelerated in recent years, particularly following the COVID-19 pandemic [2], and, in this context, AI-based tools have emerged as practical solutions to manage job offers, screen candidates, conduct interviews, and support recruitment decisions more efficiently [2, 4].

Organizations face large volumes of applications that are difficult to process using traditional methods alone [5, 6]. To address these challenges, AI-driven recruitment systems are used to automate resume screening, recommend candidate profiles, and support recruiters throughout the recruitment process [7–9]. These systems rely on techniques such as Natural Language Processing (NLP) and machine learning to analyze large

A. Bernasconi et al. (Eds.): RCIS 2026, LNBIP 597, pp. 102–114, 2027.
https://doi.org/10.1007/978-3-032-35043-5_9

amounts of structured and unstructured data. Despite their potential benefits, the use of AI in recruitment raises important ethical and organizational concerns. Several studies highlight risks related to algorithmic bias, transparency, and fairness, particularly when automated systems influence employment-related decisions [3, 10–12]. These concerns are critical in recruitment contexts, where biased or opaque systems may negatively affect candidates' trust and perceptions of legitimacy.

Although existing research has explored AI applications in recruitment and addressed ethical implications, the literature remains fragmented. Many studies focus on specific tools or technologies, while fewer adopt a holistic view of AI integration across the different phases of the recruitment process [13–15]. Moreover, empirical investigations of user perceptions, considering both candidates and HR professionals, remain limited. This study addresses these gaps by adopting a phase-dependent and user-centered perspective on AI-driven recruitment. The central research question of this work is: **What are users' perceptions of the use of Artificial Intelligence across the recruitment process?** To answer this question, the study is structured around four sub-questions: *(RQ1) Which phases of the recruitment process are impacted by AI? (RQ2) Which AI techniques are used across these phases? (RQ3) How do users perceive AI-based recruitment systems? (RQ4) What challenges and limitations are associated with the use of AI in recruitment?*

Beyond performance considerations, the adoption of AI-based recruitment systems depends on user trust, which is not only a technical property but also a socio-organizational construct. In recruitment, trust relates to how users perceive fairness, transparency, and accountability when automated systems influence employment outcomes [3, 17]. From an information systems perspective, AI-based recruitment tools can be understood as decision-support systems that reshape responsibilities and control in organizational decision-making. As a result, acceptance is likely to vary depending on whether AI outputs are interpretable, whether users can adjust parameters, and whether human oversight is maintained throughout the process [3, 18]. This perspective motivates our stage-dependent approach: the same organization may accept automation for high-volume, standardized activities, while rejecting automation in stages where contextual judgment and responsibility are central. Addressing trust therefore requires both technical measures and governance arrangements [10, 17].

Methodologically, this research combines a structured literature review with quantitative and qualitative empirical approaches. The literature review enables a systematic identification of AI techniques applied to recruitment, while surveys and semi-structured interviews provide insights into user perceptions and experiences. We propose a structured mapping of AI techniques across the main phases of the recruitment process, provide empirical evidence on user perceptions of AI-based recruitment systems, and highlight key challenges related to bias, transparency, and human oversight, contributing to ongoing discussions on responsible AI adoption in recruitment [3, 12].

This paper is structured as follows. Section 2 provides background and some related work. Section 3 describes the research methodology. Section 4 presents the results. Section 5 discusses these findings and Sect. 6 concludes the paper.

2 Background and Related Works

Background in Recruitment. Recruitment is increasingly supported by digital and AI-based systems across multiple stages of the hiring process, reflecting a broader transition from traditional recruitment to e-recruitment and AI-driven practices [3, 4, 14, 30, 31]. This evolution has been widely documented in earlier studies on the digitalization of recruitment processes [30, 31]. AI-based recruitment systems rely on several complementary techniques, including NLP, machine learning, generative AI, and computer vision. NLP is widely used for resume parsing, information extraction, and candidate matching [7, 19–21]. Machine learning supports candidate ranking and predictive evaluation based on historical data [8–10, 22–24], although it may reproduce existing biases if not carefully monitored [3, 10, 12]. Generative AI is mainly used to assist candidate interaction and content generation, such as drafting job descriptions or answering applicant queries [25, 26], but remains limited in evaluative tasks due to reliability concerns [3, 27]. Computer vision is primarily applied in video interviews to analyze behavioral and non-verbal signals [13, 27–29], although its interpretability and validity remain debated [13, 29].

Related Works. AI-based recruitment systems can be conceptualized as socio-technical decision-support systems operating in a high-stakes domain. Their adoption depends not only on technical performance but also on key perceptual constructs studied in information systems research: trust, fairness, transparency, acceptability. Trust refers to the willingness of users to rely on AI-generated outputs under conditions of uncertainty [16]. In recruitment, trust is closely linked to perceived fairness and transparency. Fairness refers to the absence of systematic bias in algorithmic decision-making [3, 12]. The literature shows that AI systems may reproduce or amplify existing biases when trained on historical data [3, 10]. Transparency relates to the extent to which users can understand how AI systems generate outputs. Limited transparency reduces the ability to interpret or challenge decisions and negatively affects trust [17]. These constructs are interdependent. A lack of transparency may undermine fairness perceptions, which in turn reduces trust and acceptance. Conversely, explainability and human oversight can reinforce trust and accountability [17, 18]. AI-based recruitment systems can therefore be analyzed as decision-support systems, where acceptance depends on how control and responsibility are distributed between human and automated actors.

3 Research Methodology

The study combines a structured literature review, a quantitative survey, and semi-structured interviews to address the four research questions.

Structured Literature Review. A structured literature review was conducted to identify how AI techniques are applied across the different phases of the recruitment process and to address the first two research questions. The literature search was conducted using the SCOPUS database, which aggregates publications from major scientific publishers such as Elsevier, ACM, Springer, IEEE, and Wiley. The following search string was applied: *(TITLE-ABS-KEY("Artificial Intelligence") OR TITLE-ABS-KEY("AI")) AND*

(TITLE-ABS-KEY("selection") OR TITLE-ABS-KEY("job matching") OR TITLE-ABS-KEY("CV")) AND (TITLE-ABS-KEY("recruitment") OR TITLE-ABS-KEY("hiring")) AND PUBYEAR > 2022. A total of 236 papers were initially identified. After title and abstract screening, irrelevant studies were excluded based on lack of focus on recruitment or AI applications. Following full-text analysis, 35 papers were retained. Reasons for exclusion included lack of relevance, insufficient methodological detail, or non-peer-reviewed sources. Two additional studies were identified through snowballing, resulting in a final corpus of 37 articles. The screening process was conducted by two researchers. Each article was independently evaluated based on predefined inclusion and exclusion criteria. Disagreements were discussed and resolved through consensus. To ensure consistency, classification rules were defined prior to the analysis phase.

Quantitative Study. The survey follows a descriptive and exploratory research design and does not aim to test causal relationships. To address user perceptions of AI-based recruitment systems, a quantitative survey was conducted. The survey targeted both job candidates and HR professionals and aimed to capture perceptions related to usefulness, trust, fairness, and acceptance of AI in recruitment processes. The questionnaire contained 22 closed-ended questions with Likert-scale items adapted from prior research on trust, fairness, and technology acceptance [16, 18]. The collected data were analyzed using descriptive statistics to identify general trends in user perceptions. The questionnaire was disseminated via several digital channels (LinkedIn, Discord, university mailing lists), enabling access to a broad and diverse population. The target participants were individuals currently seeking employment or having recent job search experience, with a particular focus on candidates from technical and scientific fields. The findings should be interpreted as exploratory, given the convenience sampling strategy and the predominance of young respondents with technical or scientific backgrounds [16, 18]. In total, 110 valid responses were collected. Data collection took place in France between March and April 2025. The sample consists primarily of respondents aged between 20 and 30, with a majority having a technical or scientific background.

Qualitative Study. To complement the quantitative findings, semi-structured interviews were conducted with selected participants. Interview transcripts were analyzed thematically to identify recurring perceptions, benefits, and concerns. Four semi-structured interviews were conducted with two HR professionals and two technical professionals from medium-sized organizations. All interviewees were actively involved in recruitment activities and had direct experience with at least some form of digital or AI-supported hiring practice. Interviews were semi-structured and allowed follow-up probes. With consent, interviews were recorded, transcribed, and systematically analyzed to identify recurring perceptions, benefits, and concerns. Although limited in number, the interviews are used as an exploratory complement to the survey and aim to provide contextual insights rather than generalizable findings. This qualitative component provided contextual insights into real recruitment practices.

3.1 Validity and Limitations

Several limitations should be acknowledged. The empirical studies were conducted on a limited sample, which may affect the generalizability of the findings. In addition, the rapid evolution of AI technologies may limit the temporal validity of the results. Nevertheless, the combination of literature review, quantitative survey, and qualitative interviews enhances the robustness of the study by enabling methodological triangulation.

4 Results

4.1 Phases of the Recruitment Process Impacted by AI (RQ1)

Table 1 shows the phases impacted by AI in Recruitment. AI is used mainly in candidate screening (11 papers), followed by interviews (6 papers), job needs and job offers (5 papers), and final decision-making (4 papers). This distribution shows that AI adoption is strongest in high-volume and standardized phases, while later stages remain more human-driven. Five papers reported AI use across multiple phases.

Table 1. Phases impacted by AI.

Phases	Number	References
Job needs & Job offer	5	[7, 19, 20, 22, 26]
Candidate Screening	11	[7–10, 15, 19, 21–25]
Interviews	6	[7, 13, 19, 27–29]
Final Decision-Making	4	[23, 24, 32, 33]

4.2 AI Techniques Used in Recruitment Processes (RQ2)

The literature review identifies four main categories of AI techniques in recruitment: NLP, generative AI, decision-oriented AI, and computer vision. Their distribution across recruitment phases is summarized in Table 2.

Table 2. Synthesized Table – AI Techniques Across Recruitment Phases.

Category	Type of AI technique	Job needs & job offer	Candidate screening	Interview	Final decision
NLP	Job description analysis & skill extraction	[7, 19, 20, 23, 26]	[7, 19, 21, 23, 24]	[7, 19]	
	Resume parsing & linguistic matching		[7–10, 21, 22, 24, 25]		

(*continued*)

Table 2. *(continued)*

Category	Type of AI technique	Job needs & job offer	Candidate screening	Interview	Final decision
	Text-based candidate communication		[7, 23]	[7]	
Generative AI	Chatbots & conversational agents	[7, 20]	[7, 23]		
	Automated Q&A/guidance systems	[19, 26]	[7, 21]		
Decision-oriented AI	Candidate scoring & ranking		[8–10, 19, 22, 24, 25]		[23, 24]
	Recommendation & predictive systems		[21, 24]	[27, 29]	[23, 32] [24, 33]
Computer Vision	Facial expression analysis			[13, 19, 27]	
	Behavioral & video analysis			[7, 28, 29]	

Figure 1 complements Table 2 by mapping recruitment phases to AI functional roles, from automation to evaluative support.

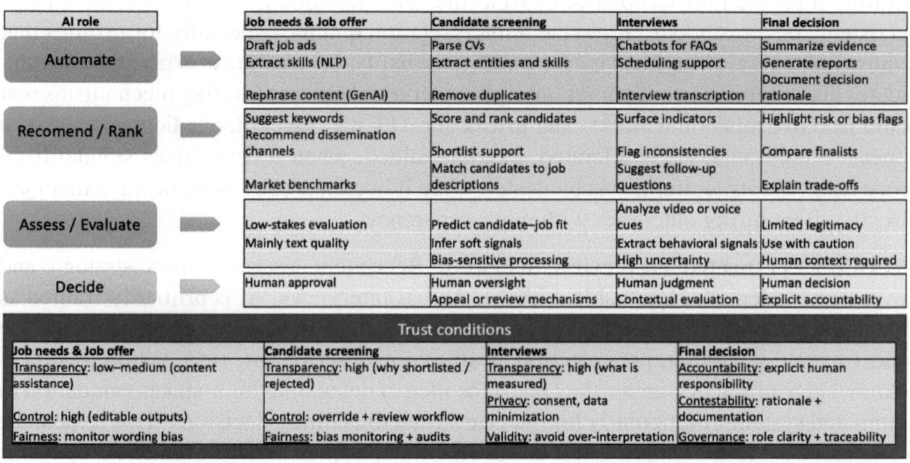

Fig. 1. Roles of Artificial Intelligence across recruitment stages.

Figure 1 shows that AI is mainly used to automate and rank in early stages, while later stages allow only limited evaluative support and keep decision authority human.

This supports a hybrid model in which AI improves efficiency without replacing human judgment in high-stakes decisions.

4.3 User Perceptions of AI-Based Recruitment Systems (RQ3)

User perceptions differ between candidates and recruiters.

Candidate Perspective (Survey Results). The survey indicates that exposure to AI in recruitment is already common among respondents, especially within the 20–30 age group, who represent the majority of the sample. Candidates most often reported encountering AI in standardized early-stage tools, notably online questionnaires and automated resume screening. By contrast, more advanced uses, like AI-supported interviews or automated pre-selection, were reported less frequently. This distribution is consistent with the literature showing that AI adoption is strongest in high-volume, standardized phases such as screening and matching [7, 19].

In terms of affective reactions, respondents reported a predominantly neutral stance. Most candidates described themselves as indifferent toward AI-based recruitment. This pattern suggests that AI does not trigger systematic rejection, but it also fails to generate strong positive engagement, an observation aligned with prior research emphasizing cautious acceptance of AI-enabled HR tools [2, 6].

Despite this emotional neutrality, candidates clearly recognized practical benefits. Speed was widely perceived as a key advantage, and many respondents judged AI as at least as effective as humans for screening resumes and cover letters. These findings align with studies highlighting efficiency, scalability, and workload reduction as central drivers of acceptance for AI in recruitment [2, 7, 34]. However, perceived effectiveness does not fully translate into acceptance of AI as the sole evaluator. Survey responses point to persistent concerns about fairness and interpretation quality, especially for profiles that require context (e.g., atypical career paths, originality, or non-linear experiences). Candidates also expressed concerns about rigid criteria and opaque filtering mechanisms that could lead to exclusion without clear justification [3, 35]. Overall, candidate acceptance appears conditional: AI is tolerated, or even valued, when it accelerates standardized processing, but its legitimacy is contested when it becomes evaluative in ways that may directly affect hiring outcomes without transparency.

Recruiter Perspective (Interview Insights). Recruiters describe a more strategic and governance-oriented relationship with AI. Across interviews, AI is primarily framed as a decision-support tool used to reduce workload and improve consistency in repetitive tasks (resume screening, pre-selection, drafting job descriptions, etc.). While recruiters acknowledge efficiency gains, they resist the idea of delegating high-stakes evaluation to automated systems, particularly for interviews and final hiring decisions. This reluctance is explained by both technical limits and professional responsibility: recruiters emphasize that motivation, authenticity, interpersonal dynamics, and cultural fit remain difficult to assess reliably through automated indicators [6, 13].

A central condition for recruiter trust is transparency combined with controllability. Interviewees report higher acceptance when they can understand the basis of AI recommendations, adjust parameters, and override outputs. Conversely, limited explainability is perceived as a threat to accountability, internally (loss of decision control) and externally (difficulty justifying outcomes to candidates) [3, 36]. This echoes broader work on trust in AI, where explainability and human oversight are repeatedly identified as key determinants of acceptance in decision-making [16, 17].

Recruiters also report an emerging challenge linked to candidates' own use of generative AI during remote tests or interviews. Interviewees describe situations where AI-assisted answers create inconsistencies between written performance and candidates' ability to justify reasoning in follow-up discussions. This trend complicates evaluation and reinforces the need for clear rules and explicit communication about acceptable AI use during selection stages [19, 27].

Synthesis: Role-Dependent Acceptance Conditions. Taken together, these results confirm that perceptions of AI-based recruitment are not uniform. Candidates primarily evaluate AI through its perceived fairness, transparency, and personal impact, while recruiters prioritize governance, accountability, and decision control [3, 16, 17]. Both groups express higher acceptance when AI is embedded in hybrid recruitment models, where automation improves efficiency in standardized steps, but human judgment remains central in interpretive and decision-critical stages [3, 13, 18]. These findings should be interpreted as indicative rather than statistically generalizable, given the exploratory nature of the study.

4.4 Challenges and Limitations of AI in Recruitment (RQ4)

The results identify several challenges and limitations associated with the use of Artificial Intelligence in recruitment processes. One of the most frequently reported challenges concerns *algorithmic bias*. Survey respondents and interviewees expressed concerns that AI-based recruitment systems may reproduce or amplify biases embedded in historical recruitment data, potentially affecting fairness in candidate evaluation [3, 35]. Prior research highlights that biased representations and stereotypes may also emerge from AI-generated outputs, reinforcing discriminatory patterns [37].

Transparency emerged as another major challenge reported by participants. Users indicated difficulties in understanding how AI systems evaluate candidates, rank profiles, or generate recommendations. This lack of transparency was identified across multiple recruitment phases, particularly during candidate screening and AI-supported interviews, and was reported as negatively influencing user acceptance [3]. Existing research further emphasizes that limited explainability of AI models constitutes a key barrier to trust and accountability in AI-driven decision-making systems [36]. Participants' concerns about transparency also point to a practical requirement: AI outputs must be explainable to remain contestable and accountable in recruitment settings [3, 17, 36]. This includes the ability to communicate why a profile was recommended or rejected, and which criteria had the strongest influence. In practice, explainable AI techniques and monitoring tools can support this requirement by providing interpretable factors and bias indicators that are meaningful for HR stakeholders, rather than only for technical teams [10, 17]. Such mechanisms do not eliminate the need for human judgment, but they can reduce opacity.

Closely related to transparency, *trust* in AI-based recruitment systems was reported as a significant challenge. The results indicate that user trust varies depending on the perceived role of AI within the recruitment process. Lower levels of trust were reported when AI systems were perceived as operating autonomously, whereas higher acceptance was associated with AI being used as a decision-support tool rather than as a replacement for human judgment [18].

In addition, *data privacy* concerns were reported, particularly in relation to AI-supported video interviews and the analysis of behavioral or biometric data. Participants expressed apprehension regarding the collection, storage, and potential reuse of such sensitive data during recruitment processes [4].

Finally, participants highlighted limitations related to the *adaptability* of AI systems. AI-based tools were reported as less effective when applied to complex or atypical candidate profiles, reinforcing the perceived need for human oversight to contextualize AI-generated outputs and ensure informed recruitment decisions [2, 6].

5 Discussion

The results show a clear asymmetry in the perception and acceptance of AI-based recruitment systems between candidates and recruiters. While both groups recognize the operational benefits of AI, their expectations and concerns differ significantly depending on their role in the recruitment process.

From the candidates' perspective, the findings suggest a form of pragmatic acceptance. Most candidates, particularly younger ones, are frequently exposed to AI-based recruitment tools and perceive them as efficient, especially in terms of speed and application processing. However, this acceptance remains largely neutral rather than enthusiastic, as reflected by the high proportion of respondents expressing indifference toward AI. This neutrality appears to stem from limited transparency regarding how AI systems operate and evaluate candidates, which constrains the development of trust.

Recruiters exhibit a more instrumental and controlled acceptance of AI. Interviews indicate that recruiters primarily value AI for its ability to automate repetitive tasks and support early-stage decision-making, while firmly rejecting its use as an autonomous decision-maker. This conditional acceptance reflects a strong emphasis on human oversight, accountability, and ethical responsibility.

This divergence in perceptions points to a critical issue in AI-enabled recruitment: trust emerges from both system properties and organizational arrangements. While candidates assess AI systems based on perceived fairness and transparency, recruiters evaluate them in relation to governance, explainability, and their capacity to remain in control of decision-making processes. These findings align with prior research emphasizing that transparency and explainability are key determinants of trust in AI-supported human resource management systems [3, 16, 17].

From an information systems viewpoint, these results suggest that AI-based recruitment should be designed and governed as a decision-support capability rather than as an autonomous decision-maker. Trust is strengthened when users can understand the rationale of outputs (transparency), influence how the system operates (controllability), and keep a clear allocation of responsibility (accountability) [16, 17]. This is particularly critical in recruitment because decisions are consequential, contestable, and closely tied to fairness perceptions [3, 36]. Based on our findings, three practical principles can be derived for responsible adoption. First, organizations should implement "human-in-the-loop" arrangements that explicitly define where AI is allowed to recommend, where it can rank, and where it cannot decide [3, 18]. Second, systems should provide actionable explanations aligned with HR reasoning, allowing users to justify outcomes and challenge questionable recommendations [17]. Third, organizations should monitor bias and data drift over time and document mitigation measures through regular audits, especially when models are trained on historical hiring data [3, 10]. These principles support a hybrid model in which efficiency gains do not come at the expense of legitimacy and user trust.

These findings contribute to research on decision-support systems by showing that AI-based recruitment tools should not be analyzed as standalone technologies, but as socio-technical capabilities embedded in organizational decision-making structures. The results highlight a phase- and role-dependent configuration of AI, where the same system may act as an automation mechanism, a recommendation engine, or an evaluative support depending on the recruitment stage and the level of responsibility involved. This contribution extends prior work on AI-supported decision-making by emphasizing that acceptance and trust are not driven solely by technical performance, but by how decision authority, accountability, and human oversight are distributed between AI systems and human actors across process stages.

The results suggest that current AI implementations are better suited to standardized and high-volume recruitment tasks than to stages requiring contextual interpretation and interpersonal judgment. The persistent reliance on human recruiters in later stages reflects not only technical limitations but also organizational and ethical considerations. Hybrid recruitment models, combining AI efficiency with human judgment, represent today the most acceptable and sustainable approach.

Emerging risks related to AI in recruitment should also be considered. These include the use of deepfake technologies or synthetic identities in video interviews, the manipulation of training data (data poisoning), and privacy attacks targeting applicant data. These risks extend concerns related to bias, transparency, and accountability [3, 10, 17]. They highlight the need for governance mechanisms such as audit procedures, usage policies, and clear accountability structures.

To conclude, we highlight that successful AI adoption in recruitment depends less on technological maturity alone than on how AI systems are integrated into existing practices, communicated to users, and governed within organizations. Addressing transparency, bias mitigation, and human control emerges as a central challenge for fostering trust and long-term acceptance of AI-based recruitment systems.

6 Conclusion

This paper studies the use of Artificial Intelligence across the recruitment process, with a particular focus on the phases affected, the techniques employed, and users' perceptions of AI-based recruitment systems. By combining a structured literature review with quantitative and qualitative empirical methods, the study provides a comprehensive and phase-dependent view of AI adoption in recruitment.

One key contribution of this research lies in the identification of a dual perception framework. The results reveal a pragmatic acceptance of AI among candidates, driven primarily by efficiency gains, alongside a more conditional and governance-oriented acceptance among recruiters. This distinction shows that acceptance of AI in recruitment is not uniform and depends strongly on users' roles, expectations, and levels of control over decision-making processes. The findings further demonstrate that AI adoption in recruitment remains phase-dependent. AI techniques are predominantly applied to early-stage, standardized tasks such as resume screening and job matching, while later stages continue to rely heavily on human judgment. Hybrid recruitment models in which AI should act as a support tool rather than an autonomous decision-maker.

From a theoretical perspective, this study contributes to the growing body of research on AI in human resource management by emphasizing the central role of trust, transparency, and human oversight in shaping user acceptance. Practically, the results suggest that organizations should prioritize explainability, clear communication, and ethical governance when deploying AI-based recruitement systems. This work highlights a concrete socio-technical requirement for AI adoption in recruitment: aligning system capabilities with governance mechanisms to preserve trust and accountability [3, 16, 17]. For organizations, this implies that deployment decisions should be phase-dependent and accompanied by clear communication to candidates about where AI is used and how it influences outcomes. This also suggests that evaluating AI recruitment systems should not be limited to accuracy or efficiency metrics, but should include user-facing criteria such as explainability and perceived fairness, which directly shape acceptance conditions [17, 36].

Despite limitations related to sample size and the exploratory nature of the empirical study, this research offers valuable insights into the socio-technical challenges of AI-enabled recruitment. Future research could extend this work by examining longitudinal adoption patterns, exploring cross-cultural differences, or assessing the impact of explainable AI mechanisms on trust and acceptance in recruitment contexts.

References

1. Madhavi, K., Kaveri, S.: AI integration in HR processes. In: Balancing Automation and Human-Machine Interaction in Business, pp. 55–72. IGI Global, Hershey, PA (2024)
2. Rajapriya, S., Subrahmanyam, M.: A conceptual study on perception towards the implementation of artificial intelligence in the recruitment and selection process in MNC companies. Balancing Automation and Human-Machine Interaction in Modern Marketing, pp. 1–18. IGI Global, Hershey, PA (2024)
3. Rigotti, L., Fosch-Villaronga, E.: Fairness, AI and recruitment. Comput. Law Secur. Rev. **52**, 105966 (2024)
4. Ben Ayed, H.: Recruitment in the era of web 4.0: what place for artificial intelligence in recruitment? Qubahan Acad. J. **4**(2), 45–58 (2024)
5. Kumar, A., Singh, V.: The impact of artificial intelligence in recruitment and selection processes in IT companies. IEEE Conf. Emerg. Comput. Appl. (2024)
6. Sharma, S., Bansal, P.: Should i adopt AI during talent acquisition? Evidence from HR professionals of Indian IT organisations. J. Organ. Effectiveness: People Performance **11**(1), 67–85 (2024)

7. Trziszka, M.: Artificial intelligence in employee recruitment. In: European Conference on Knowledge Management (ECKM), pp. 1123–1131 (2023)
8. Patil, S., Mehta, P., Kulkarni, V.: AI-based multimodal resume ranking web application for large scale job recruitment. IEEE Int. Conf. Intell. Data Anal. Process. (IDAP), 1–6. IEEE (2024)
9. Sharma, R., Verma, A.: An ML-based resume screening and ranking system. IEEE Int. Conf. Smart Comput. 1–5. IEEE (2024)
10. Albaroudi, E., Mansouri, T., Alameer, A.: A comprehensive review of AI techniques for addressing algorithmic bias in job hiring. AI **5**(1), 383–404 (2024)
11. Caliskan, A., Bryson, J., Narayanan, A.: AI generates covertly racist decisions about people based on their dialect. Nature **630**, 146–151 (2024)
12. Raghavan, M., Barocas, S., Kleinberg, J., Levy, K.: Algorithmic inclusion: shaping the predictive algorithms of artificial intelligence in hiring. Hum. Resour. Manag. J. **33**(2), 312–329 (2023)
13. Rigotti, L., Fosch-Villaronga, E.: Artificial intelligence employment interviews: examining limitations, biases, and perceptions. Computer **57**(6), 78–87. IEEE (2024)
14. Melliani, S., El Kharbouchi, A.: Impact of E-recruitment on recruiter-candidate engagement. SA J. Hum. Resour. Manag. **19**(3), 1–12 (2024)
15. Hosseini, S., Ahmadi, M.: Exploring the trends of artificial intelligence in recruitment: A bibliometric study. Int. J. Supply Oper. Manag. **11**(2) (2024)
16. Choung, H., David, P., Ross, A.: Trust in AI and its role in the acceptance of AI technologies. Int. J. Hum.-Comput. Interact. **39**(5), 1017–1032 (2023)
17. Thalpage, N.S.: Unlocking the black box: explainable artificial intelligence (XAI) for trust and transparency in AI systems. AI Ethics **4**(1), 33–48 (2023)
18. Vrontis, D., Christofi, M., Pereira, V., Tarba, S.: The adoption of artificial intelligence in employee recruitment: the influence of contextual factors. Int. J. Hum. Resour. Manag. **34**(12), 2368–2393 (2023)
19. Silva, R., Gomez, P.: Exploring the role of chatbots in the recruitment process in Latin America. Revista de Gestão Social e Ambiental **18**(1), 1–20 (2024)
20. Martínez, H., López, D.: Intent classification methods for human resources chatbots. Procesamiento del Lenguaje Natural **73**, 55–62 (2024)
21. Chen, Y., Zhou, L.: NLP-based AI-driven resume screening solution for efficient candidate selection. Lecture Notes in Computer Science, vol. 14567, pp. 401–415. Springer, Cham (2024)
22. Lopez, M., Garcia, J.: Enhancing the hiring process: a predictive system for soft skills assessment. Decis. Making Appl. Manag. **3**, 1–14 (2024)
23. Wang, J., Li, H.: Intelligent recruitment decision support combining Transformer model and knowledge graph. In: Proceedings of the ACM Conference on Intelligent Systems (2024)
24. Liu, Z., Wang, Y.: Zero-shot recommendation AI models for efficient job-candidate matching in recruitment process. Appl. Sci. **14**(6), 2601 (2024)
25. Santos, R., Pereira, L.: Examining the potential for use of ChatGPT in recruitment for hotel businesses. In: Artificial Intelligence Applications in Tourism, IGI Global (2024)
26. Nguyen, T., Brown, K.: Leveraging generative artificial intelligence for human resource management: the AI job description assignment. J. Manag. Educ. **48**(4), 567–589 (2024)
27. Jarrahi, M.H., Sutherland, W.: ChatGPT vs human expertise in the context of IT recruitment. Expert Syst. Appl. **242**, 125868 (2024)
28. Biswas, S., Jung, J.-Y., Unnam, A., Yadav, K., Gupta, S., Gadiraju, U.: "Hi. I'm Molly, Your Virtual Interviewer!" Exploring the Impact of Race and Gender in AI-Powered Virtual Interview Experiences. AAAI Conference on Human Computation and Crowdsourcing (HCOMP), vol. 12(1), pp. 12–22. AAAI Press (2024)

29. Basch, J.M., Melchers, K.G.: The effects of medium and sequence on personality trait assessments in face-to-face and videoconference selection interviews. Hum. Resour. Manag. J. **31**(3), 451–468 (2021)
30. Johansson, J., Herranen, S.: The application of Artificial Intelligence (AI) in Human Resource Management: Current state of AI and its impact on the traditional recruitment process. Bachelor thesis, Jönköping University (2019)
31. Sills, M.: E-recruitment: A comparison with traditional recruitment and the influences of social media. Bachelor thesis, Helsinki Metropolia University of Applied Sciences (2014)
32. Khan, M., Ali, S.: Exploring evolving HR and recruitment strategies in the age of artificial intelligence. Multi. Sci. J. **7**(2), 101–118 (2024)
33. Gupta, S.: Recruitment and Staffing in Educational Sectors via Explainable AI and Blockchain. Sustainable Development Using Private AI, pp. 123–138. CRC Press, Boca Raton (2024)
34. Alnsour, Y., Haddad, R.: The impact of implementing AI in recruitment on human resource management efficiency and organizational development effectiveness. J. Infrastruct. Policy Dev. **8**(8), 1–15 (2024)
35. Zhang, Y., Liu, X.: From bias to brilliance: the impact of artificial intelligence usage on recruitment biases in China. IEEE Transactions on Engineering Management (2024)
36. Kim, J., Park, S.: The impact of artificial intelligence replacing humans in making human resource management decisions on fairness: a case of resume screening. Sustainability **16**(9), 3840 (2024)
37. Johnson, K., Taylor, S.: Discriminative and exploitive stereotypes: artificial intelligence generated images of aged care nurses and the impacts on recruitment and retention. Nurs. Inq. **31**(1), e12651 (2024)

Querying Structured Data Through Natural Language Using Language Models

Valentin Micu-Hontan[1], Andrei-Alexandru Bunea[1], Nikolaos Dimitrios Tantaroudas[2], and Dan-Matei Popovici[1(✉)]

[1] National University of Science and Technology POLITEHNICA Bucharest, Bucharest, Romania
{valentin.micu,andrei.bunea,matei.popovici}@upb.ro
[2] Institute of Communication and Computer Systems (ICCS), Athens, Greece
nikolaos.tantaroudas@iccs.gr

Abstract. This paper presents an open-source methodology for allowing users to query structured, non-textual datasets through natural language. Unlike Retrieval-Augmented Generation (RAG), which struggles with numerical and highly structured information, our approach trains an LLM to generate executable queries. To support this capability, we introduce a principled pipeline for synthetic training-data generation, producing diverse questionanswer pairs that capture both user intent and the semantics of the underlying dataset. We fine-tune a compact model—DeepSeek R1-Distill-8B—using QLoRA with 4-bit quantization, making the system suitable for deployment on commodity hardware. We evaluate our approach on a dataset describing accessibility to essential services across Durangaldea, Spain. The fine-tuned model achieves high accuracy across monolingual, multilingual, and unseen-location scenarios, demonstrating both robust generalization and reliable query generation. Our results highlight that small, domain-specific models can achieve high-precision for this task without relying on large proprietary LLMs, making this methodology suitable for resource-constrained environments and adaptable to broader multi-dataset systems.

Keywords: Natural Language querying · tool-using LLMs · model fine-tuning · Large Language Models

1 Introduction

LLM reasoning is constrained by training data patterns, a limitation often addressed via Retrieval-Augmented Generation (RAG) [10]. In a typical RAG pipeline, a retrieval system identifies the top-k relevant documents from a corpus and injects them into the model's context to generate data-backed answers. This process relies on representing text as vector embeddings in n-dimensional space and identifies matches using ℓ_2 distance. Despite its success, RAG struggles with

A. Bernasconi et al. (Eds.): RCIS 2026, LNBIP 597, pp. 115–126, 2027.
https://doi.org/10.1007/978-3-032-35043-5_10

structured or numerical data—such as calculating travel times to hospitals or supermarkets—which cannot be reliably retrieved as unstructured text. While state-of-the-art commercial models with 100k-token context windows and protocols like MCP [2] can handle such data by ingesting entire schemas, they are hindered by high operational costs, privacy concerns, and prohibitive hardware requirements. Similarly, Text-to-SQL methods [6,18] often require massive models to maintain accuracy. We propose a fully open-source methodology for answering natural-language questions using non-textual data. Our key contributions include: (i) high accuracy using **small models** (7-8B parameters), (ii) an **almost entirely automated** workflow, and (iii) a **cost-effective** fine-tuning approach. We utilize large models like GPT-4 only for training; at deployment, the system remains independent and lightweight. We apply this methodology to a rural accessibility dataset from the FUTURAL project in Durangaldea, Spain. Currently piloted for Quality of Life assessments, our approach is compatible with efficient architectures like Tiny Recursive Models [9], enabling deployment on resource-constrained devices like laptops.

The remainder of this paper is organized as follows: Sect. 2 introduces the problem setting and outlines the overall methodology. Section 3 describes the construction of the dataset we used for training and evaluation. Section 4 illustrates the model fine-tuning procedure. Section 5 reports the evaluation metrics and experimental results. Section 6 analyzes the performance of the deployed application. Section 7 discusses limitations of the current approach and directions for future research. Section 8 reviews related work, and Sect. 9 concludes the paper.

2 Problem Setting and Methodology

The DLR dataset contains 100,000+ records of coordinates and travel times (walking, cycling, driving) to hospitals, supermarkets, and pharmacies across 240 km^2 in Durangaldea, Spain. Developed for accessibility studies, it examines the links between demographics, settlement patterns, and walkability. Our goal is to provide a scalable, deployable natural-language interface for stakeholders that can be easily extended to other datasets, decouples the model from continuously evolving data, and does not depend in its design on specific query syntax. Crucially, we prioritize a fully open, non-proprietary approach.

Because RAG fails to reliably encode numerical data, our system (Fig. 1) generates executable queries, retrieves the data, and reasons over the results. Inspired by tool-use prompting [16], this method requires detailed system prompts defining schemas and syntax. While effective for massive models, our experiments show that smaller models often produce syntactically incorrect or irrelevant queries, highlighting the need for specialized fine-tuning.

Early query-generation approaches like **Toolformer** [14] rely on datasets instrumented by replacing factual knowledge with queryanswer examples. Models learn to detect missing information, trigger external queries, and integrate results into the context. However, high-quality, domain-specific QA pairs are

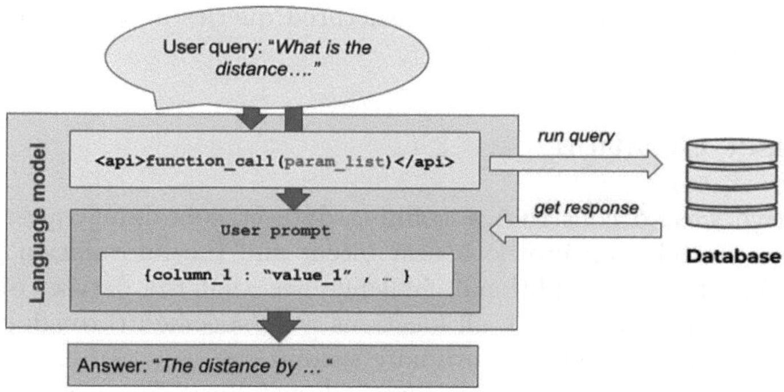

Fig. 1. Overview of the model inference and query generation process.

often unavailable for specialized databases. We address this via a **synthetic dataset creation** method designed for broad coverage and linguistic variation, ensuring robust generalization for targeted domains.

Building an expert-level data retriever with LLMs is supported by various commercial solutions; however, our solution diverges from these approaches and is guided by two key constraints: (i) it must **run on consumer-grade hardware**. In this work we focus on the NVIDIA GeForce RTX 3090, a 24-GB Ampere GPU suitable for inference and light training which is significantly more affordable than datacenter-class GPUs such as the A100; and (ii) it must rely solely on **open-source models**. We use DeepSeek R1-Distill-8B [3], a compact distilled variant of the DeepSeek-R1 family. Distillation transfers reasoning capabilities from a large model into a smaller one; in this case, the architecture is based on the 8-B Llama-3.1 Instruct model [1], fine-tuned on reasoning-oriented data. The resulting model is lightweight enough for deployment on the RTX 3090 while retaining strong problem-solving performance.

We fine-tune our model through the Quantized(Q)LoRA (Low-Rank Adaptation) [4] adapter framework. QLoRA introduces trainable low-rank matrices into selected weight layers of a frozen base model, allowing the system to learn domain-specific tasks without modifying the original model parameters. Adapters operate as lightweight, modular extensions that can be loaded or unloaded dynamically, enabling efficient specialization while preserving the underlying model's general linguistic and reasoning capabilities. This design makes training resource-efficient—only a small number of additional parameters are trained. At the same time deployment remains lightweight, since adapters require minimal memory and can be swapped in almost instant time during inference.

A central advantage of this architecture is its **extensibility**: a single base model can be paired with multiple adapters, each tailored to a different dataset or domain, thereby supporting multi-dataset systems without the need to duplicate the full model. We integrate the fine-tuned model into an application pipeline that combines geolocation services and a user-facing interface, enabling natural-language queries to be mapped onto structured queries and executed over the underlying dataset.

3 Dataset Creation

Our objective was to construct a training dataset containing questionanswer pairs that accurately capture both user intent and the information present in the underlying data. An additional challenge was that the dataset did not reference street addresses; instead, all locations were specified through geographic coordinates in the WGS-84 [8] coordinate system.

Our dataset-generation procedure proceeds as follows. For each table T available in the database, we consider all projections $\pi_{A_1,\ldots,A_i}(T)$ that capture semantically meaningful information. For every such projection, we construct a set of template questions designed to capture the types of information expressible from that attribute subset. Table 1 illustrates an example: a question instantiated from a template derived from a projection of the *Hospitals* table containing the attributes *Location* and *Distance.*

Table 1. Generating question from projections

Projection	$\pi_{\text{Location},\text{Drive\_Dist}}(Hospitals)$
Template	What is the nearest `location_type` from `location`?
Question	What is the nearest hospital from Durango?

For each identified projection, we use state of the art models (Gemini 2.5 Pro and DeepSeek R1) to generate templates that convey distinct semantic interpretations. We additionally include relevant superprojections—projections p_j whose attribute sets strictly contain those of p_i whenever they give rise to questions that reflect meaningfully different informational aspects. For instance, the question *Is the nearest hospital closer by bike or on foot?* is generated from a template p which is a superprojection of $\pi_{\text{Location}}(Hospitals)$. To ensure soundness, we manually inspect generated templates and filter-out invalid or duplicate templates. When instantiating each question template, we used the Overpass API [11] to extract a representative geographical sample consisting of 358 unique locations from the Durangaldea region. At the end of this process, we validated a total of 160 templates for the entire dataset. For every instantiated question, we also generated a corresponding correct answer, which includes the associated database query used to compute it, as illustrated in Fig. 2.

During inference, the model generates a call whenever it deems necessary. Once the call is produced, token generation is paused, the database query is invoked, and the returned value is inserted into the context. When generation resumes, the model has access to the complete callresponse pair and can use it to formulate its final answer. For this reason, the training data must contain

Question	What is the nearest hospital from Durango by drive?
Correct answer with query	The closest hospital you can find is `<API>get_closest_distance_time(category="hospital", mode="drive", location="Abadiño, Durango", metric_to_extract="distance") -> "distance": 0.402, "time": 0.537</API>` 0.402km away.

Fig. 2. Question-answer pair for model fine-tuning.

both the question and the resulting answer, ensuring the model learns how to incorporate the query output into a coherent and correct final response.

Finally, to ensure sufficient linguistic variability, we employed state-of-the-art models (Gemini 2.5 Pro and DeepSeek R1) to generate syntactic paraphrases of the questions while preserving all location-specific information from the dataset. This step enables evaluation of the model's robustness to paraphrasing and its ability to retain the parameters required for accurate querying.

The complete dataset thus created contains 44849 question-answer pairs.

4 Model Fine-Tuning

We fine-tuned the DeepSeek R1-Distill-8B model, which exhibits reasoning abilities exceeding those of typical models of comparable size. Effective query generation requires structured reasoning: recognizing when a query is necessary, identifying the required arguments, and mapping question semantics to the appropriate query. The model's strong latent reasoning capacity makes it well suited for learning these decision patterns [3]. Moreover, the model displays low variance in output structure relative to similarly sized alternatives, improving query generation reliability. The model was fine-tuned on a single NVIDIA RTX 3090 GPU using the Transformer Reinforcement Learning library, specifically the SFTTrainer component, which provides an optimized framework for supervised fine-tuning of large language models. To reduce memory consumption and enable efficient training on commodity hardware, we employed 4-bit quantization [5], which compresses model weights while preserving core representational capacity. Fine-tuning was performed using QLoRA [4] adapters, allowing us to update only a small set of low-rank adaptation parameters: approximately 10M parameters were trained, rather than the full model. Training was done with a batch size of 32 examples and a cosine learning-rate scheduler, chosen for its smooth decay properties that support stable convergence during fine-tuning. The model was trained for four epochs, each requiring several hours of compute time. To mitigate overfitting, we periodically evaluated performance on a held-out validation set and saved checkpoints corresponding to the lowest validation loss. In addition, we implemented an Early Stopping criterion that halted training when no further improvement was observed, ensuring computational efficiency and preventing unnecessary parameter updates.

Figure 3 illustrates the evolution of the evaluation loss (left) and the evaluation runtime (right) across training steps. The loss curve shows a clear downward

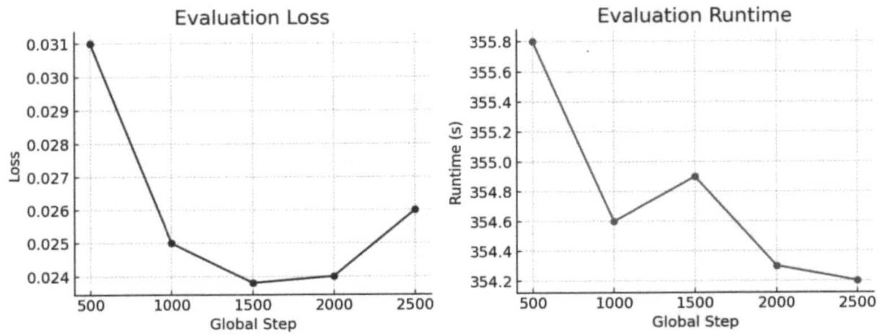

Fig. 3. Evaluation metrics during fine-tuning.

trend during the early stages of training, dropping from approximately 0.031 to 0.024 by step 1500. This indicates that the model rapidly learns the task structure during the initial epochs. After this point, the loss stabilizes, with only minor fluctuations; the slight increase observed at step 2500 suggests the onset of overfitting, consistent with the Early Stopping criterion applied during training. The evaluation runtime remains effectively constant throughout training, with variations of less than two seconds across all checkpoints. This stability reflects the fixed computational footprint of QLoRA-based fine-tuning: because only a small set of adapter parameters is updated, the computational cost of forward passes remains unchanged. The absence of runtime drift confirms that neither model degradation nor GPU resource saturation occurred during training, further validating the reliability of the training setup. The final model is publicly available[1], as well as the dataset[2] and the entire project[3], allowing anyone to reproduce the results.

5 Evaluation

Metrics. We evaluated system performance using two standard NLP metrics: ROUGE-L and BLEU-4. These metrics reflect the correctness of the generated calls. ROUGE-L utilizes the Longest Common Subsequence (LCS) to quantify the overlap between reference (R) and generated (G) tokens. By focusing on sequence order rather than strict contiguity, ROUGE-L effectively captures the preservation of key informational units—such as query arguments and factual components—even in the presence of paraphrasing. The metric is illustrated in Equation a. BLEU-4 measures similarity via contiguous n-gram overlaps (1- to 4-grams), making it highly sensitive to syntactic accuracy. This sensitivity is critical for query generation, where minor deviations in punctuation or token order can significantly degrade output validity. As shown in Equation b, the score is calculated using the geometric mean of modified n-gram precisions p_n. We assign

[1] https://doi.org/10.57967/hf/8567.
[2] https://doi.org/10.57967/hf/8566.
[3] https://github.com/valy3124/Futural_UNSTPB.

equal weights ($w_n = 1/4$) to each n-gram and apply a Brevity Penalty (BP) to penalize insufficiently long outputs. While BLEU-4 is less tolerant of paraphrasing than ROUGE-L, it serves as a rigorous proxy for the model's adherence to required syntactic structures.

$$\text{ROUGE-L} = \frac{2\,\text{LCS}_{\text{len}}}{|R|+|G|} \quad (a) \quad \text{BLEU-4} = BP \times \exp\left(\sum_{n=1}^{4} w_n \log p_n\right) \quad (b)$$

In practice, both metrics are computed for each generated response and then averaged across the evaluation set, providing different perspectives on semantic fidelity (ROUGE-L) and syntactic accuracy (BLEU-4).

Apart from BLEU-4 and ROUGE we also look at the percentage of exact matches (EM), namely the proportion of answers in which the generated query matches the reference exactly (Eq. 1), with no missing parameters, incorrect values, misspellings or wrong parameter order.

$$\text{EMA} = \frac{N_{\text{exact}}}{N_{\text{total}}} \times 100\% \qquad (1)$$

Evaluation Results. For evaluation, we reserved 2,800 queries from the training dataset as a held-out monolingual test set. In addition, we generated 500 multilingual queries using the same procedure described in Sect. 3. This evaluation design serves two purposes. First, it allows us to assess the model's ability to generalize across languages and dialects. Although DeepSeek is pretrained on a multilingual corpus, our fine-tuning was performed exclusively in English; therefore, testing multilingual generalization provides insight into how much of the original linguistic capability is retained after adaptation. Second, multilingual support is of practical importance, as real users in the Durangaldea region may interact with the application in different spoken languages.

We also evaluated the model on entirely unseen geographic locations, isolating this scenario to measure its ability to extrapolate beyond the spatial distribution encountered during training. This is essential for determining the robustness of location-dependent reasoning and query formation.

Finally, we analyzed semantic variations of questions whose locations were seen during training and found that the model exhibits best performance on these. The complete set of evaluation results is shown in Table 2.

Our model shows strong generalization to unseen locations, achieving an Exact Match of 89% and near-perfect BLEU-4 and ROUGE-L scores. This indicates that the model reliably handles novel geographic inputs and correctly forms API calls even when encountering regions not present during fine-tuning. For multilingual queries, performance varies substantially by language. Accuracy remains high for Spanish, Catalan, Galician, and French, reflecting the multilingual prior knowledge inherited from the DeepSeek pretraining. In contrast, performance on Basque is markedly lower, which is consistent with Basque's relatively limited representation in most large-scale multilingual corpora. This suggests that while the fine-tuned model preserves substantial multilingual capability, its robustness depends strongly on the linguistic coverage of the underlying

Table 2. Model accuracy

Case/Subset	Size	Exact Match (%)	BLEU-4	ROUGE-L
Unseen Locations	500	89.0	0.99	0.98
Semantic Variants	500	**94.2**	0.99	0.99
Multilingual queries				
Spanish	100	81.0	0.97	0.95
Catalan	100	86.0	0.98	0.96
Basque	100	24.0	0.81	0.71
Galician	100	93.0	0.99	0.98
French	100	64.0	0.95	0.91
Full dataset evaluation	3300	80.0	0.96	0.93
Full dataset evaluation (excluding multilingual cases)	2800	85.0	0.97	0.95

Table 3. Accuracy against state-of-the-art models

Model	Accuracy (%)	Syntax errs (%)	Location errs (%)	Other (%)
Ours	94.2	0	0	5.8
DeepSeek R1	97.2	0	0.2	2.6
GPT4	97.2	0	0.4	2.4
Gemini	98	0	2.0	0
Grok	97	0	0	3

foundation model. When evaluating the full dataset, including both monolingual and multilingual scenarios, the model achieves an Exact Match of 80%. Restricting evaluation to monolingual English queries increases performance to 85%, confirming that multilingual variability is the main source of accuracy degradation. BLEU-4 and ROUGE-L remain consistently high across settings, indicating that even when the model does not achieve exact matches, its predictions typically preserve the correct query structure and essential content. We evaluated the performance of our proposed model against several state-of-the-art (SOTA) large language models (LLMs). To ensure a fair comparison, the SOTA models were evaluated in a **zero-shot setting** (no additional training or fine-tuning), using structured prompts that detailed the API schema, dataset characteristics, and query generation logic. Comparative analysis was conducted using the *Semantic Variants* dataset subset, as this fragment represents the peak performance benchmark for our architecture. As summarized in Table 3, our model achieved a 94.2% exact match (EM) rate. Although this slightly trails the SOTA benchmark of 97.0%, the results demonstrate that our solution maintains high accuracy and remains competitive within the current landscape of large-scale generative models.

6 Application Performance

The fine-tuned model was integrated into a web application that supports both free-form natural-language queries and on-click generation of predefined question types. Model inference was performed on the same NVIDIA RTX 3090 GPU used during training. The averaged runtime breakdown in Fig. 4 indicates that model inference is the primary source of latency, contributing roughly three seconds per request. In contrast, database lookup, backend logic, and communication overhead remain comparatively small. Backend processing time also includes the application's guardrail mechanisms, which filter and validate user inputs before forwarding them to the model. These guardrails ensure that the system rejects out-of-scope or unsafe queries, thereby reducing hallucinations, preventing unsupported queries, and enforcing domain constraints. Although this contributes modestly to backend latency, it plays a critical role in maintaining reliability and user safety. The dominance of inference time suggests that end-to-end performance is chiefly limited by the model's computational footprint.

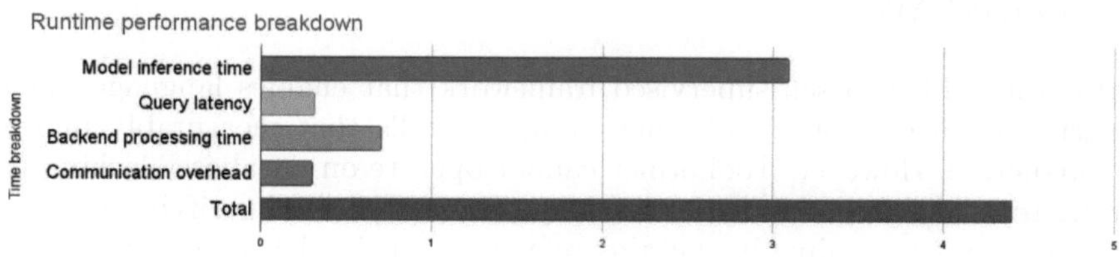

Fig. 4. Application performance.

7 Limitations and Future Work

Applicability to Broader Scenarios. Our application is currently being integrated into the FUTURAL Metasearch platform and remains to be tested against different datasets with ranging complexity. To this end, we aim for a general methodology to connect arbitrary queryable datasets to natural-language interfaces with minimal effort. To overcome the lack of supervised training data, our synthetic generation pipeline (Sect. 3) provides a systematic solution to various structured datasets. However, this approach scales poorly across numerous heterogeneous domains where manual template definition becomes impractical, where for n different datasets we might require creating 2^n combinations of different API-related question-answer pairs. For such complex scenarios, we identify ontology-based domain abstraction and LLM-driven query-space generation as promising directions for future research, and consider expanding the evaluation setting to include noisy datasets to test our system against ambiguous queries.

Application Scalability. Designed for individual request efficiency with minimal hardware, our system prioritizes simplicity over high concurrency. However, several strategies—some currently under exploration for the FUTURAL Metasearch platform—can ensure scalability as demand grows. Batching multiple queries into single forward passes improves GPU utilization and throughput during bursty workloads. Further optimizations, such as quantization, low-rank adaptation, pruning, or model distillation, can reduce memory footprints and increase per-server capacity. Finally, since many queries involve repetitive structures (e.g., travel distances or nearest locations), response caching can serve paraphrased requests directly without re-invoking the model.

Graceful Error Recovery. Currently lacking automated recovery for malformed queries, our architecture relies on standard error reporting. To enhance robustness, future iterations will explore iterative error correction: a feedback loop where execution-time diagnostics trigger re-prompting, enabling the model to refine and regenerate queries based on failure data.

8 Related Work

Toolformer [14] is a self-supervised framework that enables language models to learn tool invocation by filtering candidate calls that measurably improve text prediction. However, Toolformer cannot operate on databases lacking pre-existing textual references, as its learning process relies on verifying tool calls within a corpus that already contains references to the database's entities and relationships. **Gorilla** [12] focuses on reliable API invocation at scale by using retrieval-augmented grounding on API documentation. While Gorilla generalizes across thousands of real-world APIs, our method is intentionally lightweight and domain-specific. By utilizing a synthetic dataset-generation pipeline tailored to a fixed schema, we enable effective fine-tuning without the complexity or hardware overhead associated with large-scale retrieval systems. **ReAct** [17] interleaves natural-language reasoning with actions in a thought–action–observation loop. While ReAct is general-purpose and interactive, our method is single-step and domain-specific, generating a structured API call directly from input without iterative planning. This reduces computational costs and complexity compared to ReAct's multi-step paradigm. Other approaches like **ToolAlpaca** [15] and **Confucius** [7] rely on supervised demonstrations or iterative refinement, which require substantial annotation effort or high compute. Similarly, **GPT4Tools** [16] depends on proprietary models, and **CREATOR** [13] focuses on the broader task of tool creation. In contrast, our approach achieves high reliability in a focused domain using a principled, lightweight synthetic pipeline that avoids the need for large curated corpora or iterative interaction. Finally, the **Model Context Protocol (MCP)** [2] provides a standard for tool integration but often imposes a significant "context tax." Loading extensive JSON schemas for dozens of tools leads to context saturation and degraded accuracy. Our approach avoids this overhead by eliminating global schema loading and training the model to produce a single, domain-specific call. This results in significantly lower hardware requirements and higher precision within the targeted domain. Recent developments, such as Sphinteract [19], provide frameworks for resolving ambiguities in NL2SQL tasks, which becomes increasingly critical as dataset complexity grows. While the nature of our current dataset inherently minimizes such ambiguity, we recognize the potential of these interactive approaches and consider them a promising avenue for future system enhancements.

9 Conclusion

This paper presents a fully open-source, natural language question-answering (QA) framework optimized for accessibility data within the Durangaldea region of Spain. Our results demonstrate that the proposed system achieves performance metrics comparable to state-of-the-art (SOTA) large language models, while at the same time running on commodity hardware. It successfully bridges the gap between specialized local datasets and high-fidelity generative responses. Beyond the specific implementation, we propose a generalized methodology for the rapid deployment of QA systems across diverse tabular or relational datasets. By leveraging high levels of automation, this workflow significantly reduces the requirement for human expert intervention during the dataset preparation and model alignment phases. While the current architecture is optimized for a single-domain dataset, future research will focus on extending this system to multi-dataset environments, addressing the challenges of cross-domain query generation and data fusion to provide a more holistic analytical tool for arbitrary collections of datasets.

Acknowledgments. This paper has been funded by the European Union, via the FUTURAL project - *Empowering the FUTure through innovative Smart Solutions for rURAL areas* (HORIZON EUROPE) Project ID 101083958. Views and opinions expressed are however those of the author(s) only and do not necessarily reflect those of the European Union or the European Research Executive Agency. Neither the European Union nor the granting authority can be held responsible for them.

References

1. The llama 3 herd of models. Technical report, Meta (2024). https://ai.meta.com/research/publications/the-llama-3-herd-of-models/
2. Anthropic: Introducing the model context protocol (2024). https://www.anthropic.com/news/model-context-protocol. Accessed 30 Mar 2025
3. DeepSeek-AI: Deepseek-r1: incentivizing reasoning capability in LLMs via reinforcement learning. Nature **645**, 633–638 (2025). https://doi.org/10.1038/s41586-025-09422-z
4. Dettmers, T., Pagnoni, A., Holtzman, A., Zettlemoyer, L.: Qlora: efficient finetuning of quantized LLMs. In: Proceedings of the 37th International Conference on Neural Information Processing Systems. NIPS '23, Curran Associates Inc, Red Hook, NY, USA (2023)

5. Frantar, E., Ashkboos, S., Hoefler, T., Alistarh, D.: GPTQ: accurate post-training quantization for generative pretrained transformers. arXiv preprint arXiv:2210.17323 (2022), https://arxiv.org/abs/2210.17323
6. Gao, D., et al.: Text-to-SQL empowered by large language models: a benchmark evaluation (2023). https://arxiv.org/abs/2308.15363
7. Gao, S., et al.: Confucius: iterative tool learning from introspection feedback by easy-to-difficult curriculum (2023). https://arxiv.org/abs/2308.14034
8. Imagery, N., Agency, M.: Department of defense world geodetic system 1984: its definition and relationships with local geodetic systems. Technical report TR8350.2, NIMA (2000)
9. Jolicoeur-Martineau, A.: Less is more: recursive reasoning with tiny networks (2025). https://arxiv.org/abs/2510.04871
10. Lewis, P., et al.: Retrieval-augmented generation for knowledge-intensive NLP tasks. In: Larochelle, H., Ranzato, M., Hadsell, R., Balcan, M.F., Lin, H.T. (eds.) Advances in Neural Information Processing Systems, vol. 33, pp. 9459–9474. Curran Associates, Inc. (2020)
11. Olbricht, R.: The overpass API (2012). https://wiki.openstreetmap.org/wiki/Overpass_API. Accessed 11 Dec 2025
12. Patil, S.G., Zhang, T., Wang, X., Gonzalez, J.E.: Gorilla: large language model connected with massive APIs. In: Globerson, A., et al. (eds.) Advances in Neural Information Processing Systems, vol. 37, pp. 126544–126565. Curran Associates, Inc. (2024). https://doi.org/10.52202/079017-4020
13. Qian, C., Han, C., Fung, Y.R., Qin, Y., Liu, Z., Ji, H.: Creator: tool creation for disentangling abstract and concrete reasoning of large language models (2024). https://arxiv.org/abs/2305.14318
14. Schick, T., et al.: Toolformer: language models can teach themselves to use tools. In: Advances in Neural Information Processing Systems (NeurIPS), vol. 36 (2023)
15. Tang, Q., et al.: Toolalpaca: generalized tool learning for language models with 3000 simulated cases (2023). https://arxiv.org/abs/2306.05301
16. Yang, R., et al.: Gpt4tools: teaching large language model to use tools via self-instruction (2023). https://arxiv.org/abs/2305.18752
17. Yao, S., et al.: React: synergizing reasoning and acting in language models. In: International Conference on Learning Representations (ICLR) (2023)
18. Zeng, Y., et al.: Recparser: a recursive semantic parsing framework for text-to-sql task. In: Bessiere, C. (ed.) Proceedings of the Twenty-Ninth International Joint Conference on Artificial Intelligence, IJCAI-20, pp. 3644–3650. International Joint Conferences on Artificial Intelligence Organization (2020). https://doi.org/10.24963/ijcai.2020/504, main track
19. Zhao, F., Deep, S., Psallidas, F., Floratou, A., Agrawal, D., El Abbadi, A.: Sphinteract: resolving ambiguities in NL2SQL through user interaction. Proc. VLDB Endow. **18**(4), 1145–1158 (2024). https://doi.org/10.14778/3717755.3717772

Towards a Meta-model for Complex Performance Indicators: A Systematic Literature Review and an Illustration in the Cardiology Domain

Benito Giunta[1,2](✉), Corentin Burnay[1,2], and Alexandre Faulkner[1,2]

[1] NaDI, Namur Digital Institute, MInDIT Research Center, UNamur, Namur, Belgium
{benito.giunta,corentin.burnay}@unamur.be, alexandre.faulkner@student.uclouvain.be

[2] Faculté de Médecine, Université Catholique de Louvain (UC Louvain), Louvain-la-Neuve, Belgium

Abstract. Complex Performance Indicators (CPIs) aggregate heterogeneous data streams into multidimensional constructs supporting strategic decision-making, yet their engineering remains fundamentally challenging. Organizations face information overload, governance opacity, and extended development cycles that existing Self-Service Business Intelligence platforms fail to address. Model-Driven Engineering (MDE) promises accessible CPI implementation through intuitive abstractions, but current frameworks remain fragmented and incomplete. This work-in-progress presents a Systematic Literature Review identifying 18 MDE frameworks for indicator modeling and extracting a 31-features inventory across three dimensions: Language Expressiveness, Business User Empowerment, and MDE Capabilities. Analysis reveals that no framework achieves more than 70% feature coverage. The findings justify developing a novel Domain-Specific Modeling Language, with the inventory serving as specification blueprint for future meta-model construction.

Keywords: Complex performance indicator · Model-driven engineering · Self-service business intelligence · Systematic Literature Review

1 Introduction and Motivation

Contemporary organizations face unprecedented challenges in extracting actionable insights from exponentially growing, heterogeneous data ecosystems [1,29]. Traditional Key Performance Indicators (KPIs) prove insufficient when decision-makers require composite, context-aware measurements that integrate multiple data sources, temporal dimensions, and stakeholder perspectives [39]. Complex

A. Bernasconi et al. (Eds.): RCIS 2026, LNBIP 597, pp. 127–139, 2027.
https://doi.org/10.1007/978-3-032-35043-5_11

Performance Indicators (CPIs) emerge as an opportunity, representing multidimensional constructs that aggregate heterogeneous data streams, embed domain-specific semantics, and support dynamic reconfiguration to reflect evolving organizational priorities [11]. The conceptual distinction between KPIs and CPIs extends beyond mere compositional complexity [11]. While KPIs typically measure single organizational aspects through straightforward calculations, CPIs embody what we term the *CPI Onion* architecture [20]: a layered structure comprising (1) a mathematical core defining aggregation logic and computational dependencies, (2) a technological layer specifying heterogeneous data sources, computational infrastructure and resources, and (3) a decisional layer linking indicators to strategic goals, governance policies, and stakeholder roles. With CPI, we refer to any indicators, key or not, which is a mathematical composition, for which development efforts are not only dedicated to the mathematical modeling, but also the appropriate technical setup and decisional utility. Their multi-layered nature demands engineering approaches that transcend conventional Business Intelligence (BI) tooling, which remains predominantly code-centric and inaccessible to domain experts lacking technical expertise [2].

1.1 Challenges in CPI Engineering

Implementing CPIs aligned with decision-makers' actual requirements remains fundamentally difficult due to three interrelated challenges. *Challenge 1: Information Overload and Data Heterogeneity.* Organizations generate data from varied and highly heterogeneous sources [29], creating cognitive overload for decision-makers [37]. CPIs must synthesize disparate data formats, temporal granularities, and quality profiles while remaining comprehensible – a requirement that demands rich *language expressiveness* capable of modeling complex data integration, normalization, and aggregation logic. *Challenge 2: Strategic Alignment and Governance.* CPIs are inherently intricate artifacts whose construction risks lack of transparency, poor mathematical soundness, and misleading interpretation [7,11,26]. Decision-makers hold subjective expectations regarding visualization and interaction, and may distrust data systems they do not properly understand, preferring past experience over actual data [40,53]. Effective CPI systems must ensure *business user empowerment* by explicitly linking indicators to strategic goals, governance policies, and stakeholder roles, while providing transparent construction processes that build trust and adoption [9,60]. *Challenge 3: System Execution and Automation.* The business environment changes constantly and rapidly, yet traditional BI development cycles impose significant lead-times between need identification and CPI implementation [36]. Self-Service Business Intelligence (SSBI) promises to empower business users to produce their own BI outputs with minimal technical knowledge [2], but current SSBI offerings provide no support for CPI production. This vision requires *MDE capabilities* that enable non-technical users to specify CPIs using intuitive models while automatically generating executable implementations through predefined transformation rules [5,12,23].

1.2 Research Context and Objectives

The ultimate goal of our multi-year research project is to design, implement, and evaluate a comprehensive Model-Driven Engineering based SSBI system for managing CPIs. This project comprises six phases: (1) theoretical feature extraction via systematic literature review, (2) empirical feature identification through practitioner focus group, (3) feature validation via Delphi studies, (4) meta-model construction and formalization, (5) Domain-Specific Modeling Language (DSML) visual notation design, and (6) system prototyping and evaluation with real-world CPIs. This forum paper reports on Phase 1, representing foundational work-in-progress that establishes the theoretical basis for subsequent phases.

1.3 HFRAI Motivating Example and Research Questions

To ground our investigation in concrete requirements, we introduce the Heart Failure Risk Assessment Indicator (HFRAI), a production-grade CPI actively used in a clinical department in Wallonia, Belgium. HFRAI evaluates patient risk by integrating clinical biomarkers, patient-reported outcomes, sociodemographic variables, and temporal patterns from heterogeneous sources (EHR databases, mobile apps, pharmacy systems). This example exposes fundamental inadequacies in current paradigms: existing Self-Service BI platforms lack formal modeling capabilities to capture CPI semantics, validate computational logic, or enforce governance constraints, while established tools do not provide support for indicator-specific concerns.

To determine whether current literature adequately addresses the three identified challenges, we conduct a Systematic Literature Review (SLR) guided by two research questions. First, *what are the existing modeling frameworks capable of handling CPIs, and what are their limitations? (RQ1)* This question systematically maps MDE frameworks for indicator modeling, critically evaluating their coverage of CPI-specific requirements across the three challenge dimensions. We assess architectural assumptions, target user populations, and support for the CPI Onion's mathematical, technological, and decisional layers, identifying gaps where existing approaches fall short of comprehensive CPI lifecycle management. Second, *what is the minimal set of theoretical modeling features required to design a comprehensive CPI meta-model? (RQ2)*. This question synthesizes fragmented contributions across disparate research communities into a unified feature inventory organized by the three requirement dimensions: Language Expressiveness features (addressing Challenge 1), Business User Empowerment features (addressing Challenge 2), and MDE Capabilities features (addressing Challenge 3). The resulting inventory serves dual purposes: as an evaluation benchmark for existing frameworks and as a specification blueprint for our envisioned DSML (guiding Phase 4 meta-model development). The remainder of this paper details our SLR methodology (Sect. 2), presents results including the landscape of 18 frameworks, 31 extracted features, and HFRAI stress-test findings (Sect. 3), and concludes with implications for future meta-model development (Sect. 4).

2 Research Methodology

2.1 Systematic Literature Review Method

We adopt Kitchenham's systematic literature review guidelines [24] for identifying, evaluating, and synthesizing primary studies in IS research, ensuring reproducibility, minimizes bias, and enabling comprehensive coverage.

Search Strategy and Protocol. Our search strategy was designed to maximize recall while maintaining precision. We developed 358 keywords organized into six semantic clusters: (1) the concepts (e.g., "Complex Performance Indicator", "KPI", "composite metric"), (2) the field (e.g., "information system") and (3) the modeling paradigms (e.g., "meta-model", "ontology", "DSL", "conceptual model"). Boolean queries combined these clusters using AND/OR operators to balance sensitivity and specificity. The full 358 search strings can be accessed here[1]. Five major databases were browsed: IEEE Xplore, ACM Digital Library, Scopus, Web of Science, and PubMed, covering publications from 2000 to 2025. The initial automated search returned 3451 papers.

Screening and Selection Process. Three-stage screening applied explicit criteria with dual independent review. Stage 1 excluded non-English publications, editorials, position papers, and irrelevant papers by manually screening the title and abstract, retaining 263 papers. Stage 2 (full-text screening) applied four conjunctive criteria: (i) the framework provides graphical notation or formal constructs for indicator design, (ii) it explicitly addresses CPI concerns including composition, data integration, goal alignment, governance, or evolution, (iii) the approach is domain-independent or generalizable beyond a single proprietary context, and (iv) sufficient technical detail is provided for feature extraction. Stage 3 (quality assessment) employed a 10-item Kitchenham checklist [24] evaluating methodological rigor, clarity of framework description, empirical grounding, and contribution novelty. Papers scoring below 5.0/10.0 were excluded. Stages 2 and 3 retained 17 distinct frameworks reported across 33 papers. Discrepancies were resolved through consensus discussion involving all authors and external validation by a business intelligence researcher. We then applied forward and backward snowballing [59] on selected papers, identifying 1 additional candidate (3 distinct papers) through citation analysis making the final set of 18 frameworks (36 distinct papers).

Feature Extraction and Synthesis. We conducted systematic feature extraction through independent parallel coding. Each author was randomly assigned a subset of frameworks to analyze, identifying features. They were documented with operational definitions to ensure consistent interpretation. A synthesis phase merged individual findings into a unified inventory through iterative consensus building. Features identified by multiple coders were standardized; unique findings underwent structured discussion until agreement. The resulting 31 features were categorized into three dimensions aligned with our research questions:

[1] https://figshare.com/articles/figure/_/21995219.

Language Expressiveness, Business User Empowerment, and MDE Capabilities. We then constructed a Framework-Feature Matrix by systematically evaluating each framework's explicit support for each feature, denoting presence (✓), absence (blank), or insufficient evidence (?). Coverage scores were computed as the ratio of supported features to total features per dimension.

2.2 HFRAI Reverse-Engineering Methodology

To evaluate framework adequacy beyond abstract checklists, we developed a stress-test protocol systematically modeling HFRAI's complete specification within each framework. The methodology comprises: (1) eliciting requirements by retro-engineering the existing technological suite (databases, Excel, Power BI) and research via document analysis and semi-structured interviews. These were consolidated into a formal specification document, iteratively revised by domain experts, and structured into 47 atomic requirements spanning mathematical, informational, and decisional layers; (2) convening a validation panel (three cardiologists) reviewing instantiated models, 43 of 47 requirements were validated and 5 new added, yielding 48 requirements; (3) instantiating the HFRAI model by linking these structured requirements to our feature inventory and reviewing it with a secondary panel (two health informaticians, one MDE researcher).

Table 1. Results Set: Identified Frameworks

ID and Name	References	Concept	Model
(1) ECoWare	[4]	KPI	ontology
(2) OWL-Q Extension	[27,28]	KPI	ontology, meta-model
(3) OIM Extension	[30]	KPI	meta-model
(4) MetricM	[18,19,56]	indicators, PI	meta-model
(5) Executable KPI Pre-Conceptual Schema (KPCS)	[52]	KPI	pre-conceptual schema
(6) General Indicator Model	[16,17]	KPI, CPI	meta-model
(7) KPI-ML	[6]	KPI	meta-model
(8) PPINOT Suite	[46–51]	indicators, PI	ontology, meta-model
(9) SemPI	[13–15]	KPI, PI	ontology, meta-model
(10) URN Extension	[44,45,54]	KPI	meta-model
(11) MathML	[8]	Indicators, KPI, PI	meta-model
(12) KPIOWL	[35]	metrics, KPI	ontology
(13) TxL	[38]	KPI	DSL
(14) General Modeling Framework (GMF)	[41–43]	indicators, PI	meta-model
(15) KPI Calculation	[61]	KPI	ontology
(16) HMM Approach	[58]	PI	meta-model
(17) SMI Ontology	[3]	indicators	ontology
(18) EM-KPI	[31–34]	KPI	ontology

3 Results and Critical Synthesis

3.1 Systematic Literature Review Results

Our systematic search identified 18 frameworks from 2003 to 2024 (Table 1), exhibiting heterogeneity in concept, modeling paradigms, and domains: 56% domain-independent, 44% domain-specific (healthcare, finance, manufacturing, public sector). Our extraction identified 31 features across three categories (Table 3); Table 2 presents framework coverage across all features. Expressiveness (dim. 1) addresses mathematical logic and informational requirements. While some features reflect a kind of consensus where most frameworks converge (e.g. indicator attributes, unit, data sources), others like categories, business rules, and operators underline lack in expressiveness. Business Alignment (dim. 2) evaluates goal linkage, stakeholder roles, and governance. Most frameworks converge on features like business goal model and business process model, while discrepancies arise with conceptual model and design interface. MDE Capabilities (dim. 3) assesses technical infrastructure. The most covered features are semantics and query language, while actual computation and graphical notation remain less covered. Aggregate analysis reveals trade-offs: average coverage is quite low at 54.2%, 53.8% and 52.5% for dimensions 1, 2 and 3 respectively.

Table 2. Framework-Feature Matrix

Feature / Framework		**(1)**	**(2)**	**(3)**	**(4)**	**(5)**	**(6)**	**(7)**	**(8)**	**(9)**	**(10)**	**(11)**	**(12)**	**(13)**	**(14)**	**(15)**	**(16)**	**(17)**	**(18)**
Language Expressiveness	01 – Indicator Attributes	✓	✓	✓	✓	✓	✓	✓	✓	✓	✓	✓	✓	✓	✓		✓	✓	
	02 – Traceability		✓	✓	✓	✓		✓	✓	✓					✓		✓	✓	✓
	03 – Categories of Indicators	✓								✓	✓			✓		✓	✓	✓	
	05 – Formula		✓	✓			✓			✓		✓	✓	✓	✓	✓		✓	✓
	06 – OLAP Operations	?	✓	✓			✓	✓		✓	✓			✓	✓	✓	✓		
	04 – Operators		✓				✓		✓	✓		✓	✓	✓		✓			
	07 – Business-Rules	✓	✓	✓									✓	✓	✓	?			
	08 – Data Source	✓	✓	✓	✓	✓	✓		✓	✓	?			✓		✓			✓
	09 – Value Type and Domain				✓	✓		✓	✓		✓		?	✓	✓	✓	✓	✓	✓
	10 – Unit				✓	✓	✓	✓	✓	✓	✓		✓		✓	✓	✓	✓	✓
	11 – Weights		✓	✓		✓				✓				?					
	12 – Limit Values	✓	✓	✓	✓	✓			✓	✓	✓		✓		✓				
	13 – Status	✓	✓	✓		✓		✓			?		✓		✓				✓
	14 – Hierarchy		✓	✓			✓					?	✓					✓	
	15 – Composition of Indicators	✓	✓					✓	✓	✓		✓		✓		✓	✓	✓	
	16 – Other Relationships		✓		✓	✓		✓	✓	✓		?	✓	✓	✓	✓			✓
Dimension 1 – Score (%)		**47**	**81**	**63**	**44**	**56**	**44**	**50**	**56**	**75**	**43**	**29**	**60**	**67**	**63**	**67**	**31**	**50**	**50**
BU Empowerment	17 – Creation of Category	✓	✓	✓	✓	✓	✓				✓	✓	✓						
	18 – Business Goal Model	✓	✓	✓	✓	✓	✓	✓	✓	✓	✓	✓	✓		✓			✓	✓
	19 – Business Process Model	✓	✓		✓				✓		✓		✓		✓				✓
	20 – Conceptual Model	✓		✓			✓											✓	
	21 – Organizational Structure Model	✓	✓	✓	✓					✓	✓	✓	✓		✓				✓
	22 – Maturity	✓	✓	✓	✓	✓	✓		✓	✓	✓		✓	✓	✓	✓		✓	✓
	23 – Design Interface								✓	✓		✓				✓	✓		
Dimension 2 – Score (%)		**86**	**71**	**71**	**71**	**43**	**57**	**14**	**57**	**57**	**71**	**57**	**71**	**14**	**57**	**29**	**43**	**43**	**57**
MDE Cap./Support	24 – Semantics		✓	✓	✓	✓	✓	?	✓	✓		✓	✓	✓		✓	✓	✓	
	25 – KPI Requirements		✓		✓	✓		✓							✓		✓	✓	
	26 – KPI Computation	✓		✓								✓		✓		✓		✓	✓
	27 – Runtime	✓		✓			✓		✓	✓				✓		✓	✓	?	?
	28 – KPI Visualisation	✓	✓		✓	✓	✓	✓	✓	✓	✓		✓		✓	✓			✓
	29 – MD Component	✓		✓	✓		✓		✓			✓			✓				
	30 – Query Language and/or Code	✓	✓	✓		✓	✓		✓	✓		✓	✓	✓	✓		✓	✓	✓
	31 – Graphical Notation	✓			✓				✓					✓		✓			
Dimension 3 – Score (%)		**75**	**50**	**63**	**63**	**50**	**63**	**29**	**75**	**50**	**13**	**50**	**38**	**63**	**50**	**63**	**50**	**57**	**43**

The standard deviation of aggregate coverage reveals a stronger convergence of dimension 1 (13.9) against dimension 2 (19.8) and dimension 3 (15.4). Overall, no framework achieves 65% of coverage on all three dimensions.

Table 3. Features Descriptions

	Feature	Description
Dim. 1: Language Expressiveness	**01 – Indicator Attributes**	The constructs of the framework support the adjunction of additional pieces of information to an indicator (e.g., indicator name, creation date, etc.)
	02 – Traceability	The framework allows direct link between the indicator and the issuing-entity (a department, a business unit, a role, a team, a person,...).
	03 – Categories	The constructs of the model support the existence of sub-categories of indicator. E.g.: The framework (1) specifies a Reliability KPI.
	04 – Operators	The constructs of the model formally define the concept of basic arithmetic operators to specify a metric or an indicator.
	05 – Formula	The constructs of the model formally define and allow to manipulate the mathematical formula that describes a metric.
	06 – OLAP Operations	The constructs of the framework support OLAP operations such as basic aggregation operators, drill down, roll-up, etc. [57].
	07 – Business-rules	The constructs of the model support the formal definition of conditional constraints that the indicators must comply with.
	08 – Data Source	The framework supports the definition of data sources to be used for the computation of each indicator.
	09 – Value Type	The constructs of the model allow manipulation of the data type (numerical, textual, etc.) and the range of values.
	10 – Unit	The constructs allow to associate and retrieve its data unit (e.g. 180°C).
	11 – Weights	The constructs of the model allow the use of weighting with indicators.
	12 – Limit Values	The constructs support the definition of lower/upper bound indicators limits.
	13 – Status	The constructs allow state labels (failed, in progress, success, etc.).
	14 – Hierarchy	The framework allows formal set-up of indicators through hierarchical parent-child relationships.
	15 – Composition	The framework permits formal and dynamic creation of indicators which are a combination of other indicators.
	16 – Other relationships	Defines links other than parent-child (relationship of lagging, dependency, etc.).
Dim. 2: BU Empowerment	**17 – Creation of Cat.**	The framework allows business users to create their own sub-categories of indicators (feature 03).
	18 – Business Goal	Mapping of strategic intent [25] and indicators for objective understanding.
	19 – Business Process	Mapping of business processes and indicators for explicit context.
	20 – Conceptual Model	Mapping between various business entities involved in the organisation and the indicators monitoring them.
	21 – Org. Structure	Mapping of hierarchical levels in the enterprise architecture and indicators.
	22 – Maturity	Captures whether the framework had been applied through a real-world case study to demonstrate feasibility.
	23 – Design Interface	User-friendly interface to exploit the Graphical Notation (feature 31).
D. 3: MDE Support	**24 – Semantics**	Specification of a semantic layer for dynamic manipulation and reasoning [21].
	25 – KPI Requirements	Supports (non-)functional requirements documentation of [10].
	26 – KPI Computation	Discusses implementation; the framework performs the value computation.
	27 – Runtime	The models are partially or totally executed.
	28 – KPI Visualisation	Exposes the value to the user with functionalities that allow KPI assessment.
	29 – MD Component	Adopts a Model-Driven perspective allowing code generation.
	30 – Query/Code	Workable formalism that supports users in specification or interaction.
	31 – Graphical Notation	A notation modeling definition or operationalization for non-technical users.

Table 4. Illustrative Example Application: HFRAI Feature Mapping

	Feature	HFRAI Instantiation
Dim. 1: Language Expressiveness	**01. Indicator Attributes**	**Metadata Definition:** Defining the indicator name (HFRAI), owner as "Cardiology Dept", and version "v2.0 (2026 Guidelines)".
	02. Traceability	**Ownership:** Linking the "Biomarker Profile" to the "Pathology Laboratory" (responsible entity) and the "Cardiac Function" to the "Imaging Unit".
	03. Categories of Indicators	**Typology:** Classifying "Troponin I" as a *Laboratory Indicator* and "6-Minute Walk" as a *Functional Indicator* to allow filtering by test type.
	04. Operators	**Arithmetic Logic:** Using multiplication for weighting (e.g., $0.35 \times Cardiac\_Score$) and addition for aggregating the four dimensions ($Cardiac + Biomarker + Risk + Functional$).
	05. Formula	**Scoring Algo:** Formally defining the calculation: $HFRAI = \sum(Dimension\_Score_i \times Weight_i)$ where $Dimension\_Score$ is normalized to 0–100.
	06. OLAP Operations	**Drill-Down:** A cardiologist viewing a low HFRAI score (52) and *drilling down* to see that the "Cardiac Function" dimension is the root cause (Score: 45) due to low LVEF.
	07. Business-rules	**Alert Logic:** "IF $HFRAI < 40$ AND $Candidate\_Status = Eligible$ THEN trigger *Transplant Evaluation Protocol.*"
	08. Data Source	**Heterogeneous Fetch:** "LVEF" is fetched from *Imaging System*, "Creatinine" from *Lab Information System*, and "Age" from *Electronic Health Record.*
	09. Value Type and Domain	**Data Types:** "LVEF" as a *Percentage* (0–100), "NT-proBNP" as *Numeric* (0–35000), and "NYHA Class" as *Ordinal* (I to IV).
	10. Unit	**Standardization:** Explicitly associating "Serum Creatinine" with mg/dL to avoid errors if a lab sends results in $\mu mol/L$ (requiring conversion).
	11. Weights	**Impact Factors:** Assigning 35% importance to "Cardiac Function" vs. 15% to "Functional Capacity" based on clinical prognosis evidence.
	12. Limit Values	**Thresholds:** Defining the normal range for "NT-proBNP" as $< 125pg/mL$ and the critical high threshold as $> 450pg/mL$.
	13. Status	**State Labels:** Mapping score ranges to status labels: 0–39 = "Very High Risk"; 40–59 = "High Risk"; 60–79 = "Moderate Risk"; 80-100 = "Low Risk".
	14. Hierarchy	**Structure:** Modeling the tree structure: $HFRAI$ (Root), $BiomarkerProfile$ (Parent), $NT - proBNP$ (Child).
	15. Composition of Indicators	**Aggregation:** The "Clinical Risk Score" is a composite formed by combining Age, Diabetes status, AFib presence, and BP.
	16. Other relationships	**Interaction Effects:** Modeling a "Synergistic Risk" relationship where the simultaneous presence of *Diabetes* AND *Kidney Dysfunction* triggers an additional penalty multiplier not captured by simple addition.
D. 2: Business User Empowerment	**17. Creation of Categories**	**User Custom.:** A user defining a new group called "Modifiable Risk Factors" (grouping BP, Activity Level, Medication Adherence) without IT intervention.
	18. Business Goal Model	**Strategic Alignment:** Explicitly linking the HFRAI to the hospital strategic goal: "Reduce 30-day Heart Failure Readmission Rate by 10%".
	19. Business Process Model	**Workflow Context:** Embedding HFRAI calculation into "Patient Discharge Workflow" (BPMN), forcing a score check before a patient can be sent home.
	20. Conceptual Model	**Entity Mapping:** Linking the "Creatinine" indicator to the $Patient$ and $Kidney$ entities in the hospital's conceptual data model.
	21. Org. Structure Model	**Role-Based Views:** Mapping the "Financial Impact" of HFRAI to the *Hospital Administrator* role, and the "Clinical Detail" to the *Cardiologist* role.
	22. Maturity	Not-applicable here, this aspect refers to the meta-analysis of the approach.
	23. Design Interface	**No-Code Editor:** A GUI slider allowing a cardiologist to adjust the weight of "Cardiac Function" from 35% to 40% and see the impact on historical data.
Dim. 3: Capabilities & Support for MDE	**24. Semantics**	**Ontology Linking:** Tagging "Troponin I" with SNOMED CT code 102657007 and LOINC code 10839-9 to ensure semantic interoperability across different hospital systems.
	25. KPI Requirements	**NFRs:** Specifying non-functional requirements: "Score must be computed within 200ms of opening patient record" and "Must handle missing Troponin values by using last known value < 24h".
	26. KPI Computation	**Execution Engine:** The backend engine running the normalization logic: $Normalized\_LVEF = (Patient\_LVEF/50) * 100$ (clamped at 100).
	27. Runtime	**Live Execution:** The model is not just a document; it actively executes every time a lab result (e.g., new Potassium level) is posted to the message bus.
	28. KPI Visualisation	**UI Components:** Rendering the HFRAI as a gauge chart with color-coded zones (Red/Yellow/Green) in the Electronic Medical Record (EMR) dashboard.
	29. MD Component	**Code Generation:** Automatically generating the SQL queries to extract "Last 12 months Hospitalizations" from the Admission-Discharge-Transfer database.
	30. Query Language/Code	**DSL:** Using a Domain Specific Language to define the indicator: `DEFINE INDICATOR HFRAI AS WEIGHTED_SUM(Cardiac, Biomarker...) WHERE Context = 'Inpatient'`.
	31. Graphical Notation	**Visual Modeling:** A visual "Tree" diagram where cardiologists can drag-and-drop a new biomarker onto the "Biomarker Profile" branch updating the model.

3.2 HFRAI Instantiation

To bridge the theoretical feature inventory and practical application, we instantiate the features using the real-world HFRAI case. It serves as a descriptive evaluation via scenarios [22] for an initial assessement of method applicability [55]. Table 4 presents the instantiation matrix. We performed a feasibility exercise, mapping the 48 validated HFRAI requirements to the 31 SLR-derived features via functional and semantic correspondence. The instantiation of the HFRAI reveals critical insights into the functional demands of modern CPI modeling (referring to features into brackets). Specifically, the mapping process exposes the inherent complexity of real-world" indicators, where simple aggregation logic (F15) is insufficient without robust data source heterogeneity (F8) and semantic interoperability (F24) to handle diverse clinical inputs like imaging and lab results. Furthermore, the exercise highlights a significant dependency between Language Expressiveness and Business User Empowerment; for instance, defining synergistic risks" (F16) is only valuable if domain experts (cardiologists here) can interactively adjust weights via a design interface (F23) without technical intervention. This practical application underscores a broader conclusion; effective CPI frameworks must move beyond static definitions to support dynamic, runtime execution (F27) and context-aware business rules (F7). The set of features seems promising in addressing the complexity while they do not stand to be complete (35 out of 48 requirements are modeled), requiring additional empirical features.

4 Conclusion, Limitations and Future Work

This work presents an SLR evaluating MDE frameworks for CPI engineering, identifying 18 frameworks and extracting a validated 31-features inventory across expressiveness, business alignment, and MDE capabilities. The framework-feature matrix reveals heterogeneous coverage: no framework supports more than 70% of features, with critical gaps in data heterogeneity, semantic annotations, and missing data handling. The HFRAI empirically validates these gaps, demonstrating that existing frameworks cannot adequately model real-world cardiology CPIs. Furthermore, while the extracted features provide a promising theoretical base, this literature-derived set remains non-exhaustive; critical practical requirements (such as uncertainty modeling, data availability constraints, and feature hierarchy) highlight a persistent gap between current research and empirical needs. However, this study acknowledges several limitations. Methodologically, the SLR is bounded by the selected search string and databases, making it inherently susceptible to publication bias and the potential omission of gray literature or domain-specific terminologies. Empirically, the HFRAI stress-test relies on a single cardiology case study; while illustrative of clinical complexity, this restricts the immediate generalizability of the framework evaluation. These findings justify developing a novel DSML tailored to CPI management. Future work will proceed through the remaining project phases. First, this baseline inventory will be empirically enriched via practitioners focus groups to capture these missing dimensions, and validated through Delphi studies. Second, the validated features will be formalized into a meta-model, strictly defining its abstract syntax within frameworks like EMF/Ecore. Finally, the meta-model will ground a novel Domain-Specific Modeling Language. By pairing this abstract syntax with a no-code/low-code concrete visual syntax, we will implement and evaluate a functional prototype with real-world CPIs.

References

1. Akter, S., Wamba, S.F., Gunasekaran, A., Dubey, R., Childe, S.J.: How to improve firm performance using big data analytics capability and business strategy alignment? Int. J. Prod. Econ. **182**, 113–131 (2016)
2. Alpar, P., Schulz, M.: Self-service business intelligence. Bus. Inf. Syst. Eng. **58**(2), 151–155 (2016)
3. de los Angeles Martin, M., Olsina, L.: Towards an ontology for software metrics and indicators as the foundation for a cataloging web system. In: 2003 First Latin American Web Congress, pp. 103–113. IEEE (2003)
4. Baresi, L., Caporuscio, M., Ghezzi, C., Guinea, S.: Model-driven management of services. In: Eighth IEEE European Conference on Web Services, pp. 147–154. IEEE (2010)
5. Bézivin, J.: In search of a basic principle for model driven engineering. Novatica J. Spec. Issue **5**(2), 21–24 (2004)
6. Brandl, D.L., Brandl, D.: KPI exchanges in smart manufacturing using KPI-ML. IFAC-PapersOnLine **51**(11), 31–35 (2018)
7. Brous, P., Janssen, M.: Trusted decision-making: data governance for creating trust in data science decision outcomes. Admin. Sci. **10**(4), 81 (2020)
8. Caputo, E., Corallo, A., Damiani, E., Passiante, G.: KPI modeling in MDA perspective. In: Meersman, R., Dillon, T., Herrero, P. (eds.) OTM 2010. LNCS, vol. 6428, pp. 384–393. Springer, Heidelberg (2010). https://doi.org/10.1007/978-3-642-16961-8_59
9. Choi, Y., et al.: Towards data-driven decision-making in government: identifying opportunities and challenges for data use and analytics (2021)
10. Chung, L., Nixon, B.A., Yu, E., Mylopoulos, J.: Non-functional Requirements in Software Engineering, vol. 5. Springer, Cham (2012). https://doi.org/10.1007/978-1-4615-5269-7
11. Commission, J.R.C.E.: Handbook on Constructing Composite Indicators: Methodology and User Guide. OECD Publishing (2008)
12. De Castro, V., Mesa, J.M.V., Herrmann, E., Marcos, E.: A model driven approach for the alignment of business and information systems models. In: Mexican International Conference on Computer Science, pp. 33–43. IEEE (2008)
13. Diamantini, C., Genga, L., Potena, D., Storti, E.: Collaborative building of an ontology of key performance indicators. In: Meersman, R., et al. (eds.) OTM 2014. LNCS, vol. 8841, pp. 148–165. Springer, Heidelberg (2014). https://doi.org/10.1007/978-3-662-45563-0_9
14. Diamantini, C., Potena, D., Storti, E.: SemPI: a semantic framework for the collaborative construction and maintenance of a shared dictionary of performance indicators. Futur. Gener. Comput. Syst. **54**, 352–365 (2016)

15. Diamantini, C., Potena, D., Storti, E., Zhang, H.: An ontology-based data exploration tool for key performance indicators. In: Meersman, R. et al. (ed.) OTM 2014. LNCS, vol. 8841, pp. 727–744. Springer, Heidelberg (2014). https://doi.org/10.1007/978-3-662-45563-0_45
16. Elfouly, M., Kutzner, T., Kolbe, T.H.: General indicator modeling for decision support based on 3D city and landscape models using model driven engineering. In: Peer Reviewed Proceedings of Digital Landscape Architecture 2015 at Anhalt University of Applied Sciences (2015)
17. Elfouly, M., Labetski, A.: Flood damage cost estimation in 3d based on an indicator modelling framework. Geomat. Nat. Haz. Risk **11**(1), 1129–1153 (2020)
18. Frank, U., Heise, D., Kattenstroth, H.: Use of a domain specific modeling language for realizing versatile dashboards. In: Proceedings of the 9th OOPSLA Workshop on Domain-specific Modeling (DSM), vol. 4. Citeseer (2009)
19. Frank, U., Heise, D., Kattenstroth, H., Schauer, H.: Designing and utilising business indicator systems within enterprise models–outline of a method. In: Modellierung Betrieblicher Informationssysteme (MobIS 2008) (2008)
20. Giunta, B.: Towards the foundations for model-driven engineering of complex performance indicators: designing a systematic, user-centric, and unified CPI management framework (2025)
21. Harel, D., Rumpe, B.: Meaningful modeling: what's the semantics of "semantics"? Computer **37**(10), 64–72 (2004)
22. Hevner, A.R., March, S.T., Park, J., Ram, S.: Design science in information systems research. MIS Q., 75–105 (2004)
23. Kent, S.: Model driven engineering. In: Butler, M., Petre, L., Sere, K. (eds.) IFM 2002. LNCS, vol. 2335, pp. 286–298. Springer, Heidelberg (2002). https://doi.org/10.1007/3-540-47884-1_16
24. Kitchenham, B.: Procedures for performing systematic reviews. Keele, UK, Keele University **33**(2004), 1–26 (2004)
25. Kitsios, F., Kamariotou, M.: Business strategy modelling based on enterprise architecture: a state of the art review. Bus. Process. Manag. J. (2018)
26. Klous, S., Wielaard, N.: Remove fear and conquer resistance. In: We are Big Data, pp. 17–41. Springer, Cham (2016)
27. Kritikos, K., Plexousakis, D., Woitch, R.: A Flexible Semantic KPI measurement system. In: Ferguson, D., Muñoz, V.M., Cardoso, J., Helfert, M., Pahl, C. (eds.) CLOSER 2017. CCIS, vol. 864, pp. 237–261. Springer, Cham (2018). https://doi.org/10.1007/978-3-319-94959-8_13
28. Kritikos, K., Plexousakis, D., Woitsch, R.: Towards semantic KPI measurement. In: CLOSER, pp. 63–74 (2017a)
29. LaValle, S., Lesser, E., Shockley, R., Hopkins, M.S., Kruschwitz, N.: Big data, analytics and the path from insights to value. MIT Sloan Manag. Rev. **52**(2), 21–32 (2011)
30. Letrache, K., El Beggar, O., Ramdani, M.: Modeling and creating KPIs in MDA approach. In: 4th IEEE International Colloquium on Information Science and Technology (CiSt), pp. 222–227. IEEE (2016)
31. Li, H., Hong, T.: A semantic ontology for representing and quantifying energy flexibility of buildings. Adv. Appl. Energy **8**, 100113 (2022)
32. Li, Y., García-Castro, R., Mihindukulasooriya, N., O'Donnell, J., Vega-Sánchez, S.: Enhancing energy management at district and building levels via an EM-KPI ontology. Autom. Constr. **99**, 152–167 (2019)

33. Li, Y., O'Donnell, J., García-Castro, R., Vega-Sánchez, S.: Identifying stakeholders and key performance indicators for district and building energy performance analysis. Energy Build. **155**, 1–15 (2017)
34. Ma, R., Li, Q., Zhang, B., Huang, H., Yang, C.: An ontology-driven method for urban building energy modeling. Sustain. Cities Soc. **106**, 105394 (2024)
35. del Mar Roldán-García, M., García-Nieto, J., Maté, A., Trujillo, J., Aldana-Montes, J.F.: Ontology-driven approach for KPI meta-modelling, selection and reasoning. Int. J. Inf. Manag. **58**, 102018 (2021)
36. Michalczyk, S., Nadj, M., Azarfar, D., Maedche, A., Gröger, C.: A state-of-the-art overview and future research avenues of self-service business intelligence and analytics (2020)
37. Miller, G.A.: The magical number seven, plus or minus two: some limits on our capacity for processing information. Psychol. Rev. **63**(2), 81 (1956)
38. Monahov, I., Reschenhofer, T., Matthes, F.: Design and prototypical implementation of a language empowering business users to define key performance indicators for enterprise architecture management. In: 17th IEEE International Enterprise Distributed Object Computing Conference Workshops. IEEE (2013)
39. Parmenter, D.: Key Performance Indicators: Developing, Implementing, and Using Winning KPIs. Wiley, Hoboken (2015)
40. Pink, S., Lanzeni, D., Horst, H.: Data anxieties: Finding trust in everyday digital mess. Big Data Soc. **5**(1), 2053951718756685 (2018)
41. Popova, V., Sharpanskykh, A.: Modeling organizational performance indicators. Inf. Syst. **35**(4), 505–527 (2010)
42. Popova, V., Sharpanskykh, A.: Formal modelling of organisational goals based on performance indicators. Data Knowl. Eng. **70**(4), 335–364 (2011)
43. Popova, V., Treur, J.: A specification language for organisational performance indicators. Appl. Intell. **27**(3), 291–301 (2007)
44. Pourshahid, A., et al.: Business process management with the user requirements notation. Electron. Commer. Res. **9**(4), 269–316 (2009)
45. Pourshahid, A., Richards, G., Amyot, D.: Toward a goal-oriented, business intelligence decision-making framework. In: Babin, G., Stanoevska-Slabeva, K., Kropf, P. (eds.) MCETECH 2011. LNBIP, vol. 78, pp. 100–115. Springer, Heidelberg (2011). https://doi.org/10.1007/978-3-642-20862-1_7
46. del-Río-Ortega, A., Cabanillas, C., Resinas, M., Ruiz-Cortés, A.: PPINOT tool suite: a performance management solution for process-oriented organisations. In: Basu, S., Pautasso, C., Zhang, L., Fu, X. (eds.) ICSOC 2013. LNCS, vol. 8274, pp. 675–678. Springer, Heidelberg (2013). https://doi.org/10.1007/978-3-642-45005-1_58
47. del Río-Ortega, A., Resinas, M., Cabanillas, C., Ruiz-Cortés, A.: On the definition and design-time analysis of process performance indicators. Inf. Syst. **38**(4), 470–490 (2013)
48. del Río-Ortega, A., Resinas, M., Durán, A., Bernárdez, B., Ruiz-Cortés, A., Toro, M.: Visual PPINOT: a graphical notation for process performance indicators. Bus. Inf. Syst. Eng. **61**(2), 137–161 (2019)
49. del Río-Ortega, A., Resinas, M., Durán, A., Ruiz-Cortés, A.: Using templates and linguistic patterns to define process performance indicators. Enterp. Inf. Syst. **10**(2), 159–192 (2016)
50. del Rıo-Ortega, A., Resinas, M., Ruiz-Cortés, A.: Towards modelling and tracing key performance indicators in business processes. II Taller sobre Procesos de Negocio e Ingenierıa de Servicios, PNIS (2009)

51. del-Río-Ortega, A., Resinas, M., Ruiz-Cortés, A.: Defining process performance indicators: an ontological approach. In: Meersman, R., Dillon, T., Herrero, P. (eds.) Defining process performance indicators: an ontological approach. LNCS, vol. 6426, pp. 555–572. Springer, Heidelberg (2010). https://doi.org/10.1007/978-3-642-16934-2_41
52. Rojas, L.F.C., Jaramillo, C.M.Z.: Executable pre-conceptual schemas for representing key performance indicators. In: 2013 8th Computing Colombian Conference (8CCC), pp. 1–6. IEEE (2013)
53. Sanford, C., Oh, H.: The role of user resistance in the adoption of a mobile data service. Cyberpsychol. Behav. Soc. Netw. **13**(6), 663–672 (2010)
54. Shamsaei, A., Pourshahid, A., Amyot, D.: Business process compliance tracking using key performance indicators. In: zur Muehlen, M., Su, J. (eds.) BPM 2010. LNBIP, vol. 66, pp. 73–84. Springer, Heidelberg (2011). https://doi.org/10.1007/978-3-642-20511-8_7
55. Siau, K., Rossi, M.: Evaluation techniques for systems analysis and design modelling methods-a review and comparative analysis. Inf. Syst. J. **21**(3), 249–268 (2011)
56. Strecker, S., Frank, U., Heise, D., Kattenstroth, H.: MetricM: a modeling method in support of the reflective design and use of performance measurement systems. IseB **10**(2), 241–276 (2012)
57. Vaisman, A., Zimányi, E.: Data warehouse systems. Data-Centric Systems and Applications (2014)
58. Van Der Aa, H., Leopold, H., del Río-Ortega, A., Resinas, M., Reijers, H.A.: Transforming unstructured natural language descriptions into measurable process performance indicators using hidden Markov models. Inf. Syst. **71**, 27–39 (2017)
59. Wohlin, C.: Guidelines for snowballing in systematic literature studies and a replication in software engineering. In: Proceedings of the 18th International Conference on Evaluation and Assessment in Software Engineering, pp. 1–10 (2014)
60. Yeoh, W., Koronios, A.: Critical success factors for business intelligence systems. J. Comput. Inf. Syst. **50**(3), 23–32 (2010)
61. Zhang, Y.Y., Hu, Z.Z., Lin, J.R., Zhang, J.P.: Linking data model and formula to automate KPI calculation for building performance benchmarking. Energy Rep. **7**, 1326–1337 (2021)

Mapping Capabilities with LLMs: Requirements and Feasibility Assessment of Capability Broker

Adrian Olbricht[1], Jennifer Skopac[1], Jan Lundberg[1,2], Jelena Zdravkovic[1], and Janis Stirna[1(✉)]

[1] Stockholm University, Kista, Sweden
{adrian.olbricht,jennifer.skopac,jan.lundberg,jelenaz, js}@dsv.su.se
[2] Swedish Defence University, Stockholm, Sweden

Abstract. Capabilities provide a structured and stable business-centric view of what an organization is able to do, when, and under what conditions. Capability has become a standard design element of enterprise architecture frameworks such as NAF, TOGAF, ArchiMate and many others. Despite its usefulness, clear modularity, and purpose of the capability notion, many companies struggle becoming capability-aware and being able to continuously and efficiently manage their capability portfolios. To support this process, this paper presents requirements and an early design prototype of Capability Broker – an LLM based tool for discovering capabilities from organizational documents and creating capability models. The feasibility of the tool is demonstrated with an example case.

Keywords: LLM · Enterprise Modeling · Capability · NAF · AI

1 Introduction

Strong business performance relies on a well-designed, integrated enterprise architecture (EA) to control organizational complexity. EA aims to provide a holistic view of an enterprise through aligning multiple parts of the organization, such as different actors, products and services, organizational structures, and business processes [1, p. 3]. The concepts of EA have been widely adopted over the last few decades to manage complex organizational designs. However, the way EA connects business and IT-related aspects is often perceived as inflexible [2, p. 7]. As organizations navigate dynamic business landscapes, adaptive ways of working have become fundamental in development and management. The need for flexibility in managing both internal and external context changes has therefore sparked an interest in capability-oriented thinking in organizations [2, p. 5].

The concept of capability is generally recognized in business architecture, where it refers to an organization's ability and capacity to achieve objectives within a given context, see for example [2 p. 5; 3, 4] for capability definitions. In the context of military-oriented Capability-Based Planning, Andersson et al. [4] liken the use of capabilities to building with Lego, as each capability is a piece that helps you build new solutions when

A. Bernasconi et al. (Eds.): RCIS 2026, LNBIP 597, pp. 140–152, 2027.
https://doi.org/10.1007/978-3-032-35043-5_12

needed. Furthermore, business capabilities focus on what an organization is able to do. They provide a common language and understanding of which areas of the business should be emphasized [5]. This common language and understanding of capabilities are integral parts of the strive for continuous results in the context of changing business environments [6].

Although there are strong reasons for modern organizations to adopt capabilities, the time investment required to identify, develop, and manage them may discourage organizations from fully engaging in capability management. A common starting point towards capability management is a capability map [5]. However, creating capability maps for complex organizations is a tedious task, because it requires analyzing existing business designs (e.g. enterprise architecture models, business process descriptions, service descriptions) in detail, defining and modeling each capability, relating it with similar capabilities, and finally placing it into the capability map. As a consequence, many companies do not manage to create the map, and if they do, the map often becomes static and outdated as the organization evolves. Moreover, the process of identifying new capabilities from emerging business documents and linking them to the existing capability map needs to be supported.

To this end [6] proposes Capability Broker – a tool for capability discovery and mapping supported by a Large Language Model (LLM). The proposal is motivated by the potential of LLMs to analyze large document corpora and support conceptual modelling [7], both of which are essential tasks in capability mapping.

The project follows the design science research (DSR) framework [8] and considers Capability Broker as its design artifact. According to [9], DSR is iterative and undergoes five phases - problem explication, requirements definition, design and development, demonstration, and evaluation. Currently, Capability Broker is at the stage where the definition of the initial set of requirements has been completed, and the initial feasibility of the prototype has been carried out. Therefore, the goal of the paper is *to present the initial development cycle of Capability Broker, encompassing requirements and the initial feasibility assessment of capability discovery and mapping with LLMs in the military domain.*

The motivation for the choice of this domain is the need to have (i) sufficiently complex organization's activities, (ii) the need to work with capabilities which is common the military domain, (iii) availability of reasonably elaborated descriptions of organization's activities on various levels of details such as NATO's Allied Join Publications (AJP) and rich pictures describing operations on various levels of detail, and (iv) the need to use a pre-defined way of representing capabilities, such as the NATO Architecture Framework (NAF) [3], for military organizations of NATO member states.

The rest of the paper is structured as follows. Section 2 gives a brief background to the capability mapping and the initial vision of Capability Broker. Section 3 presents requirements for Capability Broker. Section 4 presents and discusses example prompts used in the project. Discussion and concluding remarks are presented in Sect. 5.

2 Background

2.1 Capability Modeling and Mapping

A capability is the ability and capacity that enable an enterprise to achieve a business goal within a specific context [10]. Capabilities also form the core of Capability Based Management [11]. There are two general approaches to modeling capabilities: as capability maps and as part of an enterprise model, which can be seen in modeling approaches such as Capability Driven Development (CDD) [2]. Capability maps are hierarchical i.e. they include several levels of capabilities (e.g. level 1, level 2, level 3…), where each level is a refinement of the previous one. At the top level, capabilities are typically highly stable, i.e. they represent what the organization does on a high level. Lower-level capabilities on the other hand are more operational and related to the IT applications supporting them. Each capability in the map has its internal design specifying how it is delivered by processes, resources, and services to meet a goal, as well as the context in which it is applicable.

NAF [3] covers both approaches by offering viewpoints that are mostly used for capability mapping (Capability Taxonomy C1, Enterprise Vision C2, and Capability Dependencies C3) and other viewpoints for linking capabilities with means of measurement (Performance Parameters C7), and realization (Standard Processes C4, Node Types L1, Resource Types P1, Resource Structure P2, Resource Connectivity P3 to name a few).

2.2 Overview of Capability Broker

The vision for Capability Broker was initially proposed in [6]. Its architecture consists of conceptual components such as capability map, LLM component for capability discovery and searching capabilities, and capability recommender. The tool should also have a knowledge repository for storing documents and models created.

The process of using Capability Broker for capability management is illustrated at a high level in Fig. 1:

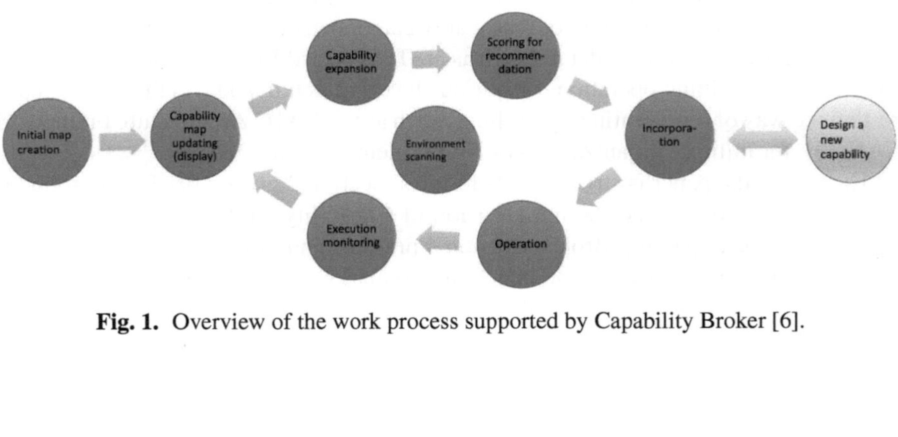

Fig. 1. Overview of the work process supported by Capability Broker [6].

Which process steps are used depends on the organization's current stage of capability adoption. If an organization has limited experience to capability-based thinking they might use the broker to create an initial map of capabilities. Meanwhile an organization that is more mature in capability-based thinking might instead use the broker to assist in implementing new capabilities and investigate the effects of the new capability on already existing capabilities.

3 Requirements for Capability Broker

Since the project follows an iterative structure, the first step is to define the requirements for creating the initial capability map and discovering related capabilities in the source documents. A complete set of requirements and other supplementary material such as models generated is available on GitHub [12].

In [6] we propose these three steps (see also Fig. 1):

Initial map creation: This task requires entering the organization's capability map into Capability Broker. This can be done in two ways. (i) If the organization (or one of its business units) does not have an explicit capability map, this task involves building an initial Capability Map in the broker. The organization can also use the LLM component to get suggestions for possible top-level capabilities by entering the business description as a prompt. Subsequent prompting can then be used to refine the level 1 capabilities into sub-capabilities at level 2 and lower levels. (ii) If the organization already possesses a capability map, it should be entered into the broker to obtain a machine-readable format and to ensure it is unified from a usage perspective.

Capability Map display: this task provides the experts with a view of existing capabilities, enabling them to browse from higher to lower level in search of a needed capability.

Capability expansion: this task enables identification of new capabilities when the capability is not found in the map. E.g., this can be done by analyzing descriptions of new technologies or business offerings from different organizational units.

Requirements for Initial Map Creation

The following requirements categories support this phase - (1) input text such as prompts and instructions, (2) file uploads for the purpose of creating a knowledge repository that focuses the LLMs response, and (3) capability map creation.

Input to Capability Broker can be in the form of text and can contain an organization description, processes, and prompts. The prompts can range from instructions to generate capability maps based on textual descriptions in attached files to instructions to identify goals and processes with the capability descriptions. The prompts can also contain several documents and instructions to combine them. These prompts should also contain instructions about the structure and the format of the output. For example, how many levels of capabilities to include, how they should be labeled, and in what format to present the output, e.g. text or PlantUML [13].

File uploads serve the purpose of creating a knowledge repository that the LLM can leverage to generate more context-aware responses. Capability Broker should support basic file upload in the following formats: PDF, JPEG, PNG, and DOCX. There should also be a limit of how many files that can be used to prevent Capability Broker's responses from becoming too unfocused. File uploads can be used for two purposes - directly in prompts and to create a knowledge base for Retrieval Augmented Generation (RAG). For the latter, the system should store domain-specific documents to provide context and use it as a knowledge base. For example, in the military domain, the base library should include the NAFv4 specification, AJPs, or other relevant external knowledge sources. The knowledge base should support using meta-models as a framework for creating and representing the output diagram, such as the NAFv4 metamodel, definitions, and examples of the modelling language. These are then used by the LLM agent as a rule set when creating the output diagram. Capability Broker should also support the user in updating (such as adding or removing files) the knowledge base (library) with additional knowledge sources to assist the LLM in identifying important relationships between capabilities, processes, and activities, among other aspects. The contents of the knowledge base should be available for viewing.

Capability map creation focuses on the functionality for creating a capability map. Capability Broker should be able to map sub-capabilities and sub-activities across different levels and accurately represent their relationships, including adding newly identified capabilities to existing capability maps. This process should enable the expert to review and refine the suggested capabilities, including correcting, complementing, and adjusting identified capabilities, activities, relationships, and structures. The expert should be able to approve, reject, rename, or modify individual suggested elements and their relationships.

Furthermore, Capability Broker should provide users with preset prompts known to yield good results and allow them to save best practice prompts.

Requirements for Capability Map Display

Capability Broker should be able to display a capability map and produce a file containing the diagram image, e.g. as PDF file, which the user can save for viewing. A key function that supports diagram creation is the ability of the LLM component to generate syntactically and semantically correct PlantUML code from which a graphical diagram is generated by another component such as a PlantUML visualization tool or a modeling tool such as Astah, which is a UML modeling tool (available from https://astah.net).

The interaction between these components should be achieved via an API and, in addition, the generated PlantUML code should be possible to copy to clipboard.

Since Capability Broker uses PlantUML tools for the purpose of displaying capability maps the UML class diagram notation is used, which imposes challenges to visually distinguish the components in the capability map. Therefore, Capability Broker should distinguish between classes that represent capabilities and those that represent activities, for example, by labelling them in different colors.

The interaction with Capability Broker is expected to be iterative. Hence, it should save the history of the current conversations' PlantUML code, so that it can be appended to the request if the user wants to make changes to a specific iteration of a diagram.

Capability Broker should also save the current conversation history of prompts and responses along with outputs for the LLM agent to use as context.

We have observed that on some occasions the PlantUML code generated contains errors. Therefore, Capability Broker needs to handle situations where the agent responds with non-compiling PlantUML code. This can be communicated to the user, prompting them to try again. Alternatively, retry behind the scenes until it works.

Requirements for Capability Expansion
Capability Broker should be able to match capabilities to other existing capabilities and processes. For example, matching newly identified capabilities to already-defined capabilities based on user input and identifying which ones support each other. It should also be able to match newly identified capabilities to an already-defined process based on user input, identifying which ones support each other.

Capability Broker should be able to re-generate the output diagram in an iterative process, taking into account the specified feedback from the user. For example, if the user wants certain classes to be renamed in the diagram, it should take previous messages into account as context when generating the response.

4 The Initial Prototype

4.1 Example Case

In the initial design cycle, the feasibility of Capability Broker was explored through a capability mapping case within the military domain. The military domain provides a relevant context since it combines organizational complexity, substantial use of capability-oriented tactical planning, and reliance on large bodies of textual documentations, e.g., manuals and processes outlined in specific publications.

At the same time, military capability development is commonly guided by established frameworks, such as NAF, which provides modeling structures for representing capabilities and their relationships. However, descriptions of organizational activities, operational functions, and emerging needs are not always expressed in capability terms, which makes identification and structuring of capabilities challenging.

The mapping case in the military domain is also motivated by the need to assess how emerging technology changes operational requirements in relation to existing capabilities. A new capability may contribute broadly by enhancing multiple sub-capabilities, thereby strengthening the overall capability. Alternatively, it may improve only a limited part of the capability structure, e.g., a specific activity or process. Being able to identify such differences is of importance when assessing a capability's operational impact, and as a result, make informed decisions regarding prioritizing development efforts, and balancing acquisition resources. From this perspective, Capability Broker is not only designed to identify and analyze new capabilities, but also to support designers' reasoning regarding where and how a new capability affects the overall capability structure.

Hence, there is a need to quickly discover capabilities from sources that are not structured according to NAF, and to align, and where relevant match, emerging capabilities with the existing capability map in order to support analysis and assessment.

4.2 Overview of the LLM Component

Regarding the LLM component of Capability Broker, the platform selected was LeChat, developed by Mistral AI. The choice of Mistral AI over other providers, such as OpenAI, was motivated by its European origin and market position. Using a European company provides greater transparency into data storage and ensures data management is compliant with GDPR. Another motivation for choosing LeChat as the LLM platform is that it is relatively unexplored for modeling purposes compared to platforms such as ChatGPT. This section provides an overview of the LLM component and explains how it fits into the overall tool architecture. Mistral Large 3 with extended reasoning (thinking) setting enabled was used in the process of prototyping the example case described Sect. 4. This version of Capability Broker calls LeChat via their API, without the extended reasoning.

The LLM component is implemented through what is commonly known as an AI Agent. In the field of Artificial Intelligence (AI), an agent is a computational entity that considers an environment when taking actions or making decisions [14]. The agent is commonly defined by a persona that describes its role and any additional tools it can use to perform its tasks [15]. The agent implemented in LeChat was defined by a set of definitions, instructions, and guardrails. Given the multiple definitions of what a capability is, it was necessary to establish a clear and consistent definition for the agent to follow. The relationships between a sub-capability and a capability, and between an activity and a capability, were also clearly specified to ensure the agent could understand the intended structure of the output. In terms of instructions, the workflow was specified to clarify both the input and output. Additionally, diagram rules were defined to control how LeChat is to structure the diagram and represent the capabilities and their relationships. To underline these guidelines, an example was provided for LeChat to follow as a reference. To enhance performance, guardrails were established to ensure LeChat produces accurate and well-structured outputs, maintains consistency, manages errors effectively, and continuously improves through user feedback. The definitions, instructions, and guardrails are presented below. It is important to note that their structure was suggested by LeChat to improve how it interprets and understands the information. See Table 1 for details.

In addition to definitions, instructions, and guardrails, external knowledge in the form of PDF documents was incorporated as a knowledge source for LeChat. This was done to reduce the chance of misinterpretations and mitigate what is commonly referred to as hallucinations in AI systems. For the purpose of this paper, the external knowledge consisted of the NAF version 4 specification. NAFv4 was selected based on its availability of documentation and relevance to the domain investigated in this paper.

The LLM components' role in the tool's overall architecture is situated in the tool's frontend, where user prompts and input files are sent to LeChat for capability identification and the construction of the capability map. The response from LeChat is then processed in the backend, where it is forwarded to a PlantUML drawing tool to generate a visual representation of the capability map, which is then presented to the user.

Table 1. Definitions, instructions, and guardrails provided to the LLM agent.

<table>
<tr><th colspan="2">Definitions</th></tr>
<tr><td colspan="2">- **Capability**: The ability to achieve a desired effect under specified standards/conditions, realized through ways and means
- **Sub-capability**: A distinct, reusable ability derived from a capability's attribute
- **Activity**: A specific action or task realizing a capability</td></tr>
<tr><th>Instructions</th><th>Guardrails</th></tr>
<tr><td>## Rules for Input
- Suggest clearer wording for ambiguous questions
- Example: If the input is ambiguous, re-phrase it for clarity
Project Workflow
1. **Input**: PDF/plain text
2. **Output**: PlantUML class diagram (no methods) with:
- Capabilities (goals + descriptions)
- Notes for context
- Activities (labeled separately)
- Sub-capabilities (if attributes represent distinct abilities)
Diagram Rules
- Label relationships clearly (e.g., "depends on")
- Validate: Every capability must have a goal/description
- The diagram should have multiple levels of sub-capabilities
Examples
Input Example
"The system must process user data and generate reports."
Output Example
"'plantuml
@startuml
class System {
+ processUserData()
+ generateReports()}
note top of System
The system must process user data and generate reports
end note
@enduml</td><td>## Input Handling
- **Clarity**: If an input is ambiguous, ask for claification
- Example: "Could you please clarify what you mean by…?"
- **Format**: Specify the preferred format for inputs
- Example: "Please provide the data in JSON format."
Output Quality
- **Accuracy**: Always verify information from reliable sources before providing an answer
- **Relevance**: Ensure that responses directly address the user's query
- **Structure**: Use clear headings and bullet points to organize information
Consistency
- **Tone and Style**: Maintain a professional and helpful tone in all responses
- **Terminology**: Use consistent terminology throughout interactions
Error Handling
- **Error Messages**: Provide clear and helpful error messages
- Example: "I'm sorry, but I couldn't understand your request. Could you please rephrase it?"
- **Fallback Mechanisms**: If the agent cannot resolve an issue, provide a fallback response
- Example: "I'm unable to assist with that request. Please contact support for further help."
Feedback Mechanism
- **User Feedback**: Encourage users to provide feedback on the agent's performance
- Example: "How was your experience with this interaction? Please let us know how we can improve."
- **Continuous Improvement**: Regularly review feedback and make necessary improvements</td></tr>
</table>

4.3 Prompting Examples

Table 2 presents examples of prompts from a conversation used for capability identification and mapping, as elaborated by using the LeChat web interface to evaluate the overall performance of the LLM.

Table 2. Examples of prompts from a conversation for capability mapping

Prompt examples of the prompts used	Comments on the response
I want to find activities and capabilities specifically relating to phase 2 in the attached file. The overarching capability should be "JointTargeting". It should have one sub-class that is Phase2	The diagram in Fig. 2 was too condensed, consisting of only two classes and thus not meaningfully representing the hierarchical structure and relationships between capabilities. However, to note, the diagram produced is in line with the prompt given to the Agent
Can you make every activity a sub-class to phase 2. Also make a differentiation if something is an activity or a capability	This diagram generated provided a significant structural improvement compared to the first diagram. The contents of the Phase2 class were made into separate classes with associations to the Phase2 class. The agent also added new capabilities for TargetDevelopment and TargetValidation
Can you check if you find any more activities that relate to phase 2? Capabilities should be represented as a sub-class of an activity if the capability relates directly to the activity	The agent re-organised the diagram. New relationships between capabilities and activities were added, e.g. it created a relationship between the activity IdentifyValidTargets and the capability TargetDevelopment
Could you check if there are any more capabilities in the document that support the different activities?	The agent identified new capabilities such as TargetSelectionStandards and linked it to the activity of IdentifyValidTargets. The agent identified five new capabilities
Could you check the document for more things that could be seen as a capability, use the following definition when checking if something could count as a capability: "A capability is the ability to achieve a desired effect under specified standards and conditions. A capability is realized through combinations of ways and means. The ability of one or more resources to deliver a specified type of effect or a specified course of action."	This rendition of the diagram was a bit of a side-step, the agent re-classified all activities as capabilities

(*continued*)

Table 2. (*continued*)

Prompt examples of the prompts used	Comments on the response
No, this isn't quite right! I still want activities to be in the diagram. I wanted to check what capabilities support different activities	The agent reverted the classification changes made in the previous rendition and added capabilities related to AdvancedTargetDevelopment. It added a new activity for IntelligenceStaffResponsibilities
I'm very happy with the diagram, but can you please remove the note and the execute() methods	The agent used class diagram notation and added operations to capabilities modeled as classes that needed to be removed

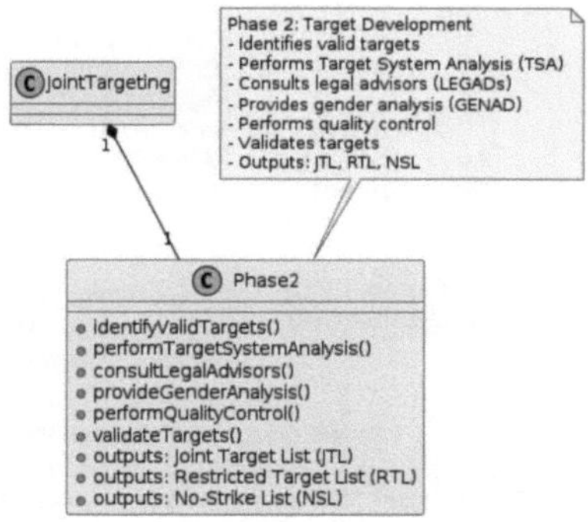

Fig. 2. The initial diagram that was produced by the agent.

In the capability identification and mapping conversation, the LLM generated PlantUML code, which was then visualized by a drawing tool, as shown in Fig. 2. The structure of this initial diagram is far from what can be expected as a capability map, however the contents of the diagram provide a good baseline. After multiple iterations of asking the agent to implement changes to the diagram, the result is shown in Fig. 3.

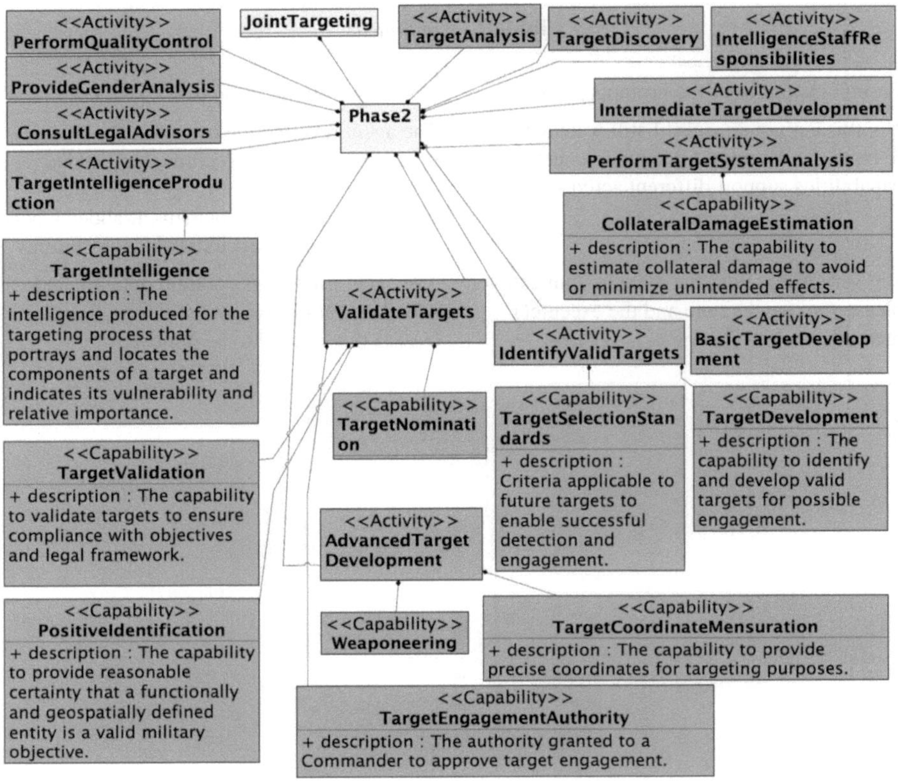

Fig. 3. The final diagram iteration.

Figure 3 shows activities of phase 2 Joint Targeting and capabilities that support the activities. The model was evaluated by one of the authors who is an expert within the military domain, and it was deemed to accurately capture the capabilities found in phase 2 of the joint targeting process. The model shown in Fig. 3 was created by importing the PlantUML code into the Astah UML modeling tool and then edited to make it fit the format of this article. Without manually editing the diagrams that are automatically generated from the PlantUML code, they tend to become excessively wide and unreadable.

During this design iteration, we tried several prompt chains and experimented with various input documents for generating a diagram. It was noted during previous attempts that that the LLM seems to perform better when given the name of the top-level capability. This structural guidance provided a more coherent response from the LLM. The iterative approach by breaking down the problem into smaller sub-tasks had a positive effect on the output, compared to the agent trying to generate the entire diagram in one go.

5 Discussion, Conclusions, and Future Work

In this idea paper we have presented the initial requirements and feasibility analysis for Capability Broker. We have observed that such a tool, using an LLM, can support capability discovery and mapping, thereby assisting the work of capability analysts. It is, however, not intended to be a replacement for human capability analysts but rather an assistant that analyses documents and proposes a capability map and/or capabilities that could be added to it. A particular impact of Capability Broker would be cases when new capabilities need to be discovered and aligned with the existing map.

The envisioned area of the use of Capability Broker is to receive capability suggestions based on domain specific data such as processes or service descriptions. The process of creating the capability map is iterative and needs an expert to prompt the agent and guide it in an appropriate direction, see Table 2 for an example of the iterative prompting process.

Concerning generalizability, Capability Broker is intended to be a domain independent tool. However, in our experience to achieve better results of capability identification, each application domain would need to be supported with some baseline documents similarly to how NATO's AJPs were used in this example case.

Concerning limitations of this study, two aspects need to be mentioned. First, the requirements have been elicited by means of action research and come mostly from one domain. Second, the outputs of the LLM have been compared with domain knowledge and documents but have not been evaluated for precision and recall due to not having an established gold standard for this problem domain. Both of these limitations will be addressed in the following design cycles of Capability Broker.

Regarding future work, we plan to elaborate and experiment further with various types of input documents such as rich text descriptions of capabilities and with other LLMs. Future versions of Capability Broker need to be able support continuous development of its recommendation logic based on feedback provided by the analyst. It shall be able to use "analyst-approved" corrections and refinements to improve future suggestions/cycles. This would also need to evolve towards a round-trip integration with modeling tools, where models from a modeling tool are improved by the LLM component and then entered back into the modeling tool.

References

1. Greefhorst, D., Proper, E.: The role of enterprise architecture. In: Architecture Principles: The Cornerstones of Enterprise Architecture, pp. 7–29. Springer (2011)
2. Sandkuhl, K., Stirna, J.: Capability Management in Digital Enterprises. Springer International Publishing (2018)
3. North Atlantic Treaty Organisation. NATO Architecture Framework, Version 4. AC/322-D(2018)0002-REV1, North Atlantic Treaty Organisation (2018)
4. Andersson, K.E., Lundberg, J., Stirna, J.: Emerging technology calls for a systemic view on military capability. In: PoEM Companion, vol. 3645, CEUR-WS.org (2024)
5. Ulrich, W., Rosen, M.: The Business Capability Map: The 'Rosetta Stone' of Business/IT Alignment. Enterprise Architecture, vol. 14/2 (2014)
6. Zdravkovic, J., Stirna, J., Tsai, C.H.: Pondering on capability brokering with LLM. In: Pufahl et al. (eds.) Intelligent Information Systems. CAiSE 2025. LNBIP 557, Springer (2025)

7. Storey V.C., et al.: Large language models for conceptual modeling: assessment and application potential. Data Knowl. Eng. **160** (2025). ISSN 0169-023X
8. Hevner, A., March, S.T., Park, J., Ram, S.: Design science in information systems research. Manag. Inf. Syst. Q. **28**(1), 75–105 (2004)
9. Johannesson, P., Perjons, E.: An Introduction to Design Science, Springer
10. Bērziša, S., et al.: Capability driven development: an approach to designing digital enterprises. Bus. Inf. Syst. Eng. **57** (2015)
11. Van Riel, J., et al.: Advancing the domain of strategy planning and implementation through enterprise architecture: a research agenda for capability-based management. Commun. Assoc. Inf. Syst. **56**, 155–166 (2025)
12. Capability Broker. https://github.com/Qp0p/Capability-Broker-diagrams-and-requirements
13. Drawing UML with PlantUML, PlantUML Language Reference Guide (Version 1.2025.0). https://plantuml.com/guide
14. Xi, Z., et al.: The rise and potential of large language model based agents: a survey. Sci. China Inf. Sci. **68**(2)
15. Masterman, T., Besen, S., Sawtell, M., Chao, A.: The landscape of emerging AI agent architectures for reasoning, planning, and tool calling: A survey. arXiv:2404.11584

Insights from Triage Processes for Fairness-Aware Process Redesign

Rachmadita Andreswari[1,2,4](✉), Stephan A. Fahrenkrog-Petersen[2,3], and Jan Mendling[1,2]

[1] Humboldt Universitat zu Berlin, Rudower Chausee 25, 12489 Berlin, Germany
{rachmadita.andre.swari,jan.mendling}@hu-berlin.de
[2] Weizenbaum Institut, Hardenbergstraße 32, 10623 Berlin, Germany
stephan.fahrenkrog@uni.li
[3] University of Liechtenstein, Fürst-Franz-Josef-Strasse, 9490 Vaduz, Liechtenstein
[4] Telkom University, Jl Telekomunikasi 1, 40257 Bandung, Indonesia

Abstract. The integration of Fairness, Accuracy, Confidentiality, and Transparency (FACT) into BPM practices is a recently increasing concern. While there are redesign heuristics, process design space related guidelines, and recommendations for general process design, research lacks an understanding of what makes a *fair* process design in order to inform design decisions. In this paper, we address this research problem by reviewing case study papers from the domain of triage processes in emergency rooms. This is a unique and specific process setting where fairness has strongly been considered during the COVID-19 pandemic. We organize our findings in a framework that integrates dimensions from social justice theory. In this way, our contribution provides a conceptual basis for fairness-aware process redesign along the BPM life cycle.

Keywords: fairness · process redesign · business process management · triage · emergency room

1 Introduction

Business processes are increasingly discussed from perspectives of ethics, responsibility, and fairness [22]. Unfair processes exhibit a major risk for companies in terms of negative press, losing customers, and litigation. The uptake of automatic decision making together with potentially biased machine learning [24] can lead to damage for members of certain types of gender or ethnical association. Examples are processes in the insurance industry where white men in the United States were offered better rates or credit approval where women got more expensive risk margins, in both cases even though other socio-economic factors were equivalent [10].

For general research on business process management (BPM), fairness is relatively new topics. New concepts related to fairness have been proposed in particular for process mining. Much of this work is engineering work, which is

A. Bernasconi et al. (Eds.): RCIS 2026, LNBIP 597, pp. 153–164, 2027.
https://doi.org/10.1007/978-3-032-35043-5_13

not yet used at a larger scale in practice. This means that we are lacking empirical insights into how organizations have been addressed concerns of fairness in their process designs. There are some exceptions though. Specific domains where processes and fairness have already been considered together provide an opportunity for developing generally applicable fairness concepts for BPM.

In this paper, we address this research problem by reflecting on lessons learnt from a specific process domain where fairness is a well established concept. More specifically, we focus on triage, the process of prioritizing patients in emergency room based on the clinical condition, the patient's prognosis, and availability of medical resources [4]. Our particular focus is the COVID-19 pandemic where extreme resource scarcity explicitly exposed fairness challenges. In this context, we review recent studies on the fairness of triage processes during the COVID-19 pandemic with the aim of identifying concepts that can be applied to any business process. In this way, we contribute a mapping of justice dimensions (from procedural, distributive and interactional justice) onto process elements and stakeholders as defined in the BPM framework [8]. This mapping provides a structured foundation for identifying potentially unfair activities across BPM contexts beyond healthcare processes.

This paper is structured as follows. Section 2 discusses the background of this work, summarizing research on fairness of triage processes contrasted with recent contributions on fairness in the area of business process management. Section 3 describes our analysis of triage processes and examining justice dimensions manifest in emergency room triage during the COVID-19 pandemic. Section 4 presents our results of mapping justice dimensions onto the BPM framework. Section 5 discusses the implications of our findings for Responsible BPM. Section 6 concludes with a summary and outlook on future research.

2 Background

In this section, we give a background of existing perspectives on fairness in BPM. Section 2.1 outlines existing studies on fairness and transparency in BPM and process mining, including the FACT framework of responsible process mining [22]. Section 2.2 highlights philosophical perspectives on fairness.

2.1 Process Design and Fairness in BPM and Process Mining

So far, BPM has approached process redesign mainly from the perspective of improving process time, cost, quality and flexibility [8]. Existing redesign heuristics and process design space approaches [17,28], do not explicitly consider fairness as a primal design objective. The dimensions of these heuristics have been later extended towards a process design space by Gross et al. [13]. These are related approaches have in common that they do not explicitly consider fairness as a primal design objective.

In BPM, notions of fairness and responsibility have been discussed mainly in the context of process mining. Here, challenges similar to those discussed in the area of data science arise, which translate into requirements of fairness, accuracy, confidentiality, and transparency (FACT) [1]. In process mining, these requirements are typically addressed with data handling and pre-processing in order to assure confidentiality, accuracy, absence of bias, and trustworthiness [1,22]. Failing to meet these requirements can entail negative consequences for stakeholders [22], not only during post-hoc process analysis, but equally during process execution. According to [27], discrimination in processes are the plausible causes of fairness issues, resulting in unequal outcomes in waiting times, re-do's, success rates, deviations, and allocation of resources [27].

2.2 Philosophical Perspectives on Fairness

The concept of fairness has long been a subject of discourses in neighboring disciplines such as law and philosophy, often in relation to notions of justice. Individuals inherently seek justice in events and decisions that affect them. Traditionally, justice has been examined within the legal domain, where it is defined objectively through formal rules and regulations. However, its scope has broadened with the incorporation of fairness, which emphasizes equality in the treatment of individuals in everyday social interactions. Fairness can thus be regarded as a factor of justice that is more closely associated with interpersonal relationships and day-to-day experiences. In the context of business processes, this implies that process execution and outcomes are consistent across cases and actors, and that any deviations are justified by objective criteria rather than irrelevant personal attributes such as gender, age, ethnicity, or social status. From this perspective, [12] conceptualizes justice as comprising three distinct dimensions: procedural justice, distributive justice, and interactional justice, each reflecting a different aspect of fairness in practice.

The first dimension is procedural justice, which refers to fairness in the processes and procedures used to make decisions [12]. This concept emphasizes adherence to rules that form the basis for fair decision-making. According to [19], procedural justice can be characterized by eight key principles: process control, decision control, consistency, bias suppression, accuracy, correctability, ethicality, and representativeness.

The second dimension is distributive justice, which concerns the perceived fairness of outcome distributions. This dimension evaluates how principles such as equity, equality, and need are applied in practice. The most widely recognized principle within distributive justice is equity, which assesses fairness based on the ratio between an individual's inputs (e.g., effort or contribution) and the outcomes received, compared to others. However, [7] extended this view by introducing two additional distribution principles: equality, where all individuals receive the same outcomes regardless of input, and need, where outcomes are distributed based on individual needs. For instance, in an organizational context, employees with above-average performance receive greater rewards (equity), all employees are granted equal leave entitlements (equality), and employees who experience workplace accidents receive compensation (need).

The third dimension is interactional justice, which is related to individuals' responses to organizational outcomes, particularly in relation to how procedures are enacted and communicated [12]. Unlike procedural justice, which focuses on the fairness of formal processes, interactional justice emphasizes the quality of interpersonal treatment individuals receive during the decision-making process [33]. According to [19], interactional justice comprises four key principles: respect, propriety, justification, and truthfulness.

2.3 Research Objective

We observe that a grounding of fairness in the three notions of justice can offer a useful basis for structuring concerns of fair process design. What is missing so far is an analysis of the repertoire of actual process design concepts that can be considered as reflecting different aspects of fairness. To this end, we conduct a review of case studies that focus on triage processes in emergency rooms during the COVID-19 pandemic [14,18,23,26,35,36]. We selected these studies based on three criteria: (1) they focus on triage processes in emergency rooms, (2) they address fairness or ethical concerns explicitly, and (3) they were conducted during or in response to the COVID-19 pandemic, a context characterized by extreme resource scarcity and high-pressure conditions that exposed inherent fairness challenges in decision-making. For each justice dimension, we examine how justice dimension manifest in triage practice and map them onto process elements and stakeholders as defined in [8].

3 Triage Processes

This section reviews ethical and fairness-driven decision-making practices found in triage processes, particularly in emergency rooms. In the following, we will first introduce the Triage process in Sect. 3.1. Next, we explain difficulties of Triage processes during the COVID-19 pandemic in multiple countries in Sect. 3.2.

3.1 What is Triage and How Does It Work?

Effective decision-making in emergency rooms (ER) is crucial for maximizing survival rates and ensuring timely medical intervention. To that end, hospitals follow strict procedures for prioritizing these cases. One of the most widely used triage processes is Simple Triage and Rapid Treatment (START) [4]. It was developed in the United States in 1983. Though being adopted worldwide, it is often implemented with modifications due to differences in social justice principles and cultural influences.

In essence, START helps classify patients based on the urgency of their condition. A key concept is the assessment of patients based on their condition using four categories: Green, Yellow, Red, and Black. The process starts with healthcare professionals evaluating the patient's Respiratory, Perfusion, and Mental status (RPM). These indicators provide a critical first assessment of the patient's condition. Respiratory function is prioritized as the primary factor, followed by blood circulation and mental state, ensuring a systematic approach to emergency assessment. Based on these evaluations, decisions are taken for how to proceed [4]: In case of Green, treatment can be delayed. In case of Yellow, treatment is urgent but can wait up to one hour. In case of Red, immediate attention is required to save the patient's life. In case of Black, the patient is deceased and does not require emergency intervention.

Triage methods generally function effectively when patient volume is manageable and sufficient healthcare personnel (nurses and doctors) are available. However, imbalances between patient numbers and medical staff often create challenges, requiring efficient decision-making to save as many lives as possible. Even in such cases, medical professionals appear to prioritize critically ill patients (Red label) well, while prioritizing non-critical cases (Yellow and Green labels) is more complex. As a first-come, first-served queue system [11,20] is often used, though it does not always align with real-world emergencies, particularly when patients have different levels of urgency. In such cases, a lottery-based priority system has been discussed as a fairer alternative, ensuring equal treatment for patients with similar conditions while minimizing subjective biases [5,11]. Several subjective factors, such as social status [6], economic background [15], age [3], and gender [3,15] may also influence triage decisions, introducing potential biases. In some regions, such as South Asia [6] and other countries [11], these factors have historically impacted patient prioritization, leading to disparities in emergency healthcare access.

Despite its structured approach, the START method is not always effective in high-pressure emergency situations. Manual assessments can be difficult to execute consistently, especially when resources are limited and rapid decisions are required. Ensuring fairness in emergency decision-making is highly challenging, particularly in overcrowded ERs with limited resources. Proper triage requires experienced healthcare professionals [3], as expertise and exposure to high-pressure situations play a critical role in making accurate and ethical decisions [11]. Establishing clear, standardized prioritization criteria based on objective health assessments [3,20] is crucial. If patient prioritization is influenced by social, cultural, or economic factors, it can lead to healthcare disparities, favoring privileged patients over others. Therefore, humanitarian values [3] and the goal of saving the maximum number of lives [3,15] must take precedence in triage assessment frameworks.

Issues that have been discussed for triage potentially exist in other business processes. Fairness concerns are particularly evident in decision-making scenarios where outcomes are influenced by actor profiles or specific process attributes, potentially leading to disparities in processing times, complexity of business process, decision point, process sequence, success process, deviation, resource allocation, ground truth and share [27].

3.2 Insights from Triage Processes in COVID Times

The challenges of triage processes become particularly visible under disaster conditions and high-pressure situations. Triage during a pandemic cannot be separated from the challenge of treating patients fairly under severe and resource-constrained conditions. As highlighted by Vinay et al. [35], the principle of fairness as an integral part of justice is applied to address issues such as age discrimination and various forms of implicit bias. Similarly, [18] emphasize that fair prioritization, based on patients' clinical conditions, is essential to provide appropriate treatment to each patient without favoring one group over another.

Furthermore, [14,23] highlight the importance of transparency in the triage process, particularly regarding patient management and decision-making. This aligns with Fairness, Accuracy, Confidentiality, and Transparency (FACT) framework in responsible process mining [22], in which transparency is necessary for making processes explainable and auditable. A transparent process facilitates the identification of deviation, such as unfairness, while simultaneously enhancing the accountability of stakeholders e.g., healthcare personnel.

The triage model used in Piacenza, Italy, as discussed by [26], highlighting that early diagnosis and dynamic resource allocation can improve process efficiency. The context-aware approach is based on available data and capacity, not just on rules. The triage study during the COVID-19 pandemic further emphasizes the importance of designing processes that are not only ethically grounded, but also applicable and responsive to real-world limitations. For example, studies such as [16,36] highlight that triage must be built on clear ethical principles and be able to adapt to evolving challenges. During the COVID-19 pandemic, overcrowding in emergency rooms increased the inaccuracies in emergency triage, resulting in both over-triage and under-triage. Additionally, the implementation of epidemic prevention measures, including advanced high-tech systems, introduced new ethical issues. Several dilemmas also emerged, such as the waste of resources for prevention, the restriction of individual patient autonomy by COVID-19 protocols, conflicts between patient interests and public interests, and inaccurate feedback from medical devices [36]. All of these issues present significant challenges for implementing triage business processes in high-pressure scenarios, reinforcing the need for ethical, adaptable, and context-aware process design.

Additionally, joint recommendations of European scientific associations, emphasize the need for a systematic, auditable, and context-sensitive framework [30]. Findings from [32] also show that continuous evaluation, real-time data integration, and process flexibility can significantly improve effectiveness and fair triage. These insights provide a strong foundation for Responsible BPM to create adaptive business processes, where process flows can be adjusted based on urgency, capacity, and stakeholder roles. They emphasize that business processes in crisis or resource-constrained situations must integrate ethical and adaptive principles, engage stakeholders, and incorporate effective governance, so that processes remain fair, transparent, and accountable.

Table 1. Comparison of COVID-19 Triage Approaches Across Selected Studies

Reference	Herreros [14]	Poggiali [26]	Orfali [23]	Jaziri [18]	Wu [36]	Vinay [35]
Input	Clinical data (age, comorbidity)	Clinical & digital assessment	Clinical assessment, Guidelines	Clinical assessment, Guidelines	Clinical assessment (Emergency Severity Index)	Clinical assessment
Country	Spain	Italy	France	U.S.	China	Swiss
Prioritization	Clinical scores (prognosis, expected benefit, potential for quick recovery)	Symptoms and Point-of-care lung US	Clinical Score (Prognosis and functional autonomy)	Clinical scores, comorbidities	Guided by maximizing benefit (prognosis)	Exclusion criteria (based on age, life expectancy, comorbidities, and frailty score)
Decision Maker	Committee triage	Emergency department team	Treating physicians and ICU teams	Triage committee (doctors, ethicists, administrators, technicians)	Emergency department team	Clinical team/ICU team
Output	Recommendation	Documentation	Transparent documentation	Ethical value guiding recommendation	Documentation	Documentation

Table 1 compares triage implementation across different countries during the COVID-19 pandemic. It illustrates how justice dimensions are operationalized in triage, prioritization, decision-making, and output processes within various contexts, providing valuable insights for designing responsible and context-aware business processes. While each country is represented by a single study, the table is intended as an illustrative overview rather than a statistically representative comparison. By understanding these international variations, organizations can more effectively tailor fairness strategies to their specific needs.

The comparison shows that most research on triage during the COVID-19 pandemic addressed a range of ethical concerns, including fair prioritization, transparency, accountability, context-awareness, and responsiveness, while also emphasizing the need to minimize deviations from established processes. This focus becomes especially important given the substantial increase in patient numbers during the pandemic, which required medical personnel to maintain resilience in order to treat all patients fairly and save more lives. Furthermore, from a practical perspective, these findings provide empirical illustrations of how triage was conducted in real-world pandemic situations.

4 Process Designs from Triage and Justice Dimensions

This section maps fairness aspects of the triage process onto the three dimensions of justice as introduced in Sect. 2.2. The eight principles of procedural justice from [19], the three principles of distributive justice from [7], and [12], and the four principles of interactional justice from [19]. Each principle is examined in relation to process elements and process stakeholders, following the BPM framework of [8]. The results of this mapping are summarized in Table 2.

Table 2. Mapping of Justice Dimensions onto Process Elements and Stakeholders

Justice Dimension	Principle	Process Element	Stakeholder	Example from Triage
Procedural	Process control	Activity	Patient, Healthcare staff	Patient arrival at triage; patient and staff discuss symptoms and medical record [31]
	Decision control	Activity	Nurse, Doctor	Vital sign assessment after patient arrival [31,34]
	Consistency	Sequence flow	Healthcare staff	Calling patient for further treatment follows the same procedure for every patient [34]
	Bias suppression	Decision point	Nurse, Doctor	Arrival mode (walking vs. ambulance) should not affect treatment received [34]
	Accuracy	Activity	Healthcare staff	Use of correct vital sign information for diagnosis and acuity determination [3,31]
	Correctability	Activity	Patient, Healthcare staff	Feedback on medication response; adjustment of treatment or re-examination [3,31]
	Representativeness	Decision point	Doctor, ER Manager	Consultation with patient, nurse, or staff before assigning triage category [34]
	Ethicality	Decision point	Doctor	Prioritization based on urgency of vital signs; resolution of ethical dilemmas (e.g., age vs. severity) [3,11]
Distributive	Equity	Decision point	Doctor, Patient	Patient accesses healthcare fairly and without socio-economic discrimination [3]
	Equality	Decision point	Doctor, Patient	All patients receive the same treatment regardless of background or contribution [20]
	Need	Decision point	Doctor, Patient	Treatment distributed based on urgency and severity of patient condition [20]
Interactional	Respect	Actor	Healthcare staff	Greeting patients kindly, listening to concerns, being mindful of language [6,9]
	Propriety	Actor	Healthcare staff	Avoiding offensive or discriminatory behavior; asking appropriate questions [21]
	Justification	Actor	Doctor	Explaining why another patient was given priority based on medical necessity [20]
	Truthfulness	Actor	Doctor, Healthcare staff	Informing patients honestly about condition, service limitations, and possible outcomes [21]

Procedural justice principles are mapped onto activities and decision points as these are the process elements where rules, consistency, and control are most directly implemented [19]. Most principles, such as process control, decision control, consistency, bias suppression, accuracy, and correctability are operationalized through structured activities carried out by medical staff. They have to ensure that decisions are grounded in vital sign assessments rather than personal judgment [31,34]. Among these, *ethicality* is the most complex as it requires balancing competing patient needs under resource constraints. For instance, when two patients have similar vital signs but different in age, the healthcare staff must consult colleagues to reach the fairest decision based on ethical standards [3,11].

Distributive justice principles are mapped onto decision points, as this is where allocation outcomes for patients are determined [12]. The three principles of equity, equality and need each represent a different basis for fair distribution. Equity ensures patients access care without socio-economic discrimination [3]; equality ensures standard treatment regardless of background of the patient [20]; and need ensures that the most urgent and vulnerable cases receive priority [20]. Overall, these principles reflect the core tension in triage between treating all patients the same and responding to individuals based on clinical severity.

All the principles of *interactional justice* are mapped onto the actor, since the quality of interpersonal treatment depends on how healthcare staff engage and interact with patients during triage [29]. While the other dimensions focus on process structure and outcomes, interactional justice focuses on the communicative dimension of fairness. For instance, how decisions are explained, how patients are addressed, and how honestly information is conveyed to patients [20,21]. These aspects are typically harder to standardize through process design alone, and are more likely to be assessed by patient feedback or survey data.

Among procedural justice dimensions, bias suppression and ethicality show the greatest challenge. Bias suppression requires that process rules be explicitly formalized so that actor attributes can not influence on decision outcome. In BPM, this can be implemented at the decision point. On the other hand, ethicality can not be fully encoded in process rules alone, it requires escalation paths and mechanisms when the issue arise. For the distributive justice principles, a process is designed equal for each cases. While need-based make the process different under the scarcity and high tension. Adjusting the process rules from equality to need is often needs adjustment by domain expert. Interactional justice dimension are the hardest to operationalize through process design because they depend on the actor behaviour rather than process itself. However, during the process design, process analyst can create standard conditions, for example, time slot for explanation and documentation of communication on the medical record.

5 Discussion

The analysis of justice, fairness and ethics principles in the triage process points to important implications for business process design. We know from research on business process design and redesign that any design choice typically comes with trade-offs. The so-called devil's quadrangle illustrates this aspect by emphasizing that an improvement in one process performance dimension of time, cost, quality, or flexibility typically entails a degradation of at least one of the other ones [8]. What we have seen as concerns of triage often implies improvements of fairness as a quality aspect with more effort in decision making and stricter conformity to explicit rules and documentation. This means that fairness and ethical design does not come for free. It can make processes slower and more costly. There is evidence that improvements in quality are possible without compromising cost and time [8]. More research is needed on discovering redesign heuristics that improve fairness without affecting other performance aspects.

Justice dimensions identified in triage are not limited to healthcare settings but extend to any business process, resource allocation, and stakeholder interaction. For instance, in recruiting processes, bias suppression maps onto the requirement that candidate selection procedures be standardized to prevent demographic characteristics from influencing hiring decisions [2]. Furthermore, distributive justice principle of equality in educational assessment processes ensures that all students receive consistent and unbiased evaluation regardless of their background [25]. Lastly, the interactional justice principle of truthfulness maps onto the obligation of banking customer service staff to provide honest and transparent information during investment consultations [37].

This study has limitations. First, the framework has not yet been empirically validated beyond the triage context, either through expert evaluation or application to non-crisis settings. The mapping is conceptually grounded but requires further testing to assess its practical applicability across domains. Furthermore, the case studies are limited to COVID-19 triage contexts which represent extreme conditions, not everyday process execution. Future work should apply the framework to non-crisis settings and evaluate it with subject matter experts across different domains.

6 Conclusion

This study demonstrates that responsible BPM, particularly fairness, can be effectively implemented within the BPM lifecycle through the emergency room triage process. By systematically mapping justice dimensions to process elements and stakeholders, this study shows that traditional process performance metrics such as time, cost, quality, and flexibility can be enriched by incorporating fairness considerations. The identified justice dimensions provide a starting point for fairness-aware process redesign. Future work could investigate how existing BPM redesign heuristics [8], such as control addition, flexible assignment, and exception handling, can be adapted to incorporate the justice dimensions identified in this study as well as the development of fairness metrics tailored specifically to BPM contexts, and exploration of the application of justice dimensions in other healthcare and critical business processes.

Acknowledgments. The research of the authors was supported by the Einstein Foundation Berlin under grant EPP-2019-524, by the Federal Ministry of Research, Technology and Space under the grant 16DII133, founded by the German Federal Ministry of Education and Research in 2017, and by Deutsche Forschungsgemeinschaft under grants 496119880 (VisualMine), 531115272 (ProImpact), SFB 1404/2 (FONDA), and Elsa Neumann Stipendium (H78027).

References

1. van der Aalst, W.M., Bichler, M., Heinzl, A.: Responsible data science. Bus. Inf. Syst. Eng. **59**, 311–313 (2017)
2. Ababneh, K.I., Hackett, R.D., Schat, A.C.: The role of attributions and fairness in understanding job applicant reactions to selection procedures and decisions. J. Bus. Psychol. **29**(1), 111–129 (2014)
3. Ahmed, S., Alsisi, R.H.: Ethical triage in public health emergency facilities: distributive justice-a decision model. Kybernetes (2024)
4. Bazyar, J., Farrokhi, M., Khankeh, H.: Triage systems in mass casualty incidents and disasters: a review study with a worldwide approach. Open Access Macedonian J. Med. Sci. **7**(3), 482 (2019)
5. Broome, J.: Fairness. In: Proceedings of the Aristotelian society, vol. 91, pp. 87–101. JSTOR (1990)
6. Chakraborty, R., Achour, N.: Setting up a just and fair ICU triage process during a pandemic: a systematic review. In: Healthcare, vol. 12, p. 146. MDPI (2024)
7. Deutsch, M.: Equity, equality, and need: what determines which value will be used as the basis of distributive justice? J. Soc. Issues **31**(3), 137–149 (1975)
8. Dumas, M., Rosa, M.L., Mendling, J., Reijers, H.A.: Fundamentals of business process management: second edition. In: Fundamentals of Business Process Management: Second Edition, pp. 1–527 (2018)
9. Efrat-Treister, D., et al.: All eyes on me: the impact of individualism vs. collectivism orientations on justice perceptions and mistreatment of frontline staff in emergency departments. Negot. Confl. Manag. Res. **17**(4) (2024)
10. Fourcade, M., Healy, K.: Classification situations: life-chances in the neoliberal era. Acc. Organ. Soc. **38**(8), 559–572 (2013)
11. Ghanbari, V., et al.: Fair prioritization of casualties in disaster triage: a qualitative study. BMC Emerg. Med. **21**, 1–9 (2021)
12. Gilliland, S.W.: The perceived fairness of selection systems: an organizational justice perspective. Acad. Manag. Rev. **18**(4), 694–734 (1993)
13. Gross, S., Stelzl, K., Grisold, T., Mendling, J., Röglinger, M., vom Brocke, J.: The business process design space for exploring process redesign alternatives. Bus. Process. Manag. J. **27**(8), 25–56 (2021)
14. Herreros, B., Gella, P., de Asua, D.R.: Triage during the COVID-19 epidemic in Spain: better and worse ethical arguments. J. Med. Ethics **46**(7), 455–458 (2020)
15. Holzer, F., Biller-Andorno, N., Baumann, H.: The role of social justice in triage revisited: a threshold conception. Med. Health Care Philos. **28**(1), 161–169 (2025)
16. Iacorossi, L., et al.: Triage protocol for allocation of critical health resources during the COVID-19 health emergency: a review. Acta Bio Medica: Atenei Parmensis **91**(4), e2020162 (2020)
17. Jansen-Vullers, M., Reijers, H.: Business process redesign in healthcare: towards a structured approach. INFOR: Inf. Syst. Oper. Res. **43**(4), 321–339 (2005)
18. Jaziri, R., Alnahdi, S.: Choosing which COVID-19 patient to save? the ethical triage and rationing dilemma. Ethics, Med. Publ. Health **15**, 100570 (2020)
19. Jess, G., Price, R.: Fairness in enforceable undertakings: comparing stakeholder voices. Saf. Sci. **94**, 1–9 (2017)
20. Lauridsen, S.: Emergency care, triage, and fairness. Bioethics **34**(5), 450–458 (2020)
21. Lee, S.H., Kim, C.W.: History-taking questions during triage in emergency medicine. Res. Lang. Soc. Interact. **55**(4), 326–349 (2022)

22. Mannhardt, F.: Responsible process mining. In: van der Aalst, W.M.P., Carmona, J. (eds.) Process Mining Handbook. Lecture Notes in Business Information Processing, vol. 448, pp. 373–401. Springer, Cham (2022). https://doi.org/10.1007/978-3-031-08848-3_12
23. Orfali, K.: What triage issues reveal: ethics in the COVID-19 pandemic in Italy and France. J. Bioethic. Inq. **17**(4), 675–679 (2020)
24. Pessach, D., Shmueli, E.: A review on fairness in machine learning. ACM Comput. Surv. (CSUR) **55**(3), 1–44 (2022)
25. Pitt, E., Winstone, N.: The impact of anonymous marking on students' perceptions of fairness, feedback and relationships with lecturers. Assess. Eval. High. Educ. **43**(7), 1183–1193 (2018)
26. Poggiali, E., Vercelli, A., Cillis, M.G., Ioannilli, E., Iannicelli, T., Magnacavallo, A.: Triage decision-making at the time of COVID-19 infection: the Piacenza strategy. Intern. Emerg. Med. **15**, 879–882 (2020)
27. Pohl, T., Qafari, M.S., van der Aalst, W.M.: Discrimination-aware process mining: a discussion. In: Montali, M., Senderovich, A., Weidlich, M. (eds.) ICPM 2022. LNBIP, vol. 468, pp. 101–113. Springer, Cham (2022). https://doi.org/10.1007/978-3-031-27815-0_8
28. Reijers, H.A., Mansar, S.L.: Best practices in business process redesign: an overview and qualitative evaluation of successful redesign heuristics. Omega **33**(4), 283–306 (2005)
29. RJ, B.: Interactional justice: communication criteria of fairness. Res. Negot. Organ. **1**, 43–55 (1986)
30. Sarmento, J.T., Pedrosa, C.L., Carvalho, A.S.: What is common and what is different: recommendations from European scientific societies for triage in the first outbreak of COVID-19. J. Med. Ethics **47**(7), e15 (2021)
31. Srinivas, S., Nazareth, R.P., Shoriat Ullah, M.: Modeling and analysis of business process reengineering strategies for improving emergency department efficiency. SIMULATION **97**(1), 3–18 (2021)
32. Turcato, G., Zaboli, A., Pfeifer, N.: The COVID-19 epidemic and reorganization of triage: an observational study. Intern. Emerg. Med. **15**, 1517–1524 (2020)
33. Tyler, T.R., Bies, R.J.: Beyond formal procedures: The interpersonal context of procedural justice. In: Carroll, J. (ed.) Applied Social Psychology and Organizational Settings, pp. 77–98. L. Erlbaum Associates, Hillsdale, NJ (1990)
34. U.S. Department of Health & Human Services: START adult triage algorithm (2024). https://chemm.hhs.gov/startadult.htm. Accessed 28 Nov 2024
35. Vinay, R., Baumann, H., Biller-Andorno, N.: Ethics of ICU triage during COVID-19. Br. Med. Bull. **138**(1), 5–15 (2021)
36. Wu, L., Kong, X.: COVID-19 pandemic: ethical issues and recommendations for emergency triage. Front. Publ. Health **11**, 1160769 (2023)
37. Yu, B., Kersten, G.E., Vahidov, R.: An experimental examination of credible information disclosure, perception of fairness, and intention to do business in online multi-bilateral negotiations: B. Yu et al. Electron. Mark. **32**(1), 217–237 (2022)

Can We Typify Health Information Seekers? Identification of Personas Among Laypeople

Mariana Dias(✉) and Carla Teixeira Lopes

INESC TEC, Faculty of Engineering, University of Porto, Rua Dr. Roberto Frias, 4200-465 Porto, Portugal
{up201606486,ctl}@fe.up.pt

Abstract. The digitalization of healthcare and the widespread use of personal health technologies generate large volumes of health data, yet individuals still struggle to manage and understand their information. This paper explores the development of personas that represent typical personal health information-seeking behaviors, motivations, and needs to support the user-centered design of personal health knowledge management systems. We adopted a mixed-method approach, conducting and analyzing an online survey that collected data on personal health information needs, purposes of access, and access-related challenges, along with using affinity diagramming to group participants by key characteristics identified through data analysis. This approach derived three personas: clinical navigator, health tracker, and care planner. These personas provide a basis for designing user-centered systems that enable individuals to access, organize, and use their personal health information.

Keywords: Survey · Personas · Personal health information needs · Personal health data · Health informatics

1 Introduction

The digitalization of healthcare and the widespread use of Internet of Things technologies have led to the continuous production of large volumes of health data from diverse sources [15]. Despite the growing availability of personal health data, it often remains fragmented across different sources, making it difficult for individuals to manage and comprehend their personal health data [11]. At the same time, recent advances in generative artificial intelligence present new opportunities to enhance personal health knowledge management by facilitating tasks such as information summarization and interpretation [6]. Effectively managing personal health information enables individuals to take a more active role in their healthcare [9]. However, designing effective personal health management systems requires a user-centered understanding of individuals' health information needs.

Persona is a widely adopted user-centered design approach for capturing user requirements, representing groups of users who share common behavioral

A. Bernasconi et al. (Eds.): RCIS 2026, LNBIP 597, pp. 165–175, 2027.
https://doi.org/10.1007/978-3-032-35043-5_14

characteristics [13]. Although personas have been applied in the design of health technologies, existing studies focus on specific population groups, such as older adults [7] or individuals with clinical conditions [16], as well as particular health information contexts, including online health communities [5], thus limiting the broader characterization of personas for personal health information seeking.

In this study, we develop personas that characterize typical profiles of individuals who interact with their personal health data. We conducted an online survey with 420 participants to determine health information needs and access-related challenges. Through data analysis and affinity diagramming, we identified different groups of individuals and used them to develop personas. The created personas serve as a user-centered foundation for the development of personal health information management systems.

2 Personal Health Information Needs

An information need arises when an individual recognizes that their existing knowledge is insufficient to satisfy a particular goal [14]. Goals can be multi-factorial and, in the health context, can include coping with health-threatening situations, participating in medical decision-making, or needing to change behaviors to prevent further complications [8]. Goals can motivate individuals to seek personal health information.

Prior work has shown that individuals differ in their preferred level of involvement in medical decision-making [3]. These differences can influence the type and level of detail of information individuals need. For instance, individuals managing serious illness may require more detailed or condition-specific health information [1]. Personal characteristics, such as age and health status, also influence information-seeking behavior [8], as, for example, individuals with closer proximity to serious conditions, such as cancer, are less likely to actively seek health information [3].

Several studies have characterized personal health information needs. Methods reported in the literature include interviews, focus groups, questionnaires, computer logs [2], and diary studies [12]. A literature review by Clarke et al. [2] on health information needs indicates that individuals commonly seek information about illnesses or medical conditions and treatment options. Other research has identified a wide range of relevant personal and health-related information, including personal data, appointments, allergies, vaccination status, medical history, diagnoses, results, therapies, and medications [1]. In diary studies of older and younger adults, older participants most often reported symptom, medication, and appointment-related questions, while younger adults frequently noted general health, symptom, and medication-related questions [12]. Among individuals with prostate cancer, the most frequently accessed information included diagnostic reports, cancer agency-dispensed medications, appointment information, diagnosis summaries, and treatment records [10].

However, most of these studies focus on health information needs within a specific electronic health record (EHR) system, rather than across broader personal information sources such as emails and Internet of Things (IoT) devices, and across diverse content such as measurements and other data not typically available in EHRs. As a result, the types of information identified in these works may not fully capture the scope of personal health information found across sources beyond EHRs.

3 Methods

We employed a mixed-method approach using a survey to collect data on individuals' health information needs and access-related challenges, followed by qualitative analysis to develop personas.

3.1 Online Survey

The survey was designed to characterize personal health information needs by identifying the types of personal health data individuals require, their motivation, and the challenges they encounter when attempting to access it. We also collected participants' age and self-reported health conditions to determine personal characteristics that may influence these needs, thereby supporting the identification of personas for personal health information access.

Study Design. We disseminated the survey through the online survey platforms SurveyCircle[1] and SurveySwap[2] in August 2025, and to the University of Porto's community via institutional email between September and mid-October 2025. The survey consisted of a semi-structured questionnaire made available in two languages, Portuguese and English, and included 17 questions. It began with a sociodemographic section collecting data on age and self-reported health conditions (3 questions). Participants were then presented with experience-based questions prompting them to recall situations (scenarios) in which they needed to access personal health information: *"Recall a situation in which you needed to access your personal health information. Then, answer the following questions."*. They were asked to detail the data required, the reason for access, whether they successfully obtained it, and any difficulties they faced (4 questions). The first three items were closed-ended and are listed in Table 1. This section could be repeated up to two additional times, allowing each participant to report a maximum of three scenarios. At the end of each section, respondents were asked if they could recall another experience: *"Do you recall another occasion when you needed to access your personal health information?"*. If they responded negatively, the survey concluded; otherwise, it proceeded to the next section.

[1] https://www.surveycircle.com/en/.
[2] https://surveyswap.io/.

Table 1. Closed-ended questions and predefined answer options.

Question	Options	Type
What information did you need to access?	Diagnostic test results | Family history of diseases | Health measurements (e.g., blood pressure, blood glucose) | Information about allergies | Vaccination records | Medical prescriptions | Medications used | Personal notes about medical appointments | Records of clinical exams performed | Records of physical activity | Scheduling of medical appointments or exams | Sleep records | Symptom records | Treatments performed | Other	Single-choice
Why did you need to consult this information?	To support decision-making | To understand the context of a diagnosis | To manage my anxiety | To identify lifestyle changes | To prepare for a medical appointment | To monitor my health | To share with my family and friends | To share with a healthcare professional | Other	Multiple-choice
Were you able to obtain the personal health information you wanted to access?	Yes, completely | Yes, partially | No	Single-choice

Ethical Considerations. The Data Protection Officer of the University of Porto approved the processing of personal data, ensuring compliance with applicable data protection legislation. Participation was anonymous, and no personally identifiable data was collected. Participants were explicitly reminded for each open-ended question not to provide any information that could disclose their identity.

Participants. We collected completed responses from 420 participants, including 275 from the University of Porto's community and 145 through online survey platforms. Participants had a mean age of 26.7 years (*SD* 11.3; range 15–75) and a median age of 23 years (interquartile range [*IQR*] 18–30), as illustrated in Fig. 1a. Respondents also indicated whether they had any health conditions: 74.3% reported no health condition, while 25.7% reported at least one. The age distribution by health condition status is presented in Fig. 1b.

Data Analysis. Qualitative data from open-ended responses on health conditions and challenges in personal health information seeking were analyzed using thematic coding. Reported health conditions were manually coded according to the International Classification of Diseases, 10th Revision (ICD-10)[3] codes.

[3] ICD-10 (2019 version) files available at: https://icdcdn.who.int/icd10/index.html.

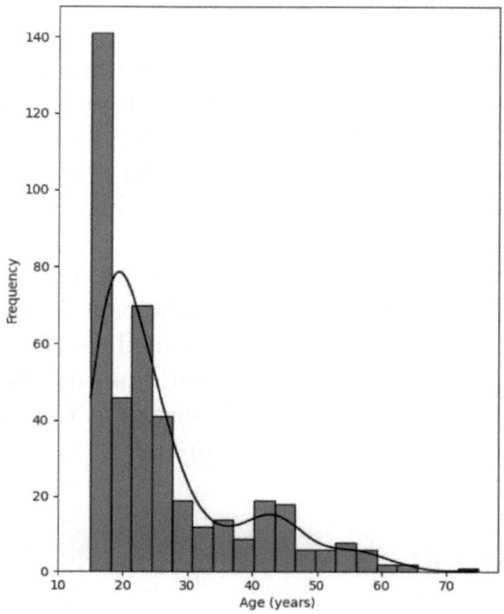

(a) Overall age distribution.

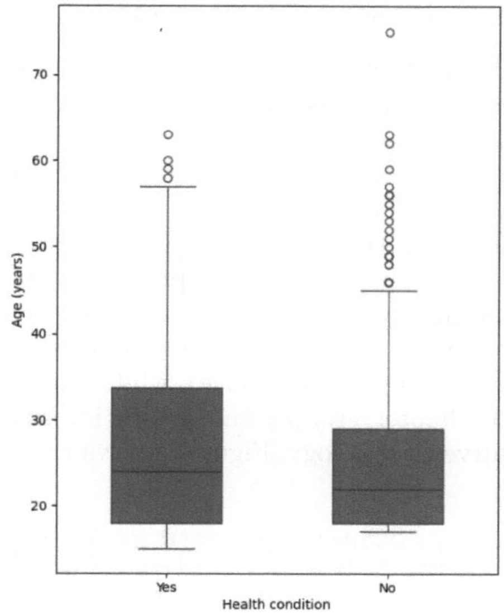

(b) Age by health condition status.

Fig. 1. Age distributions of participants.

The ICD-10 system has an alphanumeric structure in which the first three digits represent broader categories, and each subsequent digit provides increasing specificity [4]. The system is split into 22 chapters where each chapter is divided by blocks of codes. For example, chapter II Neoplasms (C00-D48) is divided into blocks such as C00-C97 ("Malignant neoplasms") and D10–D36 ("Benign neoplasms"), each containing specific codes like C50 ("Malignant neoplasm of breast"). The codes were assigned based on information with ICD10Data.com[4], an online lookup tool for ICD-10 classifications. No ambiguous cases were identified, since the ICD-10 includes sufficiently broad categories (e.g., "unspecified" or "other") that allowed all reported conditions to be coded. In instances where these categories were not applicable, conditions were assigned to the corresponding broader ICD-10 group (e.g., hernia-related conditions were coded under K40-K46, "Hernia"). Reported challenges in personal health information seeking were coded into seven thematic codes: 1) data access and permission, 2) delayed access, 3) fragmented sources, 4) health and digital literacy, 5) poor information architecture, 6) partially unavailable information, and 7) paywall barrier. Some initial codes, such as "lengthy waiting time" and "busy medical scheduling", were deemed invalid and discarded, as they referred to challenges in the healthcare system rather than barriers in accessing the information itself.

[4] https://www.icd10data.com.

3.2 Persona Development

We used affinity diagramming, a user-centered design technique, to develop personas from survey data. We did not apply standard clustering methods or factor analysis because the questionnaire included repeated measures, which would violate the independence assumptions required by those approaches. In contrast, affinity diagramming does not require independence and enables the identification of patterns across various dimensions.

Persona development progressed in two phases. In the first phase, we defined initial grouping labels using closed coding schemes based on participants' age, reported health conditions, and health information. Age was coded as older adults (aged 60 or above), while health conditions were coded according to ICD-10 chapter groups, and health information types were coded according to the survey's typology. Figure 2 shows the resulting initial label groupings.

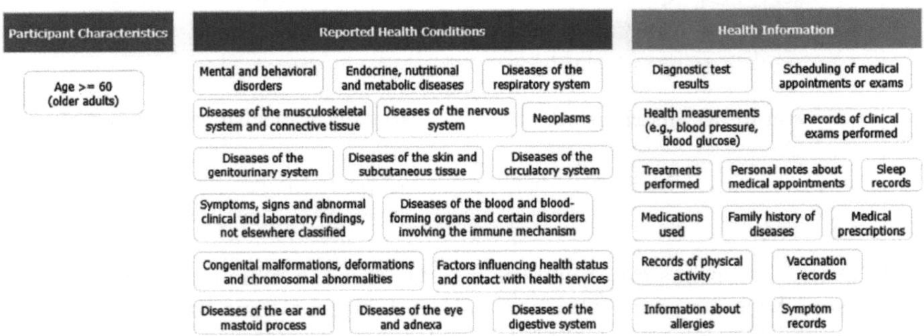

Fig. 2. Initial stage of label grouping based on closed coding.

From these groupings, we created affinity cards to support pattern identification. For groupings with few participants, we created individual participant cards, shown in Fig. 3a, each identified by a unique identifier and including the type of health information required and its purpose. For larger groupings where individual cards were not viable, we conducted descriptive analyses to characterize the data, such as identifying the most frequent purposes per health information type, and used these patterns as a basis for card creation. Figure 3b illustrates the clustering of health information types according to shared access goals.

In the second phase, we organized the affinity cards into clusters through iterative grouping, setting aside cards that did not fit into any emerging cluster. We then applied open coding to the resulting clusters, yielding three themes: treatment monitoring, preventive monitoring, and clinical event management. The clinical event management theme emerged from participants who accessed scheduling information for purposes that did not map onto either monitoring theme, leading to the creation of a different code that could capture health

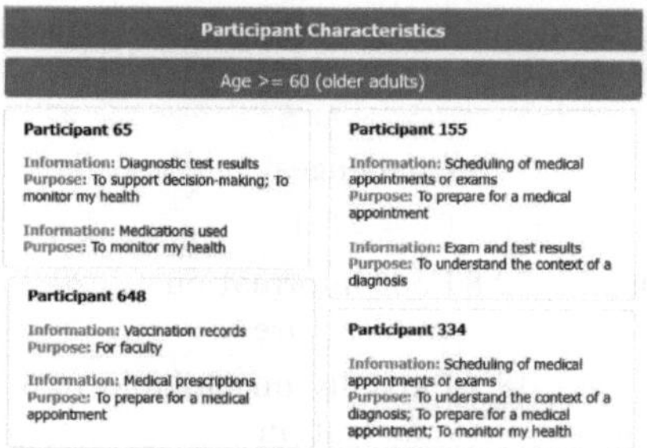

(a) Participant-based affinity cards.

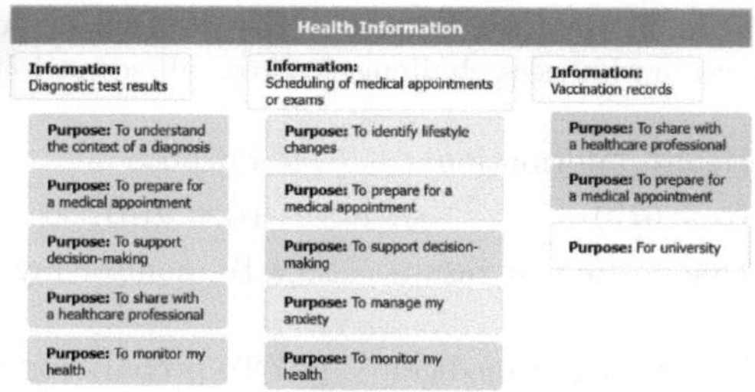

(b) Goal-based affinity cards clustered by health information.

Fig. 3. Examples of affinity cards and clusters.

information access that is administrative in nature. Figure 4 shows the refined label groupings following open coding, which formed the basis for the developed personas.

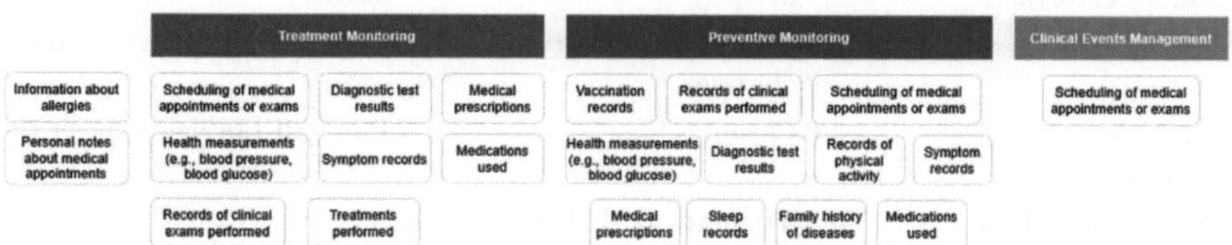

Fig. 4. Refined label grouping after open coding.

4 Results

4.1 Survey Data Analysis

All scenario responses were combined for analysis, including 610 responses for personal health information needs, 975 for access purposes, and 488 for access-related challenges. Participants also reported the degree of success in accessing personal health information: 72.6% indicated complete success, 22.5% partial success, and 4.9% inability to access information. Table 2 presents the relative frequencies of reported personal health information needs, purposes, and access challenges across all scenario responses.

The most frequently required personal health information included diagnostic test results, the scheduling of medical appointments or exams, vaccination records, and medical prescriptions. Participants most commonly accessed their health information for health monitoring, preparing for medical appointments, sharing with healthcare professionals, and understanding diagnoses.

Table 2. Relative frequency distribution of personal health information needs, purposes, and access challenges across all scenario responses.

Health information ($n = 610$)	%	Purpose ($n = 975$)	%	Challenges ($n = 211$)	%
Diagnostic test results	26.4	To monitor my health	22.8	Poor information architecture	20.9
Scheduling of medical appointments or exams	17.7	To prepare for a medical appointment	18.2	Partially unavailable information	15.6
Vaccination records	11.6	To share with a healthcare professional	16.2	Delayed access	12.8
Medical prescriptions	9.2	To understand the context of a diagnosis	14.0	Fragmented sources	12.8
Health measurements (e.g., blood pressure, blood glucose)	6.4	To support decision-making	10.9	Data access and permission	12.1
Records of clinical exams performed	6.2	To manage my anxiety	5.0	Health and digital literacy	11.8
Medications used	5.1	To identify lifestyle changes	4.7	Other	10.3
Family history of diseases	4.6	Other	4.7	Paywall barrier	3.7
Information about allergies	3.4	To share with family and friends	3.6		
Personal notes about medical appointments	2.3				
Sleep records	2.1				
Other	2.0				
Records of physical activity	1.1				
Symptom records	1.1				
Treatments performed	0.7				

We automatically aggregated reported health conditions into higher chapter-level ICD-10 categories to facilitate meaningful grouping. For example, code E28.2 (polycystic ovarian syndrome) was classified within the chapter E00–E90 (endocrine, nutritional and metabolic diseases). Chapter-level aggregation was adopted to balance granularity and generalizability, since code or block-level groupings resulted in too many clusters with few participants (e.g., with N = 151 participants, block-level groupings yielded an average of 3 participants per block, compared to 9 per chapter with chapter-level groupings). This aggregation allowed us to cluster participants by related health conditions and identify possible trends in information seeking.

Through this analysis, we identified key characteristics to guide persona development, including participants' age (e.g., older adults over 60), reported chronic health conditions (grouped by ICD-10 chapters), and the relationship between the types of personal health information most commonly accessed and the purposes for accessing them. We did not include information-seeking challenges as a basis for grouping, since most participants reported no barriers.

4.2 Personas

Based on the survey analysis and affinity diagramming, we developed three preliminary user personas to represent typical profiles of personal health information seekers. These personas were derived by grouping participants across key characteristics and identifying patterns that reflect distinct goals and personal health information-seeking behaviors. Figure 5 describes and characterizes the personas.

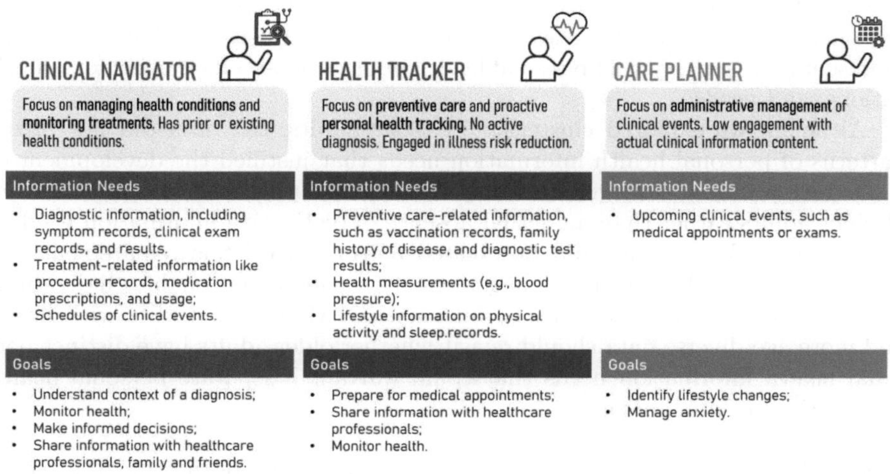

Fig. 5. Personas of personal health information seekers.

We developed the following three personas to represent typical personal health information seekers:

1. **Clinical Navigator:** user with prior or existing health conditions who is focused on condition management and treatment monitoring. They frequently access condition-specific information, including symptom records, clinical exams and results, procedure records, medication prescriptions and usage, and scheduling information related to clinical events. Their goals are to understand the context of a diagnosis, monitor health status, make informed decisions, and coordinate care with healthcare professionals, family, and friends.

2. **Health Tracker:** user with no active diagnosis and a primary focus on preventive care and proactive health monitoring. This persona frequently accesses preventive care-related information, including vaccination records, family history of disease, and diagnostic test results, as well as wellness-oriented information like health measurements, physical activity, and sleep records. Their goals are to prepare for medical appointments, share information with healthcare professionals, and monitor their health.
3. **Care Planner:** user seeking support for administrative management of clinical events. This persona accesses scheduling information for medical appointments or exams with little engagement with clinical content. Their goals focus on identifying lifestyle changes and managing anxiety.

5 Conclusion and Future Work

This work explored laypeople's personal health information needs, purposes, and challenges to develop personas that support the user-centered design of personal health knowledge management systems. Through a mixed-method approach using survey data analysis and affinity diagramming, we identified three personas: clinical navigator, health tracker, and care planner. These personas represent real-world typical personal health information-seeking behaviors, motivations, and needs.

Although we explored clustering by older adults, the data did not reveal patterns of personal health information needs that justified the development of a separate persona. However, this may reflect our data's limited representation of older adults rather than an actual absence of distinctive information needs.

Future work will build on the created personas to define requirements for a personal health knowledge management data model that enables individuals to access and organize their health information. Moreover, studies with larger and more age-diverse data should revisit whether older adults have distinct personal health information needs that could warrant a separate personal health information seeking persona.

Acknowledgment. Mariana Dias is financed by National Funds through FCT - Foundation for Science and Technology I.P., within the scope of the research grant with reference 2025.03050.BD.

References

1. Baudendistel, I., et al.: Personal electronic health records: understanding user requirements and needs in chronic cancer care. J. Med. Internet Res. **17**(5), e121 (2015). https://doi.org/10.2196/jmir.3884
2. Clarke, M.A., et al.: Health information needs, sources, and barriers of primary care patients to achieve patient-centered care: a literature review. Health Inform. J. **22**(4), 992–1016 (2016). https://doi.org/10.1177/1460458215602939

3. Degner, L.F., Sloan, J.A.: Decision making during serious illness: what role do patients really want to play? J. Clin. Epidemiol. **45**(9), 941–950 (1992). https://doi.org/10.1016/0895-4356(92)90110-9
4. Hirsch, J., et al.: ICD-10: history and context. Am. J. Neuroradiol. **37**(4), 596–599 (2016). https://doi.org/10.3174/ajnr.A4696
5. Huh, J., et al.: Personas in online health communities. J. Biomed. Inform. **63**, 212–225 (2016). https://doi.org/10.1016/j.jbi.2016.08.019
6. Kim, M., Lee, H., Jung, G.: Advancing Medical Report Interpretation with Large Language Models (2025). https://doi.org/10.2139/ssrn.5381288
7. LeRouge, C., Ma, J., Sneha, S., Tolle, K.: User profiles and personas in the design and development of consumer health technologies. Int. J. Med. Inform. **82**(11), e251–e268 (2013). https://doi.org/10.1016/j.ijmedinf.2011.03.006
8. Ormandy, P.: Defining information need in health – assimilating complex theories derived from information science. Health Expect. **14**(1), 92–104 (2011). https://doi.org/10.1111/j.1369-7625.2010.00598.x
9. Osovskaya, I., Blandford, A., Potts, H.W.: A systematic review of the effect of personal health records on patient activation. Digit. Health **11** (2025). https://doi.org/10.1177/20552076251315295
10. Pai, H., Lau, F., Barnett, J., Jones, S.: Meeting the health information needs of prostate cancer patients using personal health records. Curr. Oncol. **20**(6), e561 (2013). https://doi.org/10.3747/co.20.1584
11. Pratt, W., Unruh, K., Civan, A., Skeels, M.M.: Personal health information management. Commun. ACM **49**(1), 51–55 (2006). https://doi.org/10.1145/1107458.1107490
12. Price, M.M., Pak, R., Müller, H., Stronge, A., Breedlove, J.: Examining non-critical health information seeking: a needs analysis for personal health records. In: Proceedings of the Human Factors and Ergonomics Society Annual Meeting, vol. 53, pp. 1674–1678. SAGE Publications Sage CA, Los Angeles, CA (2009). https://doi.org/10.1177/154193120905302204
13. Pruitt, J., Adlin, T.: The Persona Lifecycle: Keeping People in Mind Throughout Product Design. Morgan Kaufmann Publishers Inc., San Francisco, CA, USA (2006)
14. Solomon, P.: Looking for information—a survey of research on information seeking, needs, and behavior. Inf. Retr. **6**, 284–288 (04 2003). https://doi.org/10.1023/A:1023944523773
15. Vijayashree, J., Jayashree, J., Chandnani, C.J., Shukla, S., Agarwal, V.: Chapter 1 The Fourth Industrial Revolution: A Paradigm Shift in Healthcare Delivery and Management, pp. 1–32. De Gruyter, Berlin, Boston (2024). https://doi.org/10.1515/9783111327853-001
16. Wang, L., et al.: Patient personas of delayed healthcare seeking among patients with breast cancer-related lymphedema: a qualitative study. Asia-Pac. J. Oncol. Nurs. 100963 (2026). https://doi.org/10.1016/j.apjon.2026.100963

Doctoral Consortium

Toward Hybrid Agile Teams: Enhancing Collaboration in Agile Projects by Exploring Traditional and AI-Based Tools for Process Optimization and Best Practices

Haoyue Liu[1,2](✉)

[1] Paris-1 Panthéon-Sorbonne University, Paris, France
Haoyue.Liu@inria.fr
[2] INRIA, Nancy, France

Abstract. Artificial intelligence (AI) is becoming an integral part of agile teams and is increasingly shaping how collaboration takes place in software development environments. AI tools are now used to support various activities such as planning, information analysis, and decision-making, which can influence team coordination and work organization. In this context, understanding how humans and AI systems collaborate within agile teams is becoming an important research issue. This PhD thesis explores how hybrid agile teams (composed of human members and AI-based agents) can work together effectively. Although some human-AI collaboration supports have been proposed in the literature, the use of process mining techniques in agile and collaborative settings remains relatively unexplored. Building on these techniques, the research examines behavioural and interaction patterns within agile workflows to identify insights that can support better team practices and improve organisational processes. Against this backdrop, the thesis proposes a human-AI hybrid collaboration support approach that incorporates process mining as a means of analysis and decision support. Particular attention is given to the organisational aspects of integrating intelligent agents into agile teams, with the goal of improving coordination mechanisms, decision-making processes, and collaboration practices within software development projects.

Keywords: Artificial intelligence · agile · software development · human-AI collaboration · Long-term collaboration · Process Mining

1 Introduction

Since the Agile Manifesto [1], agile methodologies have profoundly transformed software development by promoting iterative processes, internal and external collaboration, and adaptability to change. In 2010, Sharp and Robinson highlighted the importance of the "3Cs" within agile teams, defined through three core activities: collaboration, coordination, and communication [2], with human collaboration playing a central role. Collaborative tools such as Jira, Trello, Slack, Microsoft Teams, and Confluence support real-time communication, knowledge sharing, and work visibility, particularly in distributed development contexts.

A. Bernasconi et al. (Eds.): RCIS 2026, LNBIP 597, pp. 179–190, 2027.
https://doi.org/10.1007/978-3-032-35043-5_15

At the same time, AI is increasingly being integrated into software development environments [3]. Recent advances in natural language processing (NLP) have enabled tools that improve the quality of requirements and user stories [5], while generative AI (GenAI) systems such as ChatGPT and GitHub Copilot assist developers in producing technical artefacts, generating code, and supporting decision-making activities [6]. In parallel, machine learning (ML)-based software systems play an increasingly important role in managing project complexity, particularly in contexts where models must be continuously adapted, maintained, and integrated into the software development lifecycle [5]. These developments change work practices as well as collaboration dynamics within development teams.

The growing adoption of AI in Agile environments introduces several practical and organisational challenges. First, the large number of available AI tools makes it difficult for developers to know which one to use for a specific task. The study [3] shows that the challenge "too many tools, unclear which to use" received 73.3% of the survey votes. The situation becomes more complex because large language model (LLM)-based tools evolve quickly and new models are released frequently, which can disrupt existing workflows in the Agile process. In addition, organizational constraints, such as the mandatory use of enterprise tools (for example, Microsoft Copilot), may limit developers' choices and influence how teams collaborate with AI. Second, there is also a risk that junior developers rely too much on AI suggestions without carefully reviewing them [9]. Some teams may also stop using AI tools too quickly after early failures, without fully exploring their potential benefits. Without proper feedback loops in the development process, AI suggestions may even reinforce poor architectural decisions or existing bottlenecks. Moreover, integrating AI requires new skills within the team [8]. As AI becomes part of Agile practices, teams may need support from data scientists, AI specialists, or ML professionals. In general, team members need the right skills to work effectively with AI systems.

The introduction of AI tools also raises important questions about team structure and collaboration. This leads us to examine whether AI should be positioned within team dynamics as a "teammate" or simply as an assistant, while also considering how users communicate and collaborate with AI tools in their daily work. Unlike traditional software tools, AI systems can generate suggestions, automate tasks, and support decision-making activities. As a result, they may reshape how work is distributed within the team and raise new questions about roles and responsibilities. We therefore must consider which tasks AI could take the lead in and in which situations human developers should remain responsible for ensuring correctness, quality, and alignment with project goals.

Although previous studies [5, 6, 9] have explored the benefits of AI-supported Agile development, our study addresses the need for longitudinal assessment to better understand the long-term impact of AI adoption in Agile practices. This gap calls for approaches that can capture not only perceptions of AI use, but also observable changes in collaborative and development practices. In this research, process mining is proposed as a promising methodological lens to address this gap [7]. By analyzing event logs generated by collaborative and development tools, process mining can help identify how Agile workflows evolve, where AI-supported activities are introduced, and how these changes affect coordination, communication, and task execution over time. Combining this perspective with qualitative investigation may provide a more comprehensive understanding of human-AI collaboration in Agile environments.

This study aims to better understand how AI tools affect software development processes, how their use evolves over time. Then we propose to develop methods and recommendations to support effective human-AI collaboration in Agile teams. In this context, the main research question of this study is: **How does AI impact the agile development process over time?** This paper is structured as follows: Sect. 2 provides background. Section 3 presents the research objectives, research questions, and the proposed approach. Section 4 outlines the ongoing work. Section 5 provides a discussion of the expected contributions and highlights directions for future work.

2 Background

2.1 Agile in Software Engineering

In the context of software engineering, agile refers to a set of development methods built around short, repeating cycles, ongoing collaboration with stakeholders, and flexible responses to change. Dybå and Dingsøyr define agility as the capacity to "strip away as much of the heaviness, commonly associated with traditional software-development methodologies, as possible to promote quick response to changing environments, changes in user requirements, accelerated project deadlines" [10]. The roots of this approach can be traced back to Takeuchi and Nonaka [11], whose 1986 article in the Harvard Business Review described cross-functional, self-organizing teams that worked toward shared goals rather than fixed tasks. This organizational model foreshadowed the core principles of modern agile practice. These ideas were formalized in 2001 with the Agile Manifesto [1], which set out four core values and twelve guiding principles focused on individuals and interactions, working software, customer collaboration, and the ability to respond to change. From this foundation came a range of agile frameworks, most notably Scrum and Extreme Programming (XP), which introduced concrete team structures, roles, and iterative practices such as sprint planning, daily stand-ups, and retrospectives. A decade of empirical research reviewed by Dingsøyr et al. [12] confirmed that these frameworks clearly improved delivery speed, team cohesion, and customer satisfaction, while also pointing to ongoing challenges related to scaling agile practices and managing organizational change.

In this thesis, the term "agile" refers specifically to its use in software engineering contexts, where it shapes how development work is organized, how collaborative practices are structured, and how iterative delivery cycles are managed. Agile Project Management (APM) extends these principles to the wider coordination of software projects, covering planning, resource allocation, and stakeholder engagement within iterative frameworks [13]. This distinction matters because the research is concerned with the organizational and collaborative dimensions that agile methods bring into development teams.

2.2 Human-AI Collaboration: Definition and Scope

In this research, the term "human-AI collaboration" (HAC) refers to the set of practices, interactions, and organizational arrangements through which human team members and AI-based tools work together on software development activities. This view highlights the complementary strengths of human and machine intelligence: AI brings speed, scale, and pattern recognition, while humans provide judgment, contextual understanding, and accountability. As a result, the combination can perform better than either side acting alone [15, 17]. This complementarity is the central idea behind what Dellermann et al. [15] call hybrid intelligence, in which humans and AI jointly carry out tasks that neither could complete as effectively on its own.

At the most basic level, research on human-AI interaction has produced design principles that govern how people and AI-infused systems communicate and coordinate at the interface [14]. As AI agents gain autonomy, initiative, and the ability to adapt, the relationship shifts from interaction toward teaming, where agents take on roles traditionally held by human teammates and take part in shared coordination, cooperation, and decision-making [16, 17]. In this sense, collaboration is defined by interdependence, shared goals, and a dynamic, two-way exchange between humans and AI over time. This is what sets it apart from one-directional automation or assistance.

3 Research Design

This section presents the research objectives and gives an overview of the expected results within the overall research plan.

3.1 Research Objective

1) The first objective is to analyze the recent evolution of Agile practices in highly digitalized work environments. It examines how interaction, coordination, and collective decision-making are changing in contemporary teams.
2) The second objective is to examine how human-AI collaboration practices evolve over time in the software engineering context. It adopts a longitudinal perspective to observe how AI is introduced into processes and how AI-supported practices emerge within Agile teams.
3) The last objective is to identify and formalize best practices for hybrid collaboration between humans and AI systems in Agile contexts. It aims to define conditions that support effective and balanced human-AI collaboration in Agile teams.

3.2 Research Planning

This research is structured into successive and complementary phases (see Fig. 1.).

The first year begins with clarifying the research problem and conducting an in-depth literature review. This step helps define the scientific position of the study and formulate the research questions. Particular attention is given to current collaborative AI techniques, the challenges related to their adoption, and the ways they are integrated into the agile project lifecycle. Existing agile collaboration patterns are also identified and

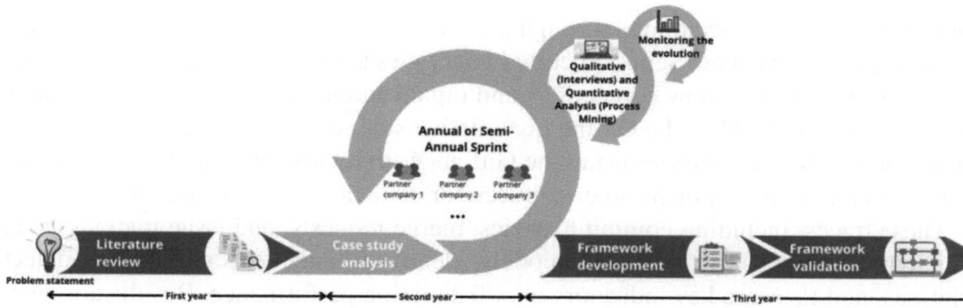

Fig. 1. Roadmap of the doctoral research project (3 years)

analyzed. To examine the changing impact of AI-supported collaboration within agile teams, the methodological approach combines qualitative data with quantitative data from digital traces. The research adopts a longitudinal design based on six-month cycles of data collection and analysis, which allows continuous monitoring of how practices evolve within the studied teams.

The second year continues the fieldwork through additional data collection cycles. The third year focuses on bringing the findings together and developing a structured support of best practices for hybrid human-AI collaboration in agile contexts. The longitudinal data collected across multiple cycles help identify patterns of change and formalize best practices based on empirical evidence. A validation phase also strengthens the robustness of the proposed support.

3.3 Research Questions

In this context, the following research questions are proposed:

(RQ1) *How does the integration of AI tools transform Agile environments in software development projects?*

Objective: This research question examines how AI tools are integrated and used within agile environments. Addressing this question provides an essential foundation for the subsequent phases of the research, particularly for identifying and observing best practices in hybrid human-AI collaboration.

How: RQ1 is addressed through a systematic literature review (SLR). The detailed review protocol is presented in the ongoing work section.

(RQ2) *What are the organizational implications of integrating intelligent agents into agile team workflows?*

Objective: This research question investigates the organizational implications of integrating intelligent agents into Agile team workflows, with a particular focus on how decision-making power is shared between human actors and automated systems, and how this distribution evolves over time. It examines the consequences of these changes for project governance, coordination practices, and the distribution of responsibilities within teams. Understanding these dynamics is essential to ensure that AI tools support collaboration and productivity without undermining Agile principles or team autonomy.

How: This question is examined through a mixed-method approach combining qualitative and quantitative data. Semi-structured interviews are conducted at different stages of the project to capture how perceptions and organizational practices evolve as teams gain experience with AI tools. On the quantitative side, event logs are collected from GitLab and GitHub repositories using the GitLab-Activity system, enabling the extraction of structured development process traces for statistical analysis and process mining. These traces, including commit histories, merge requests, and issue tracking data, are analyzed to reconstruct workflow models, identify frequent execution paths, detect deviations, and measure key indicators of collaboration performance. Results are visualized through workflow diagrams, process maps, and collaboration networks to reveal recurring behavioral patterns and interaction dynamics within teams.

***(RQ3)** What contextual factors (individual, organizational, or tool-related) influence this collaboration, and how do they affect it?*

Objective: This research question examines how contextual factors shape the effectiveness of collaborative tools, both traditional and AI-based within agile teams. It investigates organizational factors such as team size, project complexity, and members' experience, as well as individual factors including cognitive load, trust in automation, and interaction preferences. Since the impact of collaborative tools is not universal, understanding these contextual influences is essential to avoid overly general conclusions and to refine the collaboration patterns identified in RQ2.

How: This question is addressed through a qualitative research design based on semi-structured interviews and observations, which allow for an in-depth exploration of developers' perceptions, team dynamics, and the contextual conditions that shape human-AI collaboration in Agile settings.

3.4 Proposed Research Approach

The proposed research approach analyzes the practices and evolution of decision-making and operational processes within agile teams, with the objective of defining a hybrid collaborative support system. This approach supports high-performing teams composed of human actors and AI tools in making optimal decisions based on system recommendations, the specific context of each project, and the trade-offs involved in decision-making. It also guides the implementation of these decisions, especially in critical or high-pressure situations. The proposed approach consists of two main processes (see Fig. 2.).

3.4.1 Application of the Dynamic Hybrid Collaboration Board (First Process)

The first process describes the use of the dynamic collaboration board. It begins when an organizational actor, such as a project manager or team lead, expresses a need to improve collaboration practices within an agile context, particularly in relation to the use of AI tools. This process relies on multiple inputs, including different contextual dimensions (organizational, project-related, agile, team, and technological contexts), client objectives, and when available, digital traces generated by collaborative tools. These traces may include tickets and backlogs, commit histories, communication logs,

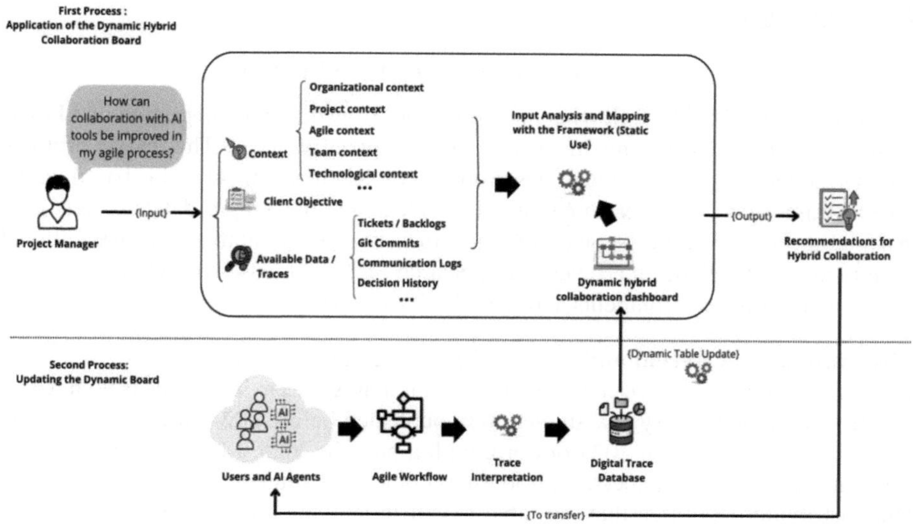

Fig. 2. Research Approach for Adaptive Hybrid Collaboration

and decision records. During this phase, the support analyzes the contextual information and produces "on-the-fly" hybrid collaboration recommendations adapted to the situation at hand.

3.4.2 Updating the Dynamic Board (Second Process)

The second process ensures the continuous evolution of the dynamic hybrid collaboration board. It is based on observing the real practices of users and AI agents within agile workflows. The interactions, decisions, and actions performed during these activities generate digital traces, which are collected and analyzed using process mining techniques before being stored in a digital trace repository [7].

Although process mining has traditionally been applied to structured business processes, its application in agile and collaborative contexts is still emerging. In agile teams, where processes are dynamic, decentralized, and context-dependent, process mining provides the opportunity to identify recurring patterns, inefficiencies, and deviations in collaborative behaviors.

When the collected traces provide sufficient informational value, the dynamic board is updated accordingly. This allows the support to remain aligned with observed practices and to continuously improve the accuracy and relevance of the recommendations it generates.

4 Ongoing Work

4.1 A Systematic Review of AI Integration in Agile Contexts

The SLR is guided by the following main question: How does the integration of AI tools transform Agile environments in software development projects? The review protocol follows the methodological guidelines proposed by Kitchenham [4] and is based on the PICOC framework. The literature search is conducted in the Scopus database, resulting in a final corpus of 32 articles. Search String 1 focuses on the limitations of the traditional agile model and includes 9 articles, while Search String 2 focuses on AI integration in agile collaborative environments and includes 23 articles.

Observation 1 - A Recent and Rapidly Emerging Research Area

The analysis reveals that research on AI integration in Agile environments is both recent and growing rapidly. Very few studies were published before 2024, while a significant increase is observed from 2024 onward, with a peak in 2025. Most publications take the form of conference papers, with journal articles remaining scarce (Fig. 3, Fig. 4). This distribution confirms that the topic is still in an exploratory phase, with limited consolidation of empirical findings.

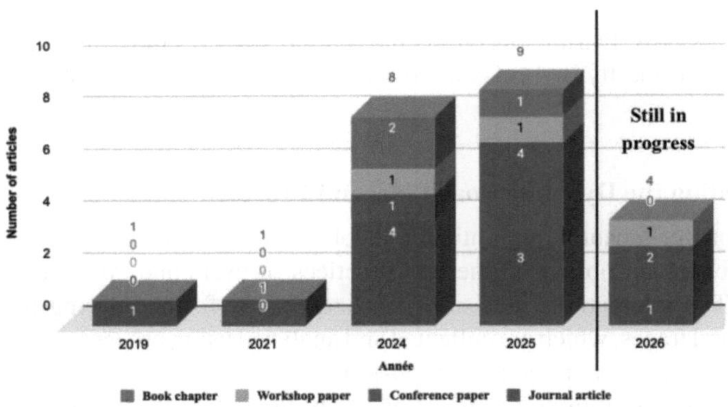

Fig. 3. Distribution of selected articles by year and by type of publication

Observation 2 - Structural Limitations of Traditional Agile Models

The review identifies recurring limitations that motivate the integration of AI into Agile workflows. These include communication and coordination challenges, particularly in distributed or large-scale contexts where stakeholder alignment becomes more difficult. Requirements management also emerges as a critical pain point, with issues such as requirements volatility and insufficient handling of non-functional requirements. More broadly, the lack of formalization and documentation practices limits knowledge sharing and shared understanding within teams. These findings position AI not as a mere productivity tool, but as a potential structural response to well-documented Agile shortcomings.

Observation 3 - AI Tools Distributed Across Three Categories of Agile Activity
AI tools identified in the corpus span five main categories: Generative AI (GenAI), Machine Learning (ML), Natural Language Processing (NLP), AI Agents and Multi-Agent Systems, and AI Decision Intelligence approaches. Their usage clusters around three dimensions (Table 1): software development activities (U1), including code generation, testing, and refactoring; Agile management and decision support (U2), encompassing backlog prioritization, sprint planning, and task allocation; and collaboration and communication (U3), covering documentation, stakeholder alignment, and knowledge sharing. GenAI tools appear prominently across all three dimensions, reflecting their versatility and growing adoption in Agile contexts.

Table 1. Synthesized Table – AI Categories, Representative Techniques, and Number of Articles per Usage (U1, U2, U3).

Category	Representative AI Techniques	U1	U2	U3
Generative AI (GenAI)	Chatbots & conversational agents (ChatGPT GPT-4, Gemini, LLM agents) AI coding assistants (Claude, GitHub Copilot)	10	8	14
Machine Learning (ML)	ML algorithms, Predictive models, Historical data analysis, Optimization algorithms	1	3	2
Natural Language Processing (NLP)	Text analysis, Semantic analysis, Ambiguity detection	0	2	2
AI Agent Tools	Cognitive agents, LLM agents, Evaluator agents, multi-agent systems	0	2	1
AI Decision Intelligence	LLM + Fuzzy BWM, AI recommendations, Hybrid decision models	0	3	1

Observation 4 - AI Integration Focused on Technical Tasks, with Limited Reach to Collaboration and Management
Examining the distribution of AI integration across Agile activities reveals a clear asymmetry. Coding is by far the most represented activity, addressed in 11 out of 23 articles, reflecting the natural accessibility of GenAI tools and their role as a first point of contact between teams and AI. Backlog management follows with 6 articles, drawing on both GenAI for requirements generation and ML models for item classification and prioritization. Testing appears in 5 articles, often alongside coding given their iterative entanglement in Agile workflows. Sprint planning is more moderately represented with 3 articles, primarily through ML-based story point estimation. What stands out, however, is the limited attention given to requirements engineering, team communication,

and project management, activities that are central to Agile practice and were identified as major sources of difficulty in the literature.

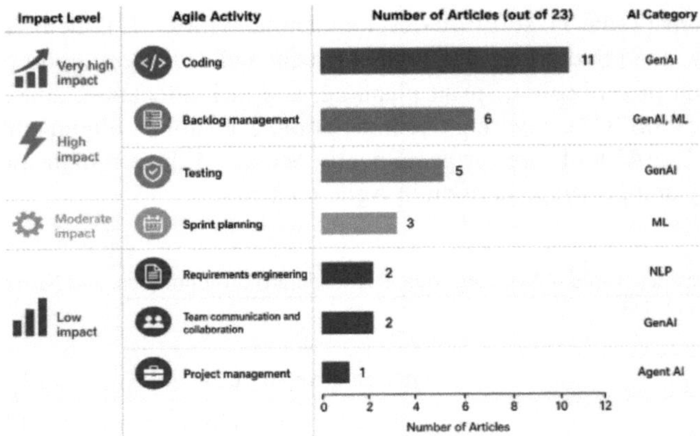

Fig. 4. Ranking of Agile Activity by Level of Impact from AI Integration.

4.2 Initial Qualitative Data Collection

As part of the first year of this doctoral research, four semi-structured interviews have been conducted with practitioners from diverse organizational contexts.

Participants and Contexts. The four participants (P1-P4) represent a deliberate diversity of organizational profiles. P1 leads a data structuration team of approximately 20 persons within a large healthcare research institution, operating under strict data sovereignty constraints that prohibit the use of cloud-based AI tools. P2 is a full-stack engineer in a 15-person digital media SME with an open, experimentation-driven culture toward AI adoption. P3 is a full-stack developer and cybersecurity lead at a maritime data startup of similar size, managing sensitive client data. P4 works within a dedicated AI Research & Development lab of a large software group of approximately 2,500 employees, whose core mission is to integrate generative AI into software products.

Agile Practices in Context. Across the four cases, no organization applies a textbook agile framework. P1 and P3 use lightweight Kanban-like workflows adapted to small team sizes and high multi-project load. P2 follows the most structured approach, with two-week sprints, planning sessions, reviews, retrospectives, and systematic peer code review. P4 was operating without formal agile rituals at the time of the interview and was transitioning to Scrum in response to growing workload. These observations confirm that agile "in practice" is considerably more fragmented and contextual than prescribed frameworks suggest.

AI Tool Integration and Adoption Patterns. The most significant variation across cases concerns the intensity of AI adoption. P1 represents the most constrained profile:

AI use remains embryonic, limited to a meeting assistant generating summaries with moderate precision, while structural constraints prevent any cloud-based processing of health data. At the opposite end, P4 uses agentic AI tools daily across the entire development lifecycle, reporting productivity gains of 10 to 20 times on specific tasks. P2 and P3 occupy an intermediate position, where coding assistants, LLM-based chat tools, and meeting transcription are actively used but with significant intra-team variation. A cross-cutting finding is that automated meeting summarization emerges as the most universally adopted AI capability, regardless of organizational size or sector.

Trust, Verification, and Collaborative Friction. None of the participants extend unconditional trust to AI-generated outputs, human verification is considered non-negotiable across all cases. The degree of trust ranges from systematic double review (P2) to a measured and quantified confidence of 70–80% (P4). A notable tension surfaces at P2 and P3: AI-generated code, being opaque and often non-idiomatic, can complicate collaborative code reading and peer review, introducing friction in agile practices centered on collective code ownership. P3 also reports a feeling of professional devaluation among some team members, a psychosocial dimension that warrants attention in further data collection.

Emerging Signals on Roles and Process. No participant reports a structural transformation of their agile process because of AI integration. AI tools accelerate specific tasks without eliminating process steps. Two emerging signals, however, deserve attention. At P4, developers increasingly position themselves as designers and verifiers rather than executors, delegating implementation to AI agents, a role shift that is nascent but explicitly acknowledged. At P2, a counterintuitive concern is raised: producing effective AI prompts requires detailed upfront specification, which may push teams toward more waterfall-like planning practices, in tension with agile principles of iterative discovery.

5 Discussion and Next Step

In this paper, we present an overview of our research program on human-AI collaboration in agile environments.

This research is currently in its exploratory phase, and an initial qualitative data collection has already started. In the next stages, additional semi-structured interviews will be conducted to gather practitioners' perceptions regarding the integration of AI tools in collaborative practices, addressing RQ2 and RQ3. We aim to reach a total of 10 to 15 interviews, selecting participants from a variety of roles, organizations, and project settings to capture a broad range of experiences and strengthen the transferability of our findings. In addition, digital trace data from development and collaboration platforms will be collected to analyze team behaviors, collaboration patterns, and workflow execution within agile projects, addressing RQ2.

Acknowledgments. This work is conducted under the supervision of Professor François Charoy (University of Lorraine, INRIA) and Professor Rebecca Deneckère (Paris 1 Panthéon-Sorbonne University, CRI). I would like to sincerely thank them for their guidance, support, and valuable feedback throughout the development of this research.

References

1. Beck, K., et al.: Manifesto for Agile Software Development (2001). https://agilemanifesto.org
2. Sharp, H., Robinson, H.: Three 'C's of agile practice: collaboration, co-ordination and communication. In: Agile software development: current research and future directions, pp. 61–85. Berlin, Heidelberg: Springer Berlin Heidelberg (2010)
3. Zhang, Z., et al.: AI and Agile Software Development: A Research Roadmap from the XP2025 Workshop. arXiv preprint arXiv:2508.20563 (2025)
4. Kitchenham, B., Charters, S.: Guidelines for Performing Systematic Literature Reviews in Software Engineering, Technical Report EBSE 2007-001, Keele University and Durham University Joint Report (2007)
5. Assres, G., Bhandari, G., Shalaginov, A., Gronli, T.M., Ghinea, G.: State-of-the-art and challenges of engineering ml-enabled software systems in the deep learning era. ACM Comput. Surv. **57**(10), 1–35 (2025)
6. Stray, V., Barbala, A., Wivestad, V.T.: Human-AI collaboration in software development: a mixed-methods study of developers' use of GitHub copilot and ChatGPT. In: Proceedings of the 33rd ACM International Conference on the Foundations of Software Engineering, pp. 1325–1332 (2025)
7. Van Der Aalst, W.: Process mining: overview and opportunities. ACM Trans. Manag. Inf. Syst. (TMIS) **3**(2), 1–17 (2012)
8. Ziyao, J.I.N.: Integrating AI into agile workflows: opportunities and challenges. Appl. Comput. Eng. **116**, 49–54 (2024)
9. Bahi, A., Ghari, J., Gahi, Y.: Integrating generative AI for advancing agile software development and mitigating project management challenges. Int. J. Adv. Comput. Sci. Appl. **15**(3) (2024)
10. Dybå, T., Dingsøyr, T.: Empirical studies of agile software development: a systematic review. Inf. Softw. Technol. **50**(9–10), 833–859 (2008)
11. Takeuchi, H., Nonaka, I.: The new new product development game. Harv. Bus. Rev. **64**(1), 137–146 (1986)
12. Dingsøyr, T., Nerur, S., Balijepally, V., Moe, N.B.: A decade of agile methodologies: towards explaining agile software development. J. Syst. Softw. **85**(6), 1213–1221 (2012)
13. Dong, H., Dacre, N., Baxter, D., Ceylan, S.: What is agile project management? Developing a new definition following a systematic literature review. Proj. Manag. J. **55**(6), 668–688 (2024)
14. Amershi, S., et al.: Guidelines for human-AI interaction. In: Proceedings of the 2019 Chi Conference on Human Factors in Computing Systems, pp. 1–13 (2019)
15. Dellermann, D., Ebel, P., Söllner, M., Leimeister, J.M.: Hybrid intelligence. Bus. Inf. Syst. Eng. **61**(5), 637–643 (2019)
16. O'neill, T., McNeese, N., Barron, A., Schelble, B.: Human–autonomy teaming: a review and analysis of the empirical literature. Hum. Factors **64**(5), 904–938 (2022)
17. Seeber, I., et al.: Machines as teammates: a research agenda on AI in team collaboration. Inf. Manag. **57**(2), 103174 (2020)

Towards a Conceptual Model for Evaluating Prospective Scenarios - A Case Study in Future Air Transport

Brieuc Danet(✉)

DTIS, ONERA, Université de Toulouse, 2 Avenue Marc Pellegrin, BP 74025, 31055 Toulouse CEDEX 4, France
brieuc.danet@onera.fr
http://www.onera.fr

Abstract. In a context of growing uncertainty, prospective scenarios have become widely used across economic and geopolitical sectors as long-term decision-support frameworks. Over the years, a broad variety of methods and scenario types have emerged, ranging from quantitative models to expert-based narrative approaches, serving purposes as diverse as risk preparedness and policy justification, yet no unified framework exists to assess scenario quality systematically.

Then a fundamental question remains: how can the quality of prospective scenarios be systematically assessed?

This thesis proposes to address this gap by developing a unified conceptual model for prospective scenarios, designed to integrate all the elements required by existing evaluation metrics and to provide a foundation for building new ones. Through a systematic review following the PRISMA2020 protocol, existing criteria and associated metrics were first catalogued.

A conceptual model will then be built, based on the scenario elements required by criteria metrics, and then validated against a corpus of existing and purpose-built scenarios. The formalism will be benchmarked against established approaches such as dynamic networks, combinatory optimization and morphological analysis.

The proposed framework will be applied to the domain of future air transport, a field rich in projections from industrial, commercial, and institutional stakeholders, to demonstrate its relevance and versatility. All scenarios created in the course of this research will be made openly available to support future work in prospective scenario evaluation.

Keywords: prospective scenarios · evaluation · criteria · metrics · conceptual models

1 Introduction

1.1 Prospective Scenarios

Faced with the uncertainties of the world to come, actors across many economic and geopolitical sectors are finding it increasingly difficult to make decisions

A. Bernasconi et al. (Eds.): RCIS 2026, LNBIP 597, pp. 191–201, 2027.
https://doi.org/10.1007/978-3-032-35043-5_16

that may prove critical. They need to rely on long-term outlooks, to collect the elements on the possible evolutions of their environment and to anticipate its transformations.

Some of them want to prepare for disruption, and be ready for critical changes that they know they can't predict. For example, national leaders must anticipate pandemics to prevent health risks to populations, as well as consider shifting geopolitical evolutions to mitigate the potential threats to security.

Other decision makers seek to assess whether their decisions will meet the objectives they have set. For example, policymakers need to ensure that environmental regulations will help reduce the future anthropogenic climate impact, and corporate leaders need to show that their decisions will deliver tangible benefits.

Finally, some projects require knowledge of future data trends to assess their accuracy. For example, regional planning authorities may rely on demographic and social shifts to anticipate the future demand for urban infrastructure (hospitals, housing, or transportation).

Prospective scenarios have been developed for this purpose, as a relevant framework for reflection. They are not predictions. Instead, each scenario provides a long-term projection built on hypotheses about how the surrounding environment might evolve over a given time horizon.

The term *scenario* as used here clearly differs from:

- the concept of *use case*, which allows a designer to place themselves in a situation in order to define the requirements to be met by the object or system being designed [5],
- the notion of *user story*, which focuses on considering problematic future situations, particularly in the field of safety (building fires, aircraft failure risks [16]).

We will rather consider in this thesis work that *scenarios provide an overall picture of the environment and highlight the interactions among several trends and events in the future* [2]. The term *prospective* refers to the idea of a long-term projection, free from any deterministic desire to forecast or predict the future situation.

1.2 Scenario Planning

A new discipline arose in the 1970s, named *scenario-planning* [20], which became active through three main prospective schools, based on different points of view and leading to different scenario development methods [2]:

1. *Probabilistic Modified Trends (PMT)*, rooted in the American school of futures studies, launched by Ted Gordon for the RAND Corporation, considers the future as the continuation of past trends. Probabilistic tools, such as the Cone of Plausibility (illustrated in Fig. 1 [22]), are used to analyse and exploit statistical data from the past, and build scenarios based on probabilistic evaluation of future events.

2. *Intuitive Logic Method (ILM)* initially developed by Herman Kahn at the Stanford Research Institute, aims at organizing structured workshops where experts base their visions of the future on intuitive reasoning rather than quantified probabilities. In this way, collaborative methods, such as the Delphi protocol [17], draw on expert opinions and collective intelligence to explore futures that could not be captured by each expert independently.
3. *La Prospective (LP)* launched by Michel Godet in France, elaborates long-term strategies by considering large sets of hypotheses and actors in the evolution of the system. Matrix-based tools, such as morphological analysis, systematically map out the relationships between key variables, named impact factors, and their evolutions to generate a wide set of possible scenarios.

Since the 1990s, these methods have diversified considerably, leading to a situation that some describe as a "methodological chaos" [7]. Commonly used by institutions, companies, and organizations of all kinds, they differ in their process, relying on hard data, on human insight, or on structural analysis to build pictures of the future.

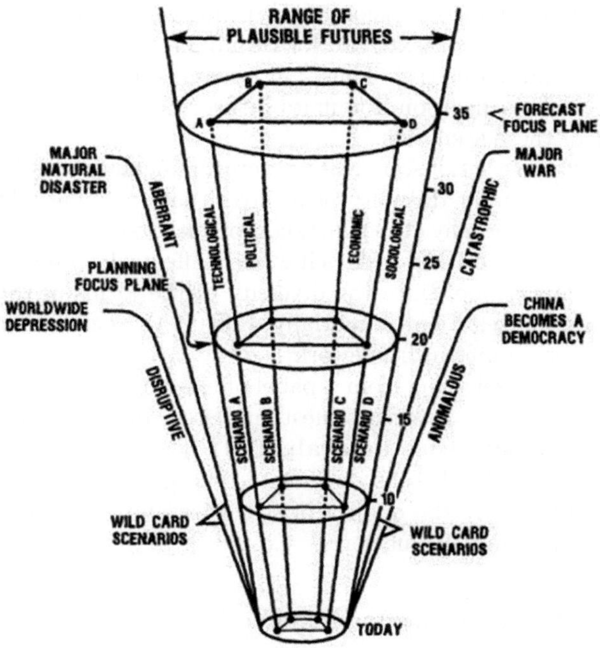

Fig. 1. Cone of Plausibility [22].

Moreover, the methods give rise to scenarios that often differ widely in form, structure, and underlying foundations. For example, some provide variable evolution graphs constructed using pseudo-deterministic simulations leveraging increasingly complex datasets; others take the form of narratives based primarily on the views of experts expressed during participatory sessions; still others present roadmaps that describe the different decision hypothesis combinations considered.

2 Research Question

Numerous applications have demonstrated the value of prospective scenarios in thinking about the future, and several reviews [7] show how widely their use has become generalized and popularized today. The library of available scenarios has grown so vast that practitioners may well prefer to select ready-made scenarios rather than resort to dedicated methods for creating new ones.

In the field of air transport, stakeholders (aircraft manufacturers, airlines and airport operators, certification and regulation bodies) have participated in the elaboration of many projections of the future [1,3,11,12], notably to demonstrate the sector's ability to meet the targets set by the Paris Agreement on climate change, and in particular carbon neutrality by 2050. Some scenarios offer a broader and more exploratory vision of the future, envisioning situations where economic or geopolitical contexts call for rethinking the role of air transport in the mobility of populations [10]. Thus, any organization linked to air transport that wishes to make a decision about its future today can select and draw on one or another of these scenarios.

In return, this large set of scenarios in literature dealing with the future of air transport presents a great diversity of future situations, where little consensus emerges. Some scenarios may even show signs of authorship bias [4]. Choosing the right scenarios is then a critical task as it greatly affects decision-making process, but also a challenging one, since the diversity of the forms of the scenarios, as illustrated before, makes them potentially difficult to analyze and exploit with a single method or a uniform framework.

Then, whether for selecting from a panel of existing scenarios or during their development process, one essential question arises: **how can the quality of these prospective scenarios be evaluated?**

This research question breaks down into two questions dealing with 1) the criteria defined to qualify the quality of scenarios and 2) the existing scenario frameworks used for their evaluation.

2.1 What Criteria Are Defined to Qualify Scenario Quality?

The literature reveals that a varying set of criteria is proposed to evaluate the overall quality of scenarios, and contribute to their validation [2]. Yet there is no consensus on the choice of criteria to use, nor even on the nature of the criteria that should be taken into account in a selection. As said before, many methods were developed to build scenarios, which have mainly been developed to fulfill expected user needs, taking one identified criterion as a main goal. For example:

- Trend-based methods [9] aim to produce *plausible* scenarios. By relying on statistical data and observable patterns, they ensure that the outlooks of the future are grounded in reality and can be reasonably expected to occur.
- Collaborative methods [17] aim to produce *creative* scenarios. By bringing together diverse perspectives and collective intelligence through group sessions and expert consultations, they push the boundaries of conventional thinking and open up a wider range of possible futures.
- Combinatory methods [13] aim to produce *consistent* scenarios. By systematically mapping the relationships between key variables, they ensure that each scenario holds together as an internally logical whole.

Other criteria can be found in the literature, each characterizing scenario quality from a different perspective. Nevertheless, while scenario development methods are generally designed to ensure the validity of their associated criteria in the scenarios they produce, the question of how to apply such criteria to existing scenarios remains largely open. What metrics could be derived from these criteria, and how could they be transposed to scenarios built through different methods?

2.2 What Frameworks Exist to Help Scenario Quality Assessment?

Indeed, some of these criteria present dedicated metrics based on specific frameworks. Notable examples include the Cross-Impact Balance Analysis [24] which uses a matrix-based formalism to identify internally *consistent* combinations, the Optimized Linear Diversity Field Anomaly Relaxation (OLDFAR) [15] which draws on combinatorial optimization to select *diverse* scenario sets, and the Dynamic Bayesian Networks which aim at representing causal dependencies between events across successive time steps to evaluate the *probability* of generated scenarios.

Furthermore, the elements of scenarios involved in these metrics can then differ considerably. For instance, the internal structure of the chosen evolution variables can be used to demonstrate *consistency* [23], while *traceability* can come from the narrative account of their development process [14]. Moreover, a single criterion may be associated with different metrics in the literature. Thus, the *plausibility* of a scenario can be linked to the overall coherence of the assumptions that construct its future situation [21], or alternatively to the causal logic of the chain of events leading up to it [6].

Nevertheless, none of these frameworks is appropriate to cover a whole set of evaluation criteria used in a scenario selection phase, as each is associated with a single criterion—and even with a specific metric of that criterion. Moreover, existing scenarios are rarely presented under a formalism that is directly suited to the metric one wishes to apply, making their evaluation against a chosen criterion all the more challenging.

3 Thesis Contribution

How, then, can one ensure that a scenario will be assessable against these various criteria and metrics?

Through this thesis, we propose to define a unified framework for prospective scenarios in the form of a conceptual model that makes it possible to account not merely for a specific criterion, but for the full set of criteria related to their evaluation.

The core insight of the doctoral work is to build the conceptual model from the scenarios elements that are referenced by existing evaluation metrics. This framework will thus enable the quality of any scenario to be assessed against any chose criterion, while also providing a foundation for building new metrics for criteria of interest that have not yet been addressed.

This formalism will be applied to the evaluation of various scenarios in order to demonstrate its relevance and versatility. Particular attention will be paid here to the case of future air transport, which is the subject of numerous projections by stakeholders across the sector, whether industrial (aircraft manufacturers), commercial (airlines), or institutional (regulatory bodies).

4 Research Program

To this end, a four-step research program was proposed:

1. The first step consisted of reviewing the existing literature on scenario quality criteria, and identifying those for which metrics already exist,
2. The second step, currently underway, aims at listing all the elements of scenarios the identified metrics draw upon.
3. The third step will build on this analysis to propose a conceptual model integrating these elements into a unified evaluation framework.
4. Finally, the fourth step will assess the accuracy and generalizability of this model across different forms of scenarios, in order to validate its applicability to the broad diversity of scenario types found in the literature.

4.1 Systematic Review of Evaluation Criteria

The first year of the doctoral work was dedicated to listing an exhaustive list of the criteria currently in use for the evaluation of prospective scenarios, through a systematic review. This review also aimed to catalogue the existing definitions of these criteria, and to identify those for which metrics already exist, potentially implemented within associated methods.

This systematic review followed the PRISMA2020 protocol [18] to ensure both the exhaustivity and the replicability of the search.

The corpus of articles was assembled in two phases. The first phase, illustrated in Fig. 2 consisted in querying the Scopus database to identify a set of publications, from which articles of interest were selected according to a set of

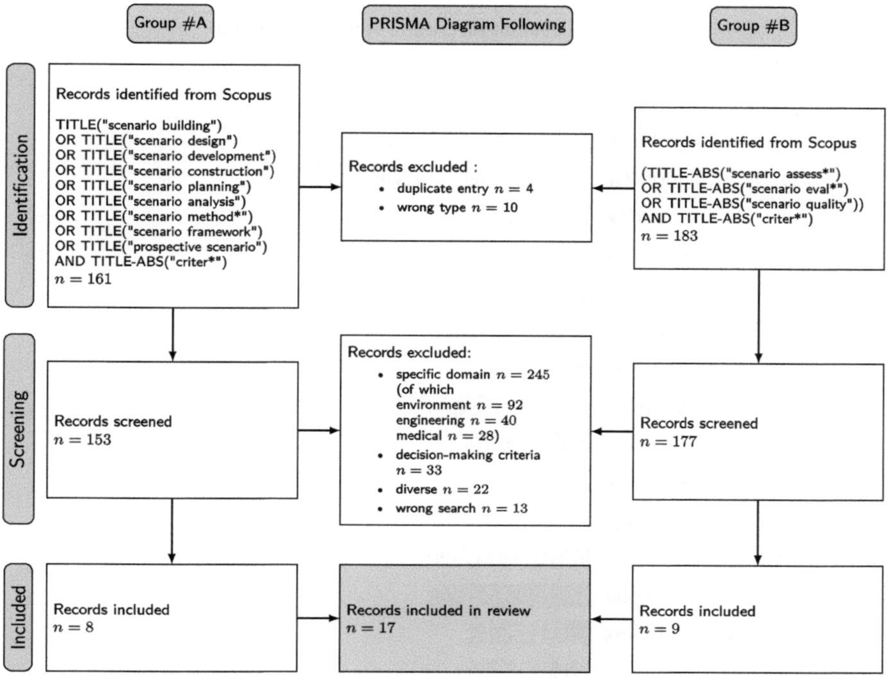

Fig. 2. Article selection following the PRISMA2020 protocol.

predefined criteria. A second selection phase, known as *"snowballing"*, was carried out to ensure that the corpus was sufficiently comprehensive to extract an exhaustive list of criteria.

An overall corpus of 39 publications was identified, and 104 criteria were extracted from them.

Since the thesis focuses on the way scenarios can be evaluated against these criteria, a meta analysis was performed to identify metrics associated with each of them.

Among the 104 criteria collected, at least one metric was found for 39 of them.

Figure 3 shows the top 20 criteria ranked by their number of associated metrics (in purple), while the number of occurrences, shown in grey, indicates how many publications reference each criterion.

For example, the *consistency* criterion appears in 23 publications, with 8 associated metrics, whereas the *coherence* criterion is referenced in 17 publications but only 3 metrics are identified for it.

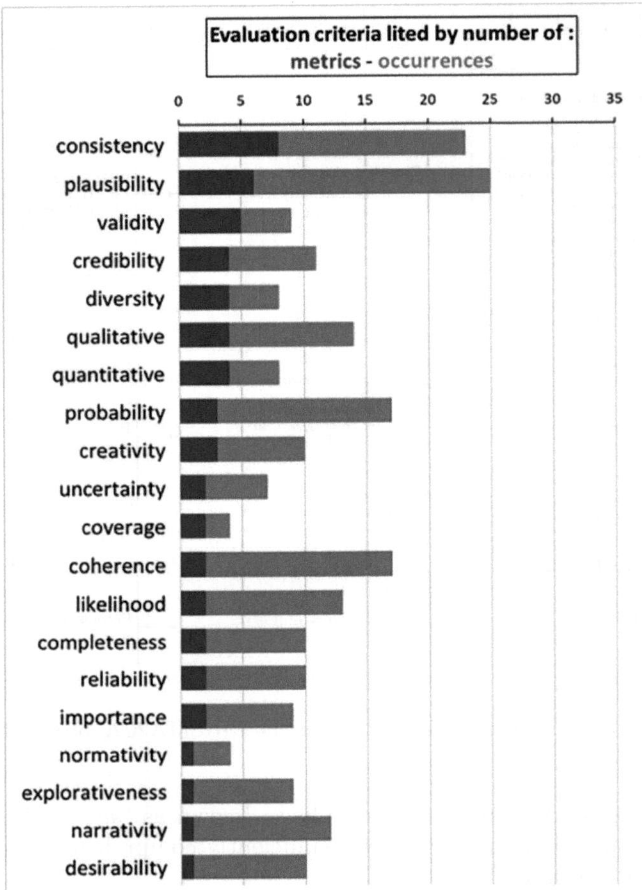

Fig. 3. Top 20 criteria listed by number of metrics. (Color figure online)

4.2 Catalogue of Scenario Elements

Ongoing step consists in identifying and analyzing each metric to list and understand which scenario elements are involved in the corresponding criterion.

Two examples illustrate how the elements of a scenario determine its assessability against a given criterion metric:

- The first concerns the *diversity* criterion. The authors define it as *"the complexity and multi-dimensionality of each scenario"*, and propose two associated metrics: *"variable type"* and *"variable control"* [19]. For a scenario to be assessable against this criterion through this metric, it must therefore explicitly provide the type of the variables used to describe the future evolution of the system, as well as the degree of control the user has over them.

- The second concerns the *traceability* criterion, defined as *"tracing or tracking results back to their production process"* [14]. Here, the associated metric relies on the *"reasons and justifications underlying the scenario design process"*. If these elements cannot be identified in a scenario, it will not be assessable through this criterion metric.

4.3 Building a Conceptual Model

The next step will be to build a conceptual model of a scenario that brings together, in a coherent manner, all the elements identified as necessary, thereby lending itself to all of the metrics previously listed.

4.4 Empirical Validation

A corpus of scenarios will then be assembled to serve as application cases. Existing scenarios will be used, whether or not they have previously been subject to evaluation. As seen before, the field of air transport will be particularly well-suited as a test case, thanks to the multitude of scenarios describing its future. Purpose-built scenarios will also be created, in order to further broaden the range of cases to be considered.

The objective here will be to assess the relevance of the proposed model by verifying that its application to the corpus of scenarios allows their quality to be evaluated against the various listed metrics.

Once the model is established, several complementary lines of work will emerge that could further strengthen the case for the proposed conceptual model. First, it can be compared with existing formalisms, such as the Dynamic Bayesian Networks [6], the Bayesian model averaging method [8], or the combinatorial models derived from Morphological Analysis [15]. The expected outcome would be to demonstrate the better versatility of the model with respect to the variety of scenarios and criteria. A further avenue consists in proposing metrics for criteria that are still too rarely, or not at all, evaluated, by leveraging the formalized elements of scenarios.

5 Open Science

The ongoing systematic review follows the principles of the PRISMA2020 protocol [18] to ensure the replicability of its results.

The scenarios used in the benchmarks will all be, either drawn from open literature and public reports, or developed as part of the thesis work, and then made available as open publications, in order to serve as application cases and test cases for future work on the evaluation of prospective scenarios.

Acknowledgments. This thesis work is supervised at ONERA by Thomas Polacsek (thomas.polacsek@onera.fr) and Anouck Chan (anouck.chan@onera.fr).

References

1. ACARE: Fly the Green Deal: Europe's vision for sustainable aviation. Technical report. ACARE (2022)
2. Amer, M., Daim, T.U., Jetter, A.: A review of scenario planning. Futures **46**, 23–40 (2013). https://doi.org/10.1016/j.futures.2012.10.003
3. ATAG, A.T.A.G.: Waypoint 2050 - An Air Transport Action Group Project. Technical report (2021)
4. Bardon, P., Massol, O., Thomas, A.: Greening aviation with sustainable aviation fuels: insights from decarbonization scenarios. J. Environ. Manag. **374** (2025). https://doi.org/10.1016/j.jenvman.2024.123943
5. Carroll, J.M. (ed.): Scenario-Based Design: Envisioning Work and Technology in System Development. Wiley, New York (1995)
6. Senni, C., Goel, C., S.: Nature scenario plausibility: a dynamic Bayesian network approach. Ecol. Econ. **236**, 108647 (2025). https://doi.org/10.1016/j.ecolecon.2025.108647
7. Cordova-Pozo, K., Rouwette, E.A.J.A.: Types of scenario planning and their effectiveness: a review of reviews. Futures **149**, 103153 (2023). https://doi.org/10.1016/j.futures.2023.103153
8. Culka, M.: Quantitative scenario design with Bayesian model averaging: constructing consistent scenarios for quantitative models exemplified for energy economics. Energy Sustain. Soc. **8**(1) (2018). https://doi.org/10.1186/s13705-018-0162-3
9. Dhami, M.K., et al.: Scenario generation and scenario quality using the cone of plausibility. Futures **142** (2022)
10. EREA: EREA Future of Aviation - the Scenarios. Technical report. Association of European Research Establishments in Aeronautics (2020)
11. EUROCONTROL: Aviation Long-Term Outlook: Flights and CO2 emissions forecast 2024 – 2050. Technical report, EUROCONTROL (2024)
12. IATA: Net Zero Roadmaps. Technical report. IATA (2023)
13. Johansen, I.: Scenario modelling with morphological analysis. Technol. Forecast. Soc. Chang. **126**, 116–125 (2018). https://doi.org/10.1016/j.techfore.2017.05.016. https://linkinghub.elsevier.com/retrieve/pii/S004016251730656X
14. Kosow, H.: New outlooks in traceability and consistency of integrated scenarios. Eur. J. Futures Res. **3**(1) (2015). https://doi.org/10.1007/s40309-015-0077-6
15. Lord, S., Helfgott, A., Vervoort, J.M.: Choosing diverse sets of plausible scenarios in multidimensional exploratory futures techniques. Futures **77**, 11–27 (2016). https://doi.org/10.1016/j.futures.2015.12.003
16. Meyer, B.: Agile roles. In: Agile!, pp. 79–87. Springer, Cham (2014). https://doi.org/10.1007/978-3-319-05155-0_5
17. Nowack, M., Endrikat, J., Guenther, E.: Review of Delphi-based scenario studies: quality and design considerations. Technol. Forecast. Soc. Chang. **78**(9), 1603–1615 (2011). https://doi.org/10.1016/j.techfore.2011.03.006
18. Page, M.J., et al.: The PRISMA 2020 statement: an updated guideline for reporting systematic reviews. Croat. Med. Assoc. School Med. **57**(4), 444–465 (2021). https://doi.org/10.21860/medflum2021264903
19. Sandoval, N., Reyna, J.L., Landis, A.E.: Internal consistency and diversity scenario development: a comparative framework to evaluate energy model scenarios. Renew. Sustain. Energy Rev. **186**, 113632 (2023). https://doi.org/10.1016/j.rser.2023.113632. https://www.sciencedirect.com/science/article/pii/S1364032123004896

20. Schoemaker, P.: When and how to use scenario planning: a heuristic approach with illustration **10**(6), 549–564 (1991). https://doi.org/10.1002/for.3980100602
21. Seeve, T., Vilkkumaa, E.: Identifying and visualizing a diverse set of plausible scenarios for strategic planning. Eur. J. Oper. Res. **298**(2), 596–610 (2022). https://doi.org/10.1016/j.ejor.2021.07.004
22. Taylor, D.C.W.: Alternative world scenarios for a new order of nations **286** (1993)
23. Thompson, R.C., Olugbara, O.O., Singh, A.: Cross-impact analysis with crowdsourcing for constructing consistent scenarios. Algorithms **19**(1) (2026). https://doi.org/10.3390/a19010041
24. Weimer-Jehle, W.: Cross-impact balances: a system-theoretical approach to cross-impact analysis. Technol. Forecast. Soc. Chang. **73**(4), 334–361 (2006). https://doi.org/10.1016/j.techfore.2005.06.005

A Bounded-Autonomy Multi-agent Architecture for Policy-Constrained Data Governance

Yasaman Hemmati[1,2,3](✉)

[1] Université Sorbonne Paris Nord, LIPN, CNRS, UMR 7030, 93430 Villetaneuse, France
yhemmati@layer-data.com

[2] LaMSN – La Maison des Sciences Numériques, Layer Data Consulting, Saint-Denis, France

[3] Layer Data Consulting, Grande Arche de la Défense, Paris, France

Abstract. Enterprise data governance teams enforce a complex ecosystem of roles covering stewardship, quality, traceability, policy, and audit, yet no technical architecture operationalizes these roles as autonomous, collaborating agents. This thesis proposes AGORA (Accountable Governance with Oversight and Role-bounded Autonomy), a governance-oriented multi-agent system (MAS) whose primary contribution is a formally verified, role-bounded enforcement mechanism coupling a Governance Role Ontology (GRO) with a Bounded Autonomy Model (BAM) to make every governance decision deterministically auditable and policy-constrained. Five role-specialized agents operate through a three-layer decision engine separating a role-specialized Large Action Model (LAM) core from a deterministic GRO rule engine and a BAM checkpoint handler—ensuring probabilistic reasoning is always constrained by deterministic enforcement. Supporting this core are a role-specialized training pipeline demonstrated on two pilot agents, and AGORABench. Phase 1 (field study and GRO construction) is complete.

Keywords: Multi-Agent Systems · Data Governance · Agentic AI · Large Action Models · Bounded Autonomy

1 Introduction

Data governance encompasses the policies, standards, roles, and operational processes that ensure data quality, security, compliance, and responsible use across the data lifecycle [2]. Contemporary data ecosystems have undergone a significant structural transformation. Organizations now operate across distributed cloud infrastructures, real-time streaming pipelines, cross-organizational exchanges, and AI-driven analytics platforms. The resulting growth in volume, velocity, and heterogeneity now exceeds the capacity of manual stewardship workflows, rule-based monitoring, and static policy enforcement mechanisms.

A. Bernasconi et al. (Eds.): RCIS 2026, LNBIP 597, pp. 202–212, 2027.
https://doi.org/10.1007/978-3-032-35043-5_17

A central challenge is the lack of a structured agent-based architecture capable of operationalizing data governance as an autonomous, continuous, policy-driven service. Existing approaches fall into two insufficient types: traditional frameworks relying on manual stewardship and periodic audits [2]; and emerging AI-powered platforms that embed automation but remain monolithic, lack coordination transparency, and are misaligned with established governance standards [9]. Neither provides a principled decomposition of governance responsibilities into collaborating agents capable of reasoning over policies, enforcing regulatory constraints deterministically, and maintaining auditable lineage across the data lifecycle [6,14]. This thesis proposes AGORA (Accountable Governance with Oversight and Role-bounded Autonomy), a governance-oriented multi-agent system integrating role specialization, ontological constraints, and bounded autonomy. The primary contribution is a formally verified, role-bounded enforcement mechanism that makes governance decisions deterministically auditable; model training and benchmarking are supporting contributions scoped to two pilot agents. This work does not aim to eliminate human oversight; rather, it reduces routine manual intervention by reserving human involvement for escalated, high-risk, or out-of-scope cases, based on explicit autonomy thresholds. Data and domain knowledge remain essential prerequisites for the success of automated workflows. The research is structured around four scientific questions:

RQ1—How can enterprise governance roles be decomposed into specialized, collaborating agents within a MAS?

RQ2—How can human authorization checkpoints be formally modeled without creating bottlenecks?

RQ3—What coordination and error-routing protocols ensure consistent, auditable, conflict-free governance behavior across the data lifecycle?

RQ4—How effective are the Stewardship and Data Quality agents at improving governance coverage, reducing manual workload, and maintaining performance against traditional and AI-powered baselines, within the full AGORA architectural framework?

2 Related Work

AGORA adopts the *native multi-agent* paradigm, in which a coordinated ecosystem of specialist agents is the primary architecture, since governance responsibilities are inherently distributed across distinct organizational roles requiring independent specialist competence [4,12].

Table 1 maps existing frameworks against six governance-specific requirements; where role specialization and deterministic enforcement refer specifically to the DAMA Guide to the Data Management Body of Knowledge (DAMA-DMBOK) [5] governance roles (steward, custodian, policy, audit) and not to generic agent roles or ad-hoc guards. No framework simultaneously satisfies all six. Four gaps follow directly: (G1) no framework decomposes the full enterprise governance role set into a native multi-agent architecture; (G2) guardrail capabilities carry no governance-domain adaptation; (G3) human-in-the-loop (HITL) checkpoints are ad hoc interruptions rather than formally modeled feedback loops; (G4) cross-boundary error routing is undefined, leaving cross-domain anomalies without a deterministic handling protocol.

Table 1. Agentic frameworks against AGORA design requirements. ✓ = fully supported; ∼ = partial; × = absent. AGORA entries reflect design claims; empirical validation is planned for Phase 4.

Framework	Gov. Role Specialization	Deterministic Enforcement	Gov. Standards Alignment	Error Taxonomy	Formal HITL	Audit Completeness
AutoGen/LangChain	×	×	×	×	×	×
MetaGPT/CrewAI	∼	×	×	×	×	×
LangGraph	×	∼	×	×	∼	∼
watsonx Orchestrate	×	∼	×	×	∼	∼
LlamaIndex/Sem. Kernel	×	×	×	×	×	×
DataGovBench agents	×	×	∼	×	×	×
AGORA (proposed)	✓	✓	✓	✓	✓	✓

To our knowledge, current frameworks treat HITL as an ad hoc interruption rather than a formally modeled architectural component [3]. EU AI Act Article 14 requires demonstrable human oversight as a built-in architectural property of high-risk AI systems, addressed through the BAM and its human authorization feedback loop in AGORA.

The DataGovBench benchmark confirms that current agents still struggle with multi-step governance workflows [9]. The L0–L5 autonomy taxonomy positions L3 as the near-term governance frontier, with safe operating region definition and formal autonomy guarantees as open challenges AGORA directly addresses.

Large Action Models. (LAMs) address large language model (LLM) hallucination in tool use by training on executable action trajectories rather than text [10]. Recent work demonstrates the effectiveness of this paradigm across GUI automation [1], REST API tool use [11,13], and multi-benchmark agentic evaluation [15,16]. To our knowledge, none of these models is trained on governance-specific action trajectories covering stewardship metadata operations, quality rule matching, lineage graph construction, policy obligation mapping, or audit log integrity verification. AGORA addresses this gap through a role-specialized supervised fine-tuning (SFT) pipeline grounded in practitioner-sourced governance trajectories.

3 AGORA: Proposed Architecture

How GRO, BAM, and the Decision Engine Relate. The three core mechanisms operate in a strict causal chain. The *GRO* defines what each agent is permitted to do: its action space, asset scope, and escalation predicates. The

BAM determines *how* each permitted action is authorized: it assigns an autonomy level and escalation threshold per role–operation class pair, and enforces the outcome deterministically. The *three-layer decision engine* operationalizes this chain within each agent at runtime: (1) the GRO Rule Engine acts as a deterministic gate before any model reasoning occurs; (2) the LAM Core handles semantic reasoning only after scope compliance is confirmed; (3) the BAM Checkpoint Handler evaluates escalation predicates and manages HITL re-entry. This separation guarantees that no probabilistic LAM output can override a deterministic governance constraint.

AGORA operationalizes enterprise data governance as a coordinated ecosystem of specialized agents. Unlike existing frameworks that provide generic orchestration, AGORA embeds governance semantics directly into agent design through the GRO, bounded autonomy constraints, and role-specific action models, Fig. 1. The architecture is organized into multiple layers: a stateless infrastructure layer handling routing and concurrency (Sect. 3.1), a stateful governance orchestration layer managing semantic delegation and human oversight (Sect. 3.2–Sect. 3.4), and an execution layer housing the five role-specialized agents (Sect. 3.3).

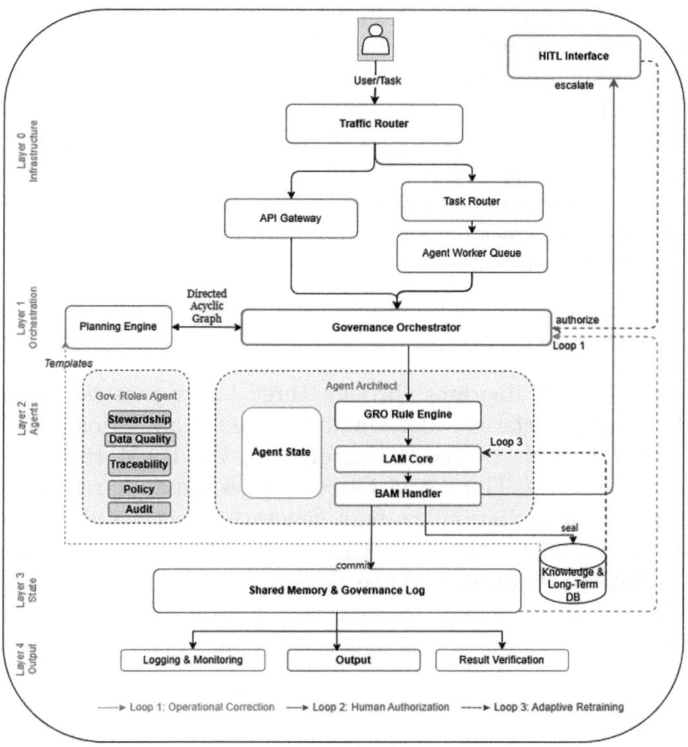

Fig. 1. AGORA architecture: layered governance pipeline with three feedback loops.

3.1 Infrastructure Layer

AGORA separates governance orchestration from infrastructure-level routing. This is a deliberate architectural requirement: if routing decisions are handled by the Governance Orchestrator, every task arrival serializes through an LLM call, creating a single-point bottleneck incompatible with concurrent governance event streams and with the auditability requirement that every task has a deterministic entry record prior to any model-based reasoning [6,14].

The layer comprises four stateless, deterministic components: **Traffic Router** (rate limiting, domain detection, LLM-free routing); **Task Router** (decomposes requests into GRO-scoped single-agent task envelopes carrying domain label, asset identifier, sensitivity tier, and timestamp); **Agent Worker Queue** (per-role buffer decoupling task arrival from agent processing rate); and **application programming interface (API) Gateway** (authenticates requests and commits a raw entry to the immutable governance log).

3.2 Governance Role Ontology (GRO)

At the core of AGORA lies the GRO, a formal ontology derived from DAMA-DMBOK [5] and instantiated through a practitioner field study. For each governance role, the GRO specifies: (i) an *action space* of permissible atomic operations; (ii) *scope constraints* over data domains, asset types, and sensitivity tiers; (iii) *accountability anchors* identifying the human role retaining ultimate accountability per action class; (iv) *escalation predicates* under which the agent must suspend autonomous action; and (v) an *error taxonomy* classifying which error types the agent may remediate autonomously, forward via shared memory, or escalate to the Orchestrator or a human. This prevents LAM-driven self-justification of incorrect decisions and guarantees deterministic error paths regardless of model confidence.

3.3 The Governance Agents

Each governance agent operates through three layers in strict sequence. The **GRO Rule Engine** acts as a deterministic gate: if an action falls outside its scope, the LAM Core is bypassed entirely and the BAM Handler receives a hard-stop signal directly. The **LAM Core** handles semantic reasoning, ambiguity resolution, and remediation strategy generation, activated only after scope compliance is confirmed, using a private vector store for retrieval-augmented context. The **BAM Checkpoint Handler** evaluates escalation predicates, constructs tamper-evident evidence records, and manages HITL re-entry. The private Short-Term Memory is scoped to the active task cycle and expires on BAM commit, never exposed externally.

The full AGORA vision defines five governance agents to operationalize the complete enterprise governance team. This thesis delivers a rigorous two-agent instantiation—the **Stewardship Agent** and the **Data Quality Agent** with the remaining three agents constituting a validated architectural design for future work.

The Stewardship Agent enriches metadata, resolves ownership conflicts, and classifies assets; GRO violations trigger hard stops with BAM logging and peer anomalies route via shared memory. The Data Quality Agent monitors completeness and consistency, applies pre-approved remediation rules. Both agents are fully trained through the SFT pipeline and evaluated on AGORABench.

3.4 Bounded Autonomy Model (BAM)

Intuitively, the BAM assigns to each agent role and operation class a policy triple that determines whether the agent may act autonomously, must notify a supervisor, or must suspend and escalate to a human. The following definitions make this precise.

Definition 1 (Governance Action). *A governance action is a tuple* $a = \langle role,\ op,\ asset,\ tier_0 \rangle$, *where* $role \in \mathcal{R}$ *is the executing agent role,* $op \in \mathcal{O}$ *is an atomic operation drawn from a finite operation set,* $asset \in \mathcal{A}$, *and* $tier_0 = tier(asset)\big|_{t_{submit}} \in \{L_1, L_2, L_3\}$ *is the declared sensitivity tier at submission time, with* $tier : \mathcal{A} \to \{L_1, L_2, L_3\}$ *and* $class : \mathcal{O} \to C$ *both total and static.*

Definition 2 (Governance State). $s(a) = \langle tier(asset),\ impact(a),\ risk(a),\ scope(r) \rangle \in \mathcal{S}$, *where* $tier(asset)$ *is the verified execution-time tier, authoritative over* $tier_0$ *if they differ;* $impact : \mathcal{O} \times \mathcal{A} \to [0,1]$; $risk : \mathcal{O} \times \mathcal{A} \to \{Low, Medium, High\}$; $scope : \mathcal{R} \to 2^{\mathcal{A}}$*—all computed deterministically by the GRO Rule Engine prior to BAM evaluation and not modifiable by the LAM Core.*

Definition 3 (BAM Triple). *For each rule* $r \in \mathcal{R}$ *and context* $c \in \mathcal{C}$, *the function* $BAM : \mathcal{R} \times \mathcal{C} \to \mathcal{A} \times \Phi \times \mathcal{E}$ *returns a triple:*

$$BAM(r, c) = \langle \alpha_{r,c},\ \varphi_{r,c},\ \varepsilon_{r,c} \rangle$$

where:

- $\alpha_{r,c} \in \{\textsc{Act\&Log},\ \textsc{Act\&Notify},\ \textsc{Await\&Escalate}\}$ *denotes the action mode.*
- $\varphi_{r,c} : \mathcal{S} \to \{\top, \bot\}$ *is the escalation condition, defined as:*

  $$\varphi_{r,c}(s(a)) = \top \iff tier(asset) = L_3 \vee impact(a) \geq \theta_{r,c} \vee risk(a) = High \vee asset \notin scope(r)$$

  where $\theta_{r,c} \in [0,1]$ *is a predefined threshold.*
- $\varepsilon_{r,c}(a, s(a))$ *defines the evidence record schema:*

  $$\varepsilon_{r,c}(a, s(a)) = \langle ts,\ a,\ s(a),\ decision,\ human\_resp \rangle$$

  where $decision \in \{\textsc{Execute},\ \textsc{Escalate},\ \textsc{Suspend}\}$, *and* $human\_resp = \bot$ *if no escalation occurs.*

The function tier remains constant during a single action lifecycle (i.e., during evaluation), but may change between submission and execution.

Definition 4 (BAM Enforcement Rule). *Let* $\langle \alpha, \varphi_{r,c}, \varepsilon_{r,c} \rangle = BAM(r,\ class(op))$. *The outcome function maps each* $(a, s(a))$ *to exactly one branch:*

$$\text{outcome}(a,\ s(a)) = \begin{cases} \textsc{Escalate}(\varepsilon_{r,c}(a, s(a))) & \varphi_{r,c}(s(a)) = \top \\ \textsc{Execute}(a) \wedge \textsc{Log}(\varepsilon_{r,c}(a, s(a))) & \varphi_{r,c}(s(a)) = \bot,\ \alpha = \textsc{Act}\,\&\,\textsc{Log} \\ \textsc{Execute}(a) \wedge \textsc{Notify}(\varepsilon_{r,c}(a, s(a))) & \varphi_{r,c}(s(a)) = \bot,\ \alpha = \textsc{Act}\,\&\,\textsc{Notify} \\ \textsc{Suspend}(a) \wedge \textsc{Request}(\varepsilon_{r,c}(a, s(a))) & \varphi_{r,c}(s(a)) = \bot,\ \alpha = \textsc{Await}\,\&\,\textsc{Escalate} \end{cases}$$

All four branches commit $\varepsilon_{r,c}(a, s(a))$ *to the immutable governance log;* $\varphi_{r,c}(s(a)) \in \{\top, \bot\}$ *and* α *ranging over three autonomy level guarantee exactly one branch applies for every* $(a, s(a))$*, with* $\varphi_{r,c}(s(a)) = \top$ *unconditionally preempting* α.

Property 1. (Escalation Completeness) For all $r \in \mathcal{R}$, $c \in \mathcal{C}$, $a \in \mathcal{R} \times \mathcal{O} \times \mathcal{A} \times \{L_1, L_2, L_3\}$ with $s(a) \in \mathcal{S}$, if $\varphi_{r,c}(s(a)) = \top$ then $\text{outcome}(a,\ s(a)) = \textsc{Escalate}(\varepsilon_{r,c}(a,\ s(a)))$. No governance action with a triggered escalation predicate executes autonomously.

This bounded autonomy produces regulatory proofs of human oversight as *runtime artifacts*, providing architectural evidence toward EU AI Act Article 14 and GDPR Article 5(2) accountability obligations. Full compliance requires additional organizational controls beyond this architecture; concurrency, temporal ordering, and identity delegation are deferred to the Phase 3 prototype.

3.5 Feedback and Self-correction Mechanisms

Three feedback loops transform AGORA into a continuously self-correcting governance infrastructure.

Loop 1—Operational Correction (Runtime). An anomaly published to shared memory triggers the Orchestrator to route a corrective task to the Policy Agent; the Quality Agent re-verifies; and the Audit Agent seals the cycle as a single evidence record.

Loop 2—Human Authorization (Runtime). When a BAM escalation predicate fires, the task is suspended and a structured decision request is posted to the HITL interface. The human decision is injected back into the LAM Core by the Orchestrator; the agent resumes and the full cycle is committed as a BAM evidence record.

Loop 3—Adaptive Training (Offline). The Audit Agent flags low-confidence decisions, overridden escalations, and human-corrected outputs, which are periodically fed back into the training pipeline, continuously updating the pilot agents' LAMs.

3.6 Orchestration Topology and Agent Training

AGORA adopts a *hybrid hierarchical* topology: the Governance Orchestrator handles strategic decisions and cross-agent routing, supported by a deterministic hierarchical task network (HTN) planning Engine producing ordered directed acyclic graphs (DAG) of single-agent sub-task. A federated extension enables independent AGORA instances to coordinate via a Governance Broker without sharing raw data [8]. Each pilot agent's LAM core is a low-rank adaptation (LoRA)-adapted model fine-tuned on role-specific governance trajectories; training details are given in Sect. 4.

4 Research Methodology

The thesis follows Design Science Research [7], a methodology appropriate for research that produces a rigorously evaluated artifact as both a scientific contribution and a practical system. Consistent with DSR protocol, the problem space and user requirements were investigated in Phase 1 before any artifact was proposed. Table 2 maps each field-study finding to the AGORA design decision it motivates, making the DSR chain of evidence explicit.

Table 2. DSR evidence chain: Phase 1 findings to AGORA design decisions.

Phase 1 Finding	Validated Problem	AGORA Design Decision
Role conflation between stewardship and custodianship	Accountability gaps across governance domains	GRO role decomposition into five specialized agents
Absence of formalized escalation paths across roles	Ad-hoc, non-auditable conflict resolution	Inter-agent error routing protocol and BAM predicates
Audit trail incompleteness in existing platforms	No tamper-evident record of governance decisions	BAM evidence record with immutable log commitment

The research is structured in four phases, of which Phase 1 is complete.

Phase 1—Field Study and GRO Construction *(completed, Jan–Apr 2026)*. Semi-structured interviews with governance practitioners at enterprise on different projects yielded a role taxonomy, action space definitions, escalation criteria, and per-role error taxonomy forming the GRO. Full findings are reported in Sect. 5.

Phase 2—Agent Training Pipeline *(Q2–Q4 2026)*. Phase 1 trajectories are converted to SFT datasets for the two pilot agents (Stewardship and Data Quality), supplemented with synthetic trajectories including deliberately injected error scenarios. Each pilot agent's LAM core is trained in one main stage: *Supervised fine-tuning* fine-tunes a base model (Mistral or LLaMA 3) with LoRA adapters on instruction–action pairs.

Phase 3—Architecture Formalization and Prototype *(Q1 2027)*. The GRO, BAM, three-layer decision engine, inter-agent error routing protocol, and three feedback loops will be formalized and implemented as a working prototype on AutoGen / LangGraph with connectors to editors APIs. A supervisor-validated scenario on data catalog governance will confirm end-to-end agent coordination, error handling, and feedback loop activation. Concurrency, temporal ordering, and identity delegation will also be addressed here.

Phase 4—Evaluation *(Q2–Q3 2027)*. The evaluation will measure HITL latency and escalation recall as indicators of human–AI interaction, and can be extended with an experiment on over-reliance and error detection when humans review AI governance decisions (Table 3).

Table 3. RQ-to-metric evaluation mapping.

RQ	Key Metric(s)	Primary Instrument
RQ1	Governance coverage: Stewardship & Data Quality roles	AGORABench (stewardship & quality domains)
RQ2	HITL latency; escalation recall	AGORABench HITL scenarios
RQ3	Error routing correctness; conflict resolution rate	AGORABench cross-domain error scenarios
RQ4	Task accuracy; workload reduction; compliance rate	DataGovBench + AGORABench

AGORA will be evaluated primarily on *DataGovBench* [9], an independently published benchmark for multi-step governance workflows providing an external validity anchor. *AGORABench* will serve as a complementary evaluation instrument covering the two pilot governance domains and error-routing scenarios not addressed by DataGovBench.

5 Preliminary Results

Phase 1 interviews with governance practitioners yielded three findings validating AGORA's core design premises: role conflation between stewardship and custodianship produces accountability gaps; the absence of a formalized escalation path motivates the inter-agent error routing protocol; and audit trail incompleteness validates the Autonomy Model's tamper-evident evidence record design.

A preliminary governance ontology prototype covering the stewardship and data quality roles has been constructed, encoding action spaces, escalation predicates, and error taxonomies. Table 4 shows four representative $BAM(r, c)$ instantiations for the Stewardship Agent, illustrating how autonomy level and escalation threshold, derived from the Phase 1 field study, vary by operation class and asset sensitivity.

Table 4. Representative BAM instantiations for the Stewardship Agent (r = Steward. $\theta_{r,c}$: impact threshold; α: autonomy level.)

Operation class c	Example op	$\theta_{r,c}$	$\alpha_{r,c}$	Escalates if
MetadataEnrich	Tag asset domain	0.8	Act & Log	tier $= L_3$ or impact ≥ 0.8
OwnershipResolve	Assign data owner	0.5	Act & Notify	tier $= L_3$ or impact ≥ 0.5
SensitivityClass	Reclassify $L_2 \rightarrow L_3$	0.3	Await & Escalate	always (tier becomes L_3)
ConflictResolve	Merge duplicate assets	0.4	Act & Notify	impact ≥ 0.4 or out of scope

Agent implementation and evaluation are planned for Phases 2 and 4, focusing on the Stewardship and Data Quality agents. The three remaining agents

(Traceability, Policy, Audit) are architecturally specified and GRO-grounded, and constitute future work beyond this thesis; the full five-agent system is intentionally ambitious and corresponds to a longer-term research agenda.

6 Conclusion

To our knowledge, AGORA is the first architecture to combine native multi-agent role decomposition grounded in DAMA-DMBOK [5], deterministic BAM enforcement, and formally modeled HITL feedback loops into a unified, policy-constrained governance system. It advances data governance from a reactive, human-centric discipline toward an intelligent, governance-by-design infrastructure. The thesis delivers four concrete contributions: (1) a GRO with per-role error taxonomy covering the full five-agent design; (2) a BAM with a three-layer decision engine providing deterministic enforcement; (3) role-specialized training rigorously evaluated on the Stewardship and Data Quality agents as the primary empirical scope; and (4) AGORABench. The GRO, BAM, and pilot agents are newly created in this thesis, while the underlying orchestration frameworks, base models, and benchmarks are reused as infrastructure and evaluation instruments. Upon completion, four artifacts will be publicly released: GRO in the Open Web Ontology Language (OWL) format, AGORABench, AGORA prototype (MIT), and anonymized field-study templates.

Supervisors. This work is conducted under the supervision of Prof. Younès Bennani (LIPN-CNRS, LaMSN – Université Sorbonne Paris Nord) and Dr. Yasmina Fedala (Layer Data Consulting).

References

1. Adept AI: Introducing ACT-1: Transforming how humans interact with computers (2022). https://www.adept.ai/blog/act-1. Accessed 16 Mar 2026
2. Agarwal, A., Amini, L., Mehta, S., Samulowitz, H., Srinivas, K.: Can AI autonomously build, operate, and use the entire data stack? (2025). https://doi.org/10.48550/arXiv.2512.07926. http://arxiv.org/abs/2512.07926
3. Ali, M.A., Dornaika, F.: Agentic AI: a comprehensive survey of architectures, applications, and future directions. Artif. Intell. Rev. **59**(1), 11 (2025). https://doi.org/10.1007/s10462-025-11422-4. http://arxiv.org/abs/2510.25445
4. Chen, K., Wang, P., Yu, Y., Zhan, X., Wang, H.: Large language model-based data science agent: a survey (2025). https://doi.org/10.48550/arXiv.2508.02744. http://arxiv.org/abs/2508.02744
5. DAMA International: The DAMA Guide to the Data Management Body of Knowledge (DAMA-DMBOK2R). Technics Publications, LLC (2024)
6. Derouiche, H., Brahmi, Z., Mazeni, H.: Agentic AI frameworks: architectures, protocols, and design challenges (2025). https://doi.org/10.48550/arXiv.2508.10146. http://arxiv.org/abs/2508.10146
7. Hevner, A.R., March, S.T., Park, J., Ram, S.: Design science in information systems research. MIS Q. **28**(1), 75–105 (2004). https://doi.org/10.2307/25148625. https://www.jstor.org/stable/25148625

8. IBM: watsonx Orchestrate: AI-powered automation for business workflows (2024). https://www.ibm.com/products/watsonx-orchestrate. Accessed 2025
9. Liu, Z., et al.: DataGovBench: benchmarking LLM agents for real-world data governance workflows (2025). https://doi.org/10.48550/arXiv.2512.04416. http://arxiv.org/abs/2512.04416
10. Moraes, H.F.: Hallucinations in LLMs: technical challenges, systemic risks and AI governance implications. International Association of Privacy Professionals (IAPP) (2025). Accessed 16 Mar 2026
11. Patil, S.G., Zhang, T., Wang, X., Gonzalez, J.E.: Gorilla: large language model connected with massive APIs (2023). https://arxiv.org/abs/2305.15334
12. Plaat, A., van Duijn, M., van Stein, N., Preuss, M., van der Putten, P., Batenburg, K.J.: Agentic large language models, a survey. J. Artif. Intell. Res. **84** (2025). https://doi.org/10.1613/jair.1.18675. http://arxiv.org/abs/2503.23037
13. Qin, Y., et al.: ToolLLM: facilitating large language models to master 16000+ real-world APIs (2023). https://arxiv.org/abs/2307.16789
14. Tran, K.T., Dao, D., Nguyen, M.D., Pham, Q.V., O'Sullivan, B., Nguyen, H.D.: Multi-agent collaboration mechanisms: a survey of LLMs (2025). https://doi.org/10.48550/arXiv.2501.06322. http://arxiv.org/abs/2501.06322
15. Zhang, J., Lan, T., Liu, Z., Hoang, T., Kokane, S., Niebles, J.C., Savarese, S., Xiong, C.: LAM-Simulator: large action model simulation for evaluating AI agents (2025). https://arxiv.org/abs/2502.11143
16. Zhang, J., et al.: xLAM: a family of large action models to empower AI agent systems. In: Proceedings of the 2025 Conference of the Nations of the Americas Chapter of the Association for Computational Linguistics: Human Language Technologies (NAACL-HLT 2025), Albuquerque, New Mexico, pp. 11583–11597. Association for Computational Linguistics (2025). https://doi.org/10.18653/v1/2025.naacl-long.578. https://aclanthology.org/2025.naacl-long.578/

Assessing Occupational Physical Workload Using IoT and Human Activity Recognition

Djalal Bouchekif(✉)

IRIT, Institut National Universitaire Champollion, University of Toulouse, Toulouse, France
djalal.bouchekif@irit.fr

Abstract. Work-related physical workload is a major concern in the healthcare sector, where professionals are frequently exposed to repetitive tasks, sustained physical effort, and demanding working conditions. Current assessment methods remain largely subjective and are often unable to support continuous monitoring in real-world environments. In this context, this PhD research aims to develop an objective framework based on wearable sensor data to identify, analyze, and characterize factors contributing to physical workload in real healthcare settings. Human Activity Recognition (HAR) represents a central component of this framework, as it enables the automatic identification of work-related activities and postures from multimodal sensor data. However, most state-of-the-art HAR approaches rely on fixed-size segmentation strategies, which limit their ability to capture variable-duration activities and complex temporal dynamics. This paper investigates these limitations through a comparative evaluation of several deep learning-based HAR architectures using both benchmark datasets and a real-world annotated dataset collected in healthcare environments. The results reveal a significant performance degradation when models are applied to real-world data, particularly for long- and variable-duration activities, highlighting critical challenges related to temporal variability and generalization. These findings emphasize the need for HAR approaches that address these limitations.

Keywords: Internet of Things · Wearable Sensors · Human Activity Recognition · Deep Learning · Physical Workload · Healthcare

1 Research Project

Work-related physical workload represents a major concern in terms of health and performance, both for workers and organizations. It is characterized by repeated physical efforts, high work intensity, constrained postures, and an increased risk of musculoskeletal disorders (MSDs), which may lead to discomfort, pain, or functional impairment [1]. Monitoring physical workload is therefore essential to prevent occupational accidents and diseases, improve

A. Bernasconi et al. (Eds.): RCIS 2026, LNBIP 597, pp. 213–222, 2027.
https://doi.org/10.1007/978-3-032-35043-5_18

ergonomics, and support the design of safer working conditions. Traditionally, physical workload assessment relies on subjective methods, such as questionnaires [2], perceived exertion scales [3], or self-reported data. Although widely used, these approaches present several limitations: they lack objectivity, do not allow continuous monitoring, and are sensitive to inter-individual variability. Moreover, they struggle to capture the complexity of real-life situations, where the duration, intensity, and nature of activities can vary significantly [4].

To overcome these limitations, more objective approaches have emerged, based on data collected from wearable sensors such as accelerometers, gyroscopes, and magnetometers, which are commonly embedded in smartphones, smartwatches, activity trackers, and other wearable devices [5]. These technologies enable continuous and reproducible measurements of movement, posture, and physiological or environmental parameters, thereby allowing a more accurate assessment of physical workload in real working conditions [6].

The objective of this PhD research is to develop and evaluate a multimodal data-driven framework for analyzing work-related physical workload in real occupational settings. The proposed framework combines objective data collected from wearable IoT sensors with subjective information obtained through questionnaires in order to characterize complex work activities, extract relevant workload indicators, and identify the main factors contributing to perceived physical strain. The expected outcome of the PhD is both a methodological framework and an evaluated artefact capable of supporting the analysis and prevention of physically demanding work situations.

More specifically, this research aims to address the following questions:

- **RQ1:** What are the most relevant multimodal data sources for assessing physical workload in real-world healthcare settings?
- **RQ2:** How can complex occupational activities be automatically recognized from irregular temporal patterns collected by wearable inertial sensors?
- **RQ3:** Which contextual factors contribute to physical workload and perceived physical strain?

Within this broader PhD research, Human Activity Recognition (HAR) constitutes a key step for transforming raw inertial signals into meaningful information about observed work situations. It enables the segmentation of data sequences, the distinction between different performed activities, and the interpretation of physical workload measurements within their execution context. This contextualization is essential, as the same level of physical effort or physiological response may have different meanings depending on the activity being performed, its duration, its repetition, and the postures involved.

In this study, we focus on the HAR component of the proposed framework and examine how existing deep learning-based HAR architectures perform when applied to complex occupational activities collected in real healthcare environments. The contributions of this study can be summarized as follows:

- We introduce and describe PEMESA HAR, a real-world healthcare dataset designed to support the analysis of physical workload.

– We provide an analysis of the limitations of existing approaches through a comparative evaluation of several deep learning-based HAR architectures, tested on both benchmark datasets and a real-world dataset collected in healthcare environments.

The remainder of this paper is organized as follows. Section 2 presents the related work. Section 3 describes the proposed methodology including the PEMESA Dataset and the assessment results. Finally, Sect. 4 discusses the limitations and outlines future research directions.

2 Related Work

Humans perform a wide range of activities in their daily lives. The structure of these activities can be hierarchical, ranging from elementary actions to complex tasks. In the literature, human activity is defined as a sequence of actions performed by one or more individuals [7]. As illustrated in Fig. 1, human activity can be divided into three distinct levels (gestures, actions, and activities) that vary in complexity and duration [8].

At the lowest level are gestures, corresponding to elementary, repetitive movements (e.g., hand movements). At an intermediate level are actions, which consist of combinations of gestures directed toward specific objectives. At the highest and most complex level are activities, which represent coordinated sets of actions forming meaningful occupational tasks or professional processes.

In practice, most existing HAR approaches primarily focus on the first two levels, namely gestures and actions, while the recognition of complex, long-term

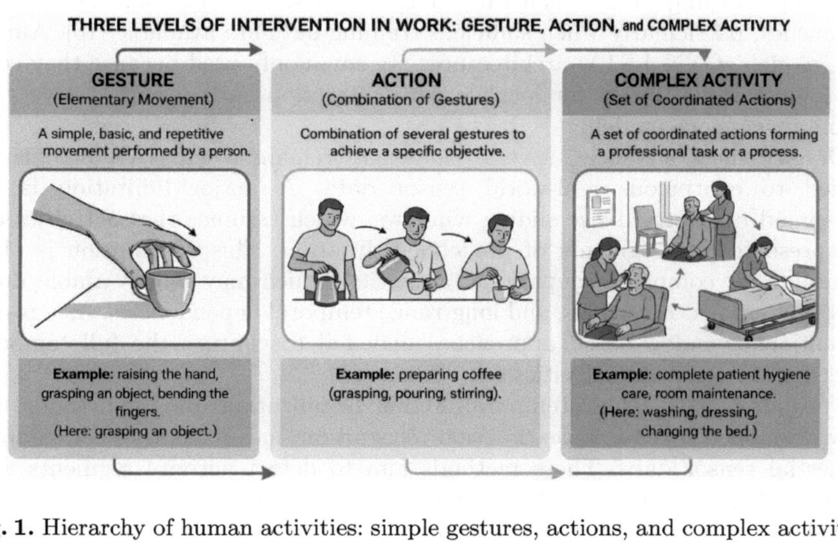

Fig. 1. Hierarchy of human activities: simple gestures, actions, and complex activities.

professional activities remains relatively less explored. In this context, occupational activities are considered the most complex category, involving extended temporal structure, multiple coordinated actions, and goal-oriented behavior.

The ultimate goal of Human Activity Recognition (HAR) is to train algorithms to accurately identify and recognize these different types of activities, using various state-of-the-art techniques.

2.1 HAR Methods

Early Human Activity Recognition (HAR) approaches mainly relied on traditional signal processing and machine learning techniques. In these methods, handcrafted features were extracted from inertial signals using time- and frequency-domain techniques such as Fourier transform, wavelet transform, or Principal Component Analysis (PCA), and then classified using supervised models such as Decision Trees (DT), Random Forests (RF), k-Nearest Neighbors (k-NN), Artificial Neural Networks (ANN), and Support Vector Machines (SVM) [9]. Although these approaches have shown good performance in controlled settings, they strongly depend on manual feature engineering and remain limited in their ability to capture complex temporal dependencies in sequential human activities.

To overcome these limitations, recent HAR research has increasingly shifted toward deep learning architectures, which can automatically learn discriminative representations from raw sensor data. Convolutional Neural Networks (CNN), Recurrent Neural Networks (RNN), Long Short-Term Memory networks (LSTM), Gated Recurrent Units (GRU), hybrid models, and attention-based architectures have been widely investigated. Comparative studies have shown that deep learning models generally outperform traditional machine learning approaches, particularly when sufficient training data are available [10]. Among these models, CNN–LSTM architectures are commonly used because they combine convolutional layers for local feature extraction with recurrent layers for temporal sequence modeling [11].

Despite these advances, several challenges remain when HAR models are applied to continuous real-world sensor data. A major limitation is the widespread use of fixed-size sliding windows, which assumes that activities can be represented by segments of predefined duration. This assumption is often unsuitable for complex occupational activities, which may have variable durations, ambiguous transitions, and long-range temporal dependencies. As a result, conventional window-based approaches may fail to capture the full temporal structure of real-world activities.

To address this issue, Temporal Action Localization (TAL) methods, initially developed for video-based action recognition, have recently been adapted to inertial sensor data. These methods aim to detect activity segments with arbitrary start and end boundaries, without relying on predefined windows, and have shown substantial performance gains, with up to a +26% improvement in F1-score over conventional HAR models on several benchmark datasets [12].

2.2 HAR Datasets

We selected four widely used benchmark datasets for HAR: UCI-HAR [13], PAMAP2 [14], WISDM [15], and Opportunity [16]. These datasets cover a range of scenarios from simple daily activities to complex multimodal environments.

UCI-HAR contains triaxial accelerometer and gyroscope data from smartphone sensors collected from 30 subjects performing six daily activities at 50 Hz. PAMAP2 includes multimodal inertial (accelerometer, gyroscope, magnetometer) and physiological (heart rate) data from 9 subjects performing 18 activities at 100 Hz. WISDM provides accelerometer data from 36 participants performing six activities at 20 Hz. Opportunity is a multimodal dataset combining wearable, object, and ambient sensors, capturing complex activities across multiple levels (gestures, actions, and high-level behaviors) with detailed annotations.

3 Research Methodology

This PhD research follows a four-phase methodological framework designed to address the research questions:

(1) the identification of relevant activities and data sources through field studies; (2) the deployment of a multimodal data acquisition protocol in real-world healthcare settings (RQ1); (3) the development of machine learning models for activity and posture recognition from time-series data (RQ2); and (4) the analysis of physical workload through the integration of objective and subjective measures (RQ3). This approach aims to ensure greater ecological validity compared to laboratory-based studies. In the following sections, the presented results focus on the third phase of the proposed methodology, corresponding to RQ2.

3.1 PEMESA HAR Dataset

Most open-source datasets commonly used in Human Activity Recognition (HAR) focus on simple activities performed in controlled laboratory conditions, which limits their transferability to real occupational settings. To address this gap, we conducted a field study in a real healthcare environment in collaboration with a project partner. This phase combined structured observations and semi-structured interviews with healthcare staff, leading to the identification of two professional profiles particularly exposed to physical workload and biomechanical constraints: *nursing assistants* and *hospital service agents (ASH)*. These observations guided the selection of the professional activities to be captured under real working conditions. As part of the broader objective of quantifying work-related physical fatigue, we designed a multimodal data collection protocol, illustrated in Fig. 2. The protocol combines objective measurements from IoT sensors, including inertial, physiological, and environmental data, with subjective self-assessments.

In this paper, we focus on inertial signals collected using a commercial smartwatch, namely the Samsung Galaxy Watch 7. The device provides triaxial accelerometer and gyroscope measurements, corresponding respectively to linear acceleration and angular velocity along three axes. Activity labels were assigned in real time using a mobile application connected to the smartwatch, allowing sensor samples to be associated with the corresponding occupational activity observed in the field. In addition to wearable sensing, environmental sensors integrated into Raspberry Pi devices were deployed at different locations in the healthcare environment to capture contextual factors such as temperature, air quality, noise, and light. Beacon devices were also used to support movement tracking within the facility. The PEMESA dataset was collected from four participants and comprises approximately 20 h of annotated recordings, corresponding to more than 4 million synchronized inertial samples. The recordings include six selected professional activity classes, as well as an additional "other" category for activities not considered in this study. The distribution of the six selected activity classes is reported in Table 1.

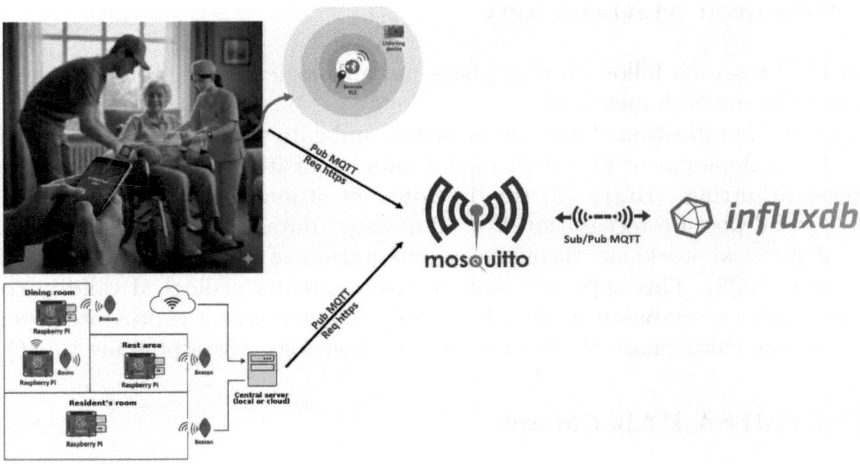

Fig. 2. Overview of the data collection protocol.

Table 1. Distribution of activities in the PEMESA dataset

Class	Number of Samples	Proportion
Assistance with bathroom bathing	354 400	42.2%
Assistance with bed bathing	220 177	26.2%
Positioning patient in wheelchair	40 177	4.8%
Bed preparation	50 427	6.0%
Cleaning the floors	24 939	3.0%
Cleaning the bathroom	150 395	17.9%

3.2 Evaluating HAR Approaches

We evaluated several state-of-the-art deep learning architectures commonly used for sensor-based Human Activity Recognition (HAR), including Convolutional Neural Networks (CNN), Long Short-Term Memory networks (LSTM), Bidirectional LSTMs (Bi-LSTM), Gated Recurrent Units (GRU), CNN-LSTM hybrids, Transformer-based models, InceptionTime, and Fully Convolutional Networks

Table 2. Performance Metrics of Deep Learning Models

Dataset	Technique	Accuracy	Precision	Recall	F1-Score
UCI-HAR	CNN	0.95	0.95	0.95	0.95
	LSTM	0.82	0.83	0.82	0.82
	Bi-LSTM	0.85	0.85	0.84	0.84
	GRU	0.82	0.83	0.82	0.82
	CNN-LSTM	0.96	0.95	0.96	0.96
	Transformers	**0.97**	**0.96**	**0.97**	**0.97**
WISDM	CNN	0.63	0.62	0.57	0.59
	LSTM	0.63	0.61	0.56	0.58
	Bi-LSTM	0.63	0.61	0.57	0.59
	GRU	0.63	0.62	0.57	0.59
	CNN-LSTM	0.70	0.69	0.67	0.66
	Transformers	**0.75**	**0.74**	**0.72**	**0.72**
PAMAP2	CNN	**0.99**	**0.99**	**0.99**	**0.99**
	LSTM	0.98	0.98	0.98	0.98
	Bi-LSTM	**0.99**	**0.99**	**0.99**	**0.99**
	GRU	0.98	0.98	0.98	0.98
	CNN-LSTM	0.97	0.97	0.96	0.96
	Transformers	0.98	0.98	0.97	0.97
OPPORTUNITY	CNN	**0.89**	**0.91**	**0.90**	**0.90**
	LSTM	0.69	0.71	0.71	0.69
	Bi-LSTM	0.79	0.81	0.81	0.80
	GRU	0.70	0.69	0.74	0.70
	CNN-LSTM	0.85	0.86	0.85	0.85
	Transformers	0.88	0.89	0.88	0.88
PEMESA HAR	CNN	0.55	0.52	0.50	0.51
	LSTM	0.52	0.50	0.48	0.49
	Bi-LSTM	0.54	0.52	0.49	0.50
	GRU	0.53	0.51	0.48	0.49
	CNN-LSTM	0.58	0.56	0.53	0.54
	Transformers	0.60	0.58	0.55	0.56
	InceptionTime	0.62	0.60	0.57	0.58
	FCN	0.60	0.58	0.55	0.56
	TAL	**0.62**	**0.62**	**0.59**	**0.60**

(FCN). We also explored Temporal Action Localization (TAL) models, which predict activity segments with explicit start and end boundaries instead of relying on fixed-window classification. The results are summarized in Table 2.

The evaluation reveals a clear performance gap between benchmark datasets and the PEMESA HAR dataset. While most architectures achieve strong results on standard HAR datasets, their performance decreases substantially when applied to real-world occupational data. This degradation highlights the limited generalization ability of current HAR models when facing distribution shifts related to users, sensor placement, devices, and environmental context.

This gap can also be explained by the nature of occupational activities, which are longer, less structured, and more heterogeneous than the activities commonly represented in benchmark datasets. In PEMESA HAR, some activities can last up to 30 min and may include multiple motion patterns and transitions. Under these conditions, fixed-size sliding windows become insufficient to capture the complete temporal structure of the activity, leading to fragmented representations and reduced recognition performance.

Overall, these findings show that the limitations observed in this study are not specific to a single model or dataset. Rather, they reflect broader challenges in sensor-based HAR, particularly regarding robustness to real-world variability, temporal modeling of long-duration activities, and cross-domain generalization. These challenges motivate the need for more adaptive and hierarchical HAR approaches for complex occupational settings.

4 Future Work

Several research directions will be explored in the next stages of this PhD.

First, the dataset will be enriched by increasing the number of participants and the diversity of observed situations. In parallel, adaptive temporal segmentation approaches, such as automatic change-point detection and dynamic windowing, will be investigated to better model occupational activities with variable durations and complex temporal structures. Hybrid deep learning architectures combining convolutional networks, recurrent models, and attention mechanisms will also be explored to improve the recognition of long, heterogeneous, and weakly structured activities in real healthcare environments.

Second, hierarchical modeling approaches will be studied to better represent complex professional activities. Rather than considering each activity as a single global label, future work will explore enriched annotation strategies enabling multi-level labeling, in which complex activities can be decomposed into elementary actions, postures, or motion primitives. This direction aims to improve both the recognition of composite activities and the interpretability of Human Activity Recognition models.

Finally, the HAR component will be integrated into the broader objective of physical workload assessment. Future work will combine activity recognition outputs with objective sensor-based indicators and subjective self-assessments, such as perceived exertion and questionnaire-based measures.

This integration will support the identification of physically demanding situations and the analysis of the contextual factors contributing to perceived physical strain. The expected outcome of this PhD is an evaluated multimodal methodological framework for analyzing work-related physical workload in real occupational settings.

This framework aims to support the interpretation of occupational activities, the identification of workload-related risk factors, and the prevention of physically demanding work situations.

Open Science

The PEMESA study received a favorable opinion from the Ethics Committee for Research of the University of Toulouse. Data processing follows GDPR-oriented safeguards, including consent, minimization, pseudonymization, restricted access, and secure storage. Given the sensitivity of occupational data, raw participant-level records will not be openly released. Reproducibility will be supported by documenting the protocol, preprocessing, models, training settings, and metrics, and by releasing code, scripts, and aggregated results.

Supervisor

This PhD work is supervised by **Imen Megdiche** (IRIT, INUC) and **Rémi Bastide** (IRIT, INUC).

Acknowledgment. This research was funded under the PEMESA project, supported by the European Union and awarded by the Occitanie Region. The author would like to thank the medical partners involved in this project, in particular the AGIR nursing home (EHPAD), for granting access to their facilities and for their support in conducting the experimental studies.

References

1. Motta, F., Varrecchia, T., Chini, G., Ranavolo, A., Galli, M.: The use of wearable systems for assessing work-related risks related to the musculoskeletal system—a systematic review. Int. J. Environ. Res. Public Health **21**(12), 1567 (2024)
2. Arsalan, A., Majid, M., Nizami, I.F., Manzoor, W., Anwar, S.M., Ryu, J.: Human stress assessment: a comprehensive review of methods using wearable sensors and non-wearable techniques (2023). arXiv preprint arXiv:2202.03033
3. Morishita, S., Tsubaki, A., Takabayashi, T., Fu, J.B.: Relationship between the rating of perceived exertion scale and the load intensity of resistance training. Strength Conditioning J. **40**(2), 94–109 (2018)
4. Matthews, G., de Winter, J., Hancock, P.: What do subjective workload scales really measure? Operational and representational solutions to divergence of workload measures. Theor. Issues Ergon. Sci. **21** (2019)

5. Naranjo, J.E., Mora, C.A., Bustamante Villagómez, D.F., Mancheno Falconi, M.G., Garcia, M.V.: Wearable sensors in industrial ergonomics: enhancing safety and productivity in industry 4.0. Sensors **25**(5), 1526 (2025)
6. Donisi, L., Cesarelli, G., Pisani, N., Ponsiglione, A.M., Ricciardi, C., Capodaglio, E.: Wearable sensors and artificial intelligence for physical ergonomics: a systematic review of literature. Diagnostics **12**(12), 3048 (2022)
7. Arshad, M.H., Bilal, M., Gani, A.: Human activity recognition: review, taxonomy and open challenges. Sensors **22**(17), 6463 (2022)
8. Ziaeefard, M., Bergevin, R.: Semantic human activity recognition: a literature review. Pattern Recogn. **48**, 2329–2345 (2015)
9. Rehman, S., Ali, A., Khan, A., Okpala, C.: Human activity recognition: a comparative study of validation methods and impact of feature extraction in wearable sensors. Algorithms **17**, 556 (2024)
10. Kumar, P., Suresh, S.: Deep-HAR: an ensemble deep learning model for recognizing the simple, complex, and heterogeneous human activities. Multimedia Tools Appl. (2023)
11. Ordóñez, F.J., Roggen, D.: Deep convolutional and LSTM recurrent neural networks for multimodal wearable activity recognition. Sensors **16**(1), 115 (2016)
12. Bock, M., Moeller, M., Van Laerhoven, K.: Temporal action localization for inertial-based human activity recognition. arXiv preprint arXiv:2311.15831 (2024)
13. Anguita, D., Ghio, A., Oneto, L., Parra, X., Reyes-Ortiz, J.L., et al.: A public domain dataset for human activity recognition using smartphones. In: ESANN, vol. 3, no. 1, pp. 3–4 (2013)
14. Reiss, A., Stricker, D.: Introducing a new benchmarked dataset for activity monitoring. In: 2012 16th International Symposium on Wearable Computers (ISWC), pp. 108–109 (2012). https://doi.org/10.1109/ISWC.2012.13
15. Kwapisz, J.R., Weiss, G.M., Moore, S.A.: Activity recognition using cell phone accelerometers. ACM SIGKDD Expl. Newsl. **12**(2), 74–82 (2011)
16. Chavarriaga, R., et al.: The opportunity challenge: a benchmark database for on-body sensor-based activity recognition. Pattern Recogn. Lett. **34**(15), 2033–2042 (2013)

Cyber Resilience by Design for Critical-Infrastructure Information Systems: A Three-Phase Framework for Systemic Risk Assessment, Management, and Recovery

Fatimah Faraji(✉)

Institute for Analytics and Data Science, School of Computer Science and Electronic Engineering, University of Essex, Colchester, UK
f.faraji@essex.ac.uk

Abstract. Critical infrastructures are increasingly exposed to cascading cyber-physical disruptions, making cyber resilience a systemic problem that must connect pre-incident assessment, incident-time management, and post-incident recovery. This doctoral research develops a Cyber Resilience-by-Design framework for software-intensive industrial CI information systems spanning three phases: (1) pre-incident resilience analysis through a dynamic multi-criteria criticality method and the Risk Impact Pathway Analysis (RIPA) risk-propagation algorithm, (2) during-incident resilience management through a Remedial Action Scheme (RAS) formulated as a Markov Decision Process (MDP) and addressed using Deep Reinforcement Learning (DRL), and (3) post-incident resilience improvement through recovery and learning mechanisms. Preliminary results currently concern Phase 1 and show improved systemic assessment by identifying pathway-centric hotspots and prioritising interventions under constrained budgets. Validation uses an industrial dependency network (21 nodes, 33 dependencies); planned evaluation uses metrics including systemic risk reduction, service continuity, decision latency, and safety-constraint violations.

Keywords: Information Science · Information Security and Risk Management · Cyber-Physical Systems · Critical Infrastructure · Systemic Risk · Design Science Research · Decision Support

1 Introduction and Problem Significance

Critical infrastructures (CIs) are foundational to modern society, sustaining economic activity, public health, safety, and national security [4,5]. From an Information Science perspective, and more specifically through Information Systems Engineering, critical infrastructures can be understood as software-intensive

A. Bernasconi et al. (Eds.): RCIS 2026, LNBIP 597, pp. 223–234, 2027.
https://doi.org/10.1007/978-3-032-35043-5_19

cyber-physical information systems whose resilience depends on how dependencies, information flows, and operational decisions are modelled and managed across the incident lifecycle. A defining characteristic is their dense web of dependencies and interdependencies—no system operates in isolation—which improves efficiency but also introduces vulnerabilities, particularly at cyber–physical interfaces [6,8,9,35]. In practice, these dependencies are realised through information-system interfaces and service dependencies (e.g., APIs, message brokers, data hubs, and identity services), so disruptions often propagate along information flows as well as physical couplings.

This interconnectedness increases exposure to systemic risk: disruptions can propagate across services and subsystems and escalate into cascading failures [10,11]. Real incidents and recent research demonstrate how cyber intrusions can traverse logical and operational dependencies to trigger widespread impacts, including attacks on industrial control environments [12,13]; such propagation is often non-linear and difficult to predict [14–16]. These realities expose the inadequacy of conventional component-centric risk analytics and decision-support approaches [17].

This work targets software-intensive industrial CI information systems where cascading disruption propagates through information-system dependencies (e.g., APIs, identity services, data hubs, message brokers) as well as cyber–physical couplings. The research is grounded in an industrial cyber–physical infrastructure case modelled as a dependency network (21 nodes, 33 dependencies) capturing data exchange, process sequencing, control signals, and computational-resource couplings, which enables concrete, pathway-level analysis and validation rather than generic resilience guidance. The graph represents the system under study, while criticality, resilience, and risk scores are assigned at node level and propagated through dependency edges. A node represents a system element that may provide, consume, process, or control a service. Depending on the modelling scope, nodes may correspond to physical assets, cyber assets, software services, data repositories, or operational functions.

In this work, *cyber resilience* is understood as the ability of a critical- infrastructure information system to prepare for, absorb, respond to, recover from, and adapt after cyber disruption across the incident lifecycle. In this sense, resilience is broader than *resistance*: resistance refers to the ability to withstand disruption at the point of impact, whereas resilience also includes response, restoration, and adaptation. Resilience is quantified as a composite score derived from preparedness, absorption, responsiveness, recoverability, and adaptability. Each dimension is assessed through measurable indicators and aggregated into a node-level resilience score used in risk propagation calculations. This distinction is important because the work addresses not only whether disruption can be resisted, but also how systemic risk can be managed and improved over time. Consequently, building resilience capabilities in CI information systems increasingly requires a resilience-by-design perspective, engineering systems that can absorb shocks, adapt to changing conditions, and recover rapidly [18]. In interdependent CI/CPS settings, this requires systemic risk assessment as an engineering capability: identifying critical assets, modelling dependencies, and quantifying how disruptions propagate across coupled networks [19–21]. Secure Tropos is used during requirements and dependency modelling to identify actors, goals, assets, trust assumptions, and interdependencies. The resulting dependency model serves as an input to the graph-based criticality and propagation analyses. We use the NIST Cybersecurity Framework (Identify–Protect–Detect–Respond–Recover) as a lifecycle structure [22,23,26], and focus on extending it with cascade-aware analytics and decision support suitable for CI information systems. These gaps motivate three research questions:

RQ1: What models, artefacts, and lifecycle mechanisms are needed for a Cyber Resilience by Design framework for *interdependent critical-infrastructure information systems* that links design-time systemic assessment, incident-time decision support, and post-incident learning?

RQ2: How can *dependency-pathway* modelling, dynamic criticality assessment, and risk-propagation analytics be represented and operationalised within critical-infrastructure information systems to support *cascade-aware* resilience across the incident lifecycle?

RQ3: What methods can provide scalable, time-critical decision support for mitigating cyber-risk propagation *along dependency pathways* in interdependent critical infrastructures under dynamic dependencies and cascading effects, while satisfying *operational safety and service-continuity constraints*?

Research Objective. The objective of this doctoral research is to design, implement, and evaluate a lifecycle framework for systemic risk assessment, incident-time decision support, and recovery-oriented learning in critical-infrastructure information systems. The objective of the framework is not only to evaluate systemic cyber risk but also to support risk-informed mitigation, recovery, and adaptation decisions across the entire incident lifecycle. For this reason, the work distinguishes between two linked but analytically different capabilities: *systemic risk assessment*, which estimates and prioritises critical assets, dependencies, and pathways; and *systemic risk management*, which selects and validates feasible interventions under operational constraints. The proposed framework integrates these capabilities through three linked phases, but does not treat diagnosis and treatment as the same task.

Expected contributions. The scientific contribution of this work is the design and phased evaluation of a lifecycle resilience engineering framework for interdependent critical-infrastructure information systems. Rather than addressing only one stage of the incident lifecycle, the work contributes linked artefacts for systemic risk assessment, incident-time decision support, and post-incident adaptation. The proposed three-phase framework operationalises Cyber Resilience Act (CRA) principles across the entire cyber resilience lifecycle, embedding security-by-design in Phase 1, supporting incident handling in Phase 2, and enabling recovery, learning, and continuous improvement in Phase 3. Concretely, this work makes five contributions.

C1 (RQ1) Lifecycle Framework [developed/integration ongoing]: A three-phase Cyber Resilience by Design lifecycle integrating design-time assessment, incident-time decision support, and post-incident model/policy update through feedback loops.

C2 (RQ2) Dynamic Criticality Modelling [completed/evidenced]: A dynamic multi-criteria criticality artefact combining organisational importance with systemic network role via scenario-dependent weighting for resilience-aware prioritisation under changing interdependencies.

C3 (RQ2) Risk Impact Pathway Analysis (RIPA) [completed/evidenced]: A dependency-aware risk-propagation algorithm combining linkage intensity, resilience, and criticality to quantify propagated risk and identify pathway-centric systemic hotspots for mitigation planning.

C4 (RQ3) Run-time Decision Support [developed/validation ongoing]: A remedial action scheme (RAS) formulated as an MDP and addressed using DRL to select time-critical mitigation actions under operational constraints and stochastic cascade dynamics.

C5 (RQ1, RQ2) Requirements and Dependency Modelling [developed/planned integration]: Secure Tropos is used as the requirements and dependency modelling backbone of the framework. Its role is not to replace the risk-propagation or decision-support mechanisms, but to capture actors, dependencies, and resilience/security requirements, and to provide traceability from design-time assumptions to run-time actions and post-incident updates across the three phases.

The remainder of this paper is organised as follows. Section 2 presents the state of the art and the research methodology underpinning this doctoral research. Section 3 describes the proposed three-phase Cyber Resilience by Design framework and summarises the preliminary results achieved to date, particularly for the pre-incident phase. Finally, Sect. 4 discusses the Future Research Directions, outlines the open science approach, and concludes the paper.

2 State of the Art and Research Methodology

We conducted a PRISMA-based systematic literature review (SLR) of 41 primary studies (2017–2025), complemented by targeted backward and forward snowballing, to examine how cybersecurity and resilience of Critical Infrastructure (CI) systems are addressed across the incident lifecycle [2,25]. The review revealed a fragmented body of knowledge spanning pre-incident risk assessment, incident-time response, and post-incident recovery, with limited integration across these phases. Furthermore, key gaps were identified in dynamic criticality assessment, dependency-aware risk propagation, incident-time decision support, and recovery-driven adaptation.

The methodological choices adopted in this work are driven by these identified gaps and by the characteristics of the problem domain rather than by any single preferred technique. Phase 1 focuses on understanding how risk propagates through interconnected infrastructures; therefore, graph-based modelling, dynamic criticality assessment, and pathway-oriented risk analysis are employed to capture dependencies, cascading effects, and systemic impact. Phase

2 addresses incident-time decision making, where operators must select mitigation actions under uncertainty, evolving cascade conditions, and operational constraints. This problem is formulated as a Markov Decision Process (MDP), enabling system states, actions, and transition dynamics to be explicitly represented. Deep Reinforcement Learning (DRL) is proposed as the decision-support mechanism because it can learn adaptive response policies for large-scale environments where traditional optimisation approaches become computationally intractable. Since large real-world incident datasets are rarely available in critical infrastructure settings, Phase 2 validation will be conducted using simulation and cyber-physical testbed environments derived from the dependency and risk models developed in Phase 1. Finally, Phase 3 extends the framework beyond response by incorporating recovery, learning, and adaptation mechanisms, enabling continuous improvement of cyber resilience throughout the incident lifecycle.

On the design-time side, recent work advances interdependency modelling and cascade analysis, including studies on cascading effects in interdependent critical infrastructures [7] and cascading disasters in urban infrastructure [10]. Governance- and framework-oriented contributions are also maturing [21,23], but they largely remain at the level of guidance rather than providing an end-to-end method that yields executable incident-time policies or specifies how models should evolve from operational evidence.

At run time, the literature offers decision-support and response mechanisms, including PRAETORIAN [27], the DYNABIC approach [31], and validation support for CI security systems [28]. However, these contributions typically address the during-incident phase in isolation and provide limited guidance on how design-time dependency and criticality models are operationalised into actionable response policies, or how incident evidence is used to update models and policies after the event.

Recovery and learning are frequently addressed through restoration and resource-allocation approaches [29,30,32,33], yet closed-loop mechanisms that feed incident observations back into risk, dependency, and policy models remain under-engineered in CI information systems.

Across these strands, existing work remains fragmented across lifecycle stages. What is still missing is an end-to-end information-science framework that integrates dependency modelling, systemic risk assessment, incident-time decision support, and post-incident learning into a closed-loop resilience capability for critical-infrastructure information systems [18,24]. In this work, this gap is examined through three dimensions: (i) Systemic Risk Assessment Methodologies, (ii) Dynamic Risk Management Strategies, and (iii) Resilience Enhancement and Recovery.

Research Methodology and Empirical Design. The empirical research instruments used in this work include a PRISMA-based systematic literature review, industrial case-based validation, simulation-based experimentation, sensitivity analysis, and comparative benchmarking against baseline approaches. Building on the limitations identified through the SLR [2], this research adopts the Design

Science Research Methodology (DSRM) to design, develop, and evaluate a Cyber Resilience-by-Design framework for critical-infrastructure information systems. The SLR provides the evidence base to derive design objectives and requirements, particularly for (i) modelling interdependencies and cascading effects, (ii) integrating cyber and physical viewpoints within CI service architectures, and (iii) supporting time-critical, scalable decision-making during incidents.

Following DSRM, the research produces and evaluates a coherent set of engineering artefacts spanning the incident lifecycle: (i) pre-incident assessment artefacts (adaptive criticality modelling and RIPA), (ii) a during-incident decision artifact (MDP/DRL-based RAS), and (iii) post-incident improvement mechanisms that enable model and policy update through feedback. The artefacts are integrated through Secure Tropos models (actors, dependencies, and resilience/security requirements), which provide a unifying representation and traceability across phases, from design-time assumptions and dependencies to run-time actions and post-incident updates.

Evaluation is performed using real-world case studies and simulation-based experimentation, benchmarked against static criticality, propagation-unaware risk scoring, and heuristic response strategies. We assess whether the artefacts enable (i) improved prioritisation of critical assets, dependencies, and high-risk pathways, measured by higher systemic risk reduction under fixed intervention budgets (Phase 1); (ii) more effective and timely mitigation under operational constraints, measured by decision latency, time-to-mitigation, cascade containment, service continuity, and safety/service constraint violations (Phase 2); and (iii) improved resilience through systematic learning, measured by evidence-driven model/policy updates that improve containment and continuity over successive incidents without increasing constraint violations (Phase 3).

Although the framework is not proposed as a compliance method for any single regulation, governance and regulatory considerations remain relevant as contextual constraints, particularly for traceability, documentation, operational accountability, and post-incident learning. In this sense, the framework can support security-by-design and resilience-by-design expectations that are increasingly emphasised in regulatory contexts such as the Cyber Resilience Act (CRA), without being limited to CRA compliance alone.

3 Proposed Framework and Preliminary Results

This section presents the preliminary outputs of the research, including the proposed three-phase Cyber Resilience by Design framework and the phase-specific results obtained so far.

3.1 Three-Phase Cyber Resilience by Design Framework

Figure 1 summarises the proposed three-phase Cyber Resilience by Design framework for critical-infrastructure information systems. At the current stage, the Phase 1 pre-incident artefacts have been developed and evaluated, drawing on the published RIPA framework [1] and extending work currently under review [2,3]. Phase 2 has been developed at the conceptual and methodological level and is under validation, while Phase 3 remains planned as the post-incident learning and adaptation layer.

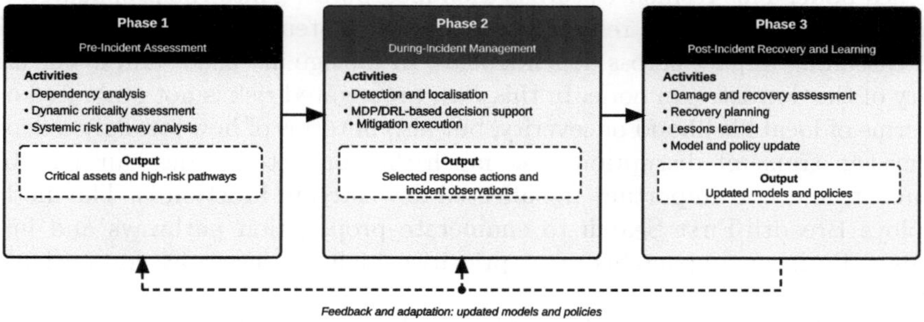

Fig. 1. Three-phase Cyber Resilience by Design framework linking pre-incident systemic risk assessment, during-incident decision support, and post-incident recovery and learning for critical-infrastructure information systems, with feedback and adaptation through updated models and policies.

The framework comprises three phases:

Phase 1 – Pre-Incident: Cyber Resilience Analysis (Completed)

This phase focuses on preparedness, robustness, and redundancy through two design-time assessment artefacts:

Dynamic Multi-Criteria Criticality Assessment [3]*:* A three-stage framework addressing RQ2. First, the Dynamic Organisational Impact Assessment evaluates context-dependent importance through a weighted combination of Operational Imperative, Contextual Security Relevance, and Resilience Capacity Modifier. Second, Adaptive Systemic Interconnectivity Analysis captures network role using Influence Centrality, Susceptibility Centrality, and Cross-Domain Betweenness Centrality (CDBC) to identify cyber–physical interface nodes. Third, Dynamic Criticality Synthesis combines organisational and systemic dimensions:

$$\mathrm{DCS}_a = (\mathrm{DOIS}_a \times W_{\mathrm{org}}) + (\mathrm{SIS}_a \times W_{\mathrm{sys}}) \quad (1)$$

where the weights adapt to context: normal operations prioritise service continuity, whereas active threats prioritise cascade prevention.

Risk Impact Pathway Analysis (RIPA) [1]*:* A propagation-aware risk assessment method addressing RQ2. Unlike traditional approaches focused only on initial risk (R_0), RIPA quantifies propagated risk through dependencies:

$$R_{\mathrm{pro}}(i \to j) = R(i) \cdot (1 - \mathrm{RE}(i)) \cdot (1 - \mathrm{RE}(j)) \cdot \mathrm{LI}(i, j) \cdot C(j) \quad (2)$$

where $R(i)$ denotes the initial risk of node i, estimated from vulnerability severity, exploitation likelihood, threat characteristics, and operational impact using

CVSS, EPSS, CAPEC-derived metrics, and asset-specific criticality information. RE denotes resilience; In the proposed framework, resilience is operationalised as a quantitative multi-dimensional capability comprising preparedness, absorption, responsiveness, recoverability, and adaptability. Each dimension is assessed through a set of measurable indicators and combined to produce a resilience score for each node. This resilience score is then incorporated into the propagated-risk calculation, where higher resilience reduces the potential for risk transmission and cascading impact across dependencies. LI linkage intensity; and C the criticality of the downstream node. In this way, propagated risk is not evaluated only in terms of local likelihood or severity, but also in terms of how strongly a dependency can transmit disruption, how much the connected nodes can absorb or reduce it, and how important the affected downstream function is. The method employs Breadth-First Search to enumerate propagation pathways and introduces a Pathway Impact Score to prioritise high-risk cascades for mitigation planning.

Phase 2 – During-Incident: Cyber Resilience Management (Developed; Validation Ongoing)

This phase focuses on absorption, resourcefulness, and adaptability through real-time decision support for RQ3's time-critical and scalable requirements. Because remediation must occur during rapid cascade propagation, the proposed Remedial Action Scheme (RAS) comprises:

Step 1 – Rapid Detection & Localisation: Infer the likely failure location from available telemetry and focus the response on the affected subgraph.

Step 2 – MDP-Based Decision Making: Cascading-failure mitigation is formulated as a Markov Decision Process $\{S, A, T, R\}$, where S denotes system states, A remedial actions, T denotes cascade-transition dynamics, and R denotes mitigation effectiveness. Deep Reinforcement Learning is used to identify actions that maximise service yield while containing propagation under nonlinear stochastic dynamics. Operational constraints are enforced through action masking, with a rule-based fallback when confidence or time budget is insufficient. The Markov assumption is applied at the state-transition level: given the current incident state and selected action, the model estimates the next state of the affected system, enabling sequential action evaluation over time. The incident-response process is modelled as a sequence of states and actions, where future states depend on the current system state and selected mitigation action.

Step 3 – Timely Execution: Execute control actions through field devices under strict time constraints.

Phase 2 uses Phase 1 outputs, including criticality rankings and high-PIS pathways, to prioritise feasible remedial actions under time constraints. Importantly, Phase 2 is not limited to identifying risky nodes or pathways; it addresses the separate problem of selecting feasible treatment actions. In this phase, reducing systemic risk may conflict with service continuity or system functionality, for example when isolation or throttling actions reduce operational performance. For this reason, mitigation is treated as a constrained decision problem rather than as pure risk minimisation, and service/safety constraints are explicitly included in the action-selection process.

Phase 3 – Post-Incident: Cyber Resilience Improvement (Planned) This phase focuses on recovery and adaptation through damage assessment, recovery planning, and continuous improvement. Lessons learned are used to update pre-incident assessment models and during-incident decision policies, addressing RQ1 and RQ2.

3.2 Pre-Incident Phase Validation

The Phase 1 artefacts are evaluated on an industrial cyber–physical infrastructure case comprising 21 nodes across three subsystems and 33 dependencies, derived from industrial system architecture and operational knowledge (data exchange, process sequencing, control signals, and computational-resource couplings). Experiments replay representative disruption scenarios to assess hotspot/ pathway rankings and the impact of budget-constrained interventions.

Finding 1 – Dynamic Criticality Adaptation: Under normal operations, the Scheduler ranked highest (DCS = 82.53). During a simulated active cyber threat, the Authentication Service priority increased by 34%, demonstrating the effect of adaptive weighting.

Finding 2 – Cyber-Physical Interface Detection: The CDBC metric identified three critical convergence points—LAMFlow controller (CDBC = 0.73), API Service gateway (CDBC = 0.68), and Authentication service (CDBC = 0.71)—that were not highlighted by traditional degree centrality.

Finding 3 – Systemic Hot Spots: Comparing R_0 with total systemic risk revealed substantial amplification, with one node increasing from 55.68% to 65.30% (17.3% relative increase). PIS analysis further identified one node appearing in 95% of top pathways. Pathway-targeted interventions achieved greater total systemic risk reduction under equal resource constraints. Sensitivity analysis showed that resilience produced approximately 2× greater risk reduction than equivalent changes in criticality or linkage intensity.

4 Future Research Directions, Open Science, and Conclusion

Future Research Directions. The next steps focus on validating Phase 2, developing Phase 3, and evaluating the integrated three-phase framework end to end. For Phase 2, the next step is the experimental validation of the Remedial Action Scheme (RAS) in a multi-domain testbed and simulation environment, with Phase 1 outputs used to prioritise mitigation actions during cascade propagation. The evaluation will focus on decision latency, systemic risk reduction, service continuity, and robustness under stochastic dynamics. For Phase 3, future work will develop post-incident learning mechanisms that use incident evidence to update dependency models, propagation parameters, criticality estimates, and response policies, and reintegrate these updates into pre-incident assessment artefacts. Finally, the full framework will be benchmarked against representative baselines, including static criticality assessment, propagation-unaware risk assessment, and heuristic incident-response strategies, using metrics such as time-to-decision, time-to-mitigation, cascade containment, service continuity, and safety-constraint violations.

A current limitation of the research is that empirical evidence is strongest for Phase 1 and is based on a single industrial case model. At this stage, generalisation should therefore be interpreted as analytical and design-oriented rather than statistical. The problem addressed by the work is intentionally broad, but the framework is developed in phased contributions so that each stage can be justified and evaluated separately. Future work will strengthen external validity through comparative scenarios, additional case contexts, and simulation/testbed-based evaluation of Phase 2 and Phase 3. **Open Science.** The research adheres to the principles of Open Science throughout its research life cycle. The research articles will be made openly accessible through open access publications, preprints, and the institutional repository of the University of Essex. To ensure reproducibility of the research results, the Python code for the proposed approaches and experiments, along with the experiment code, parameters, synthetic configurations of the dependency graph, and replication materials, will be made publicly available in a public repository such as Zenodo or GitHub upon acceptance of the research articles. The research on the validation of Phase 1 utilises real-world data provided by LAMAD. The data is subject to confidentiality and data governance agreements and cannot be made publicly available. However, if possible, anonymised and aggregated data products may be made available. Otherwise, the research aims to make the research methodology and design sufficiently clear to allow replication of the research. In addition, the research review protocol and materials prepared according to the PRISMA guidelines will be made available to facilitate independent replication and extension of the review.

Conclusion. This work develops a three-phase Cyber Resilience-by-Design framework for critical-infrastructure information systems that treats cyber resilience as a systemic lifecycle capability rather than a component-level property. At the current stage of the research, the pre-incident phase has been developed and evaluated, the during-incident phase has been developed and awaits empirical validation, and the post-incident phase remains planned as the final feedback-driven adaptation layer.

Doctoral Consortium History. This is the first time that this work has been submitted to a doctoral consortium, and it has not been presented in any former doctoral consortium at another conference.

Acknowledgments. The author thanks Prof. Haralambos Mouratidis (University of Essex) for supervision and guidance.

Disclosure of Interests. The author has no competing interests to declare that are relevant to the content of this article.

References

1. Faraji, F., Mouratidis, H., Al Zakwani, A.: Risk Impact Pathway Analysis (RIPA): a dynamic risk assessment framework for interdependent critical infrastructures. Int. J. Inf. Secur. **25**(1), 26 (2025). https://doi.org/10.1007/s10207-024-00926-x
2. Faraji, F., Mouratidis, H., Santos-Olmo, A., Islam, S., Grosado, D., Fernández-Medina, E.: Towards resilience by design: systematic review of critical infrastructure protection against systemic risks (2025). Manuscript under review
3. Faraji, F., Javadpour, A., Sangaiah, A.K., Zavieh, H.: A solution for resource allocation through complex systems in fog computing for the internet of things. Computing **106**(7), 2107–2131 (2024). https://doi.org/10.1007/s00607-023-01199-1
4. Kruse, P.M., Schmitt, H.C., Greiving, S.: Systemic criticality of infrastructure. Clim. Change **165**(1–2), 2 (2021)
5. Seppänen, H., Luokkala, P., Zhang, Z., Torkki, P., Virrantaus, K.: Critical infrastructure vulnerability. Int. J. Crit. Infrastruct. Prot. **22**, 25–38 (2018)
6. Ani, U.D., Watson, J.D.M., Nurse, J.R., Cook, A., Maples, C.: A review of critical infrastructure protection approaches: improving security through responsiveness to the dynamic modelling landscape. In: Living in the Internet of Things (IoT 2019), p. 6 (2019)
7. Abdelgawad, A., Saleh, M., Gonzalez, J.: Analysing cascading effects in interdependent critical infrastructures. In: Murayama, Y., Velev, D., Zlateva, P. (eds.) Information Technology in Disaster Risk Reduction: Third IFIP TC 5 DCITDRR International Conference, ITDRR 2018, Held at the 24th IFIP World Computer Congress, WCC 2018, Poznan, Poland, September 20–21, 2018, Revised Selected Papers 3, pp. 50–65. Springer (2020). https://doi.org/10.1007/978-3-030-32169-7_5
8. Tabansky, L., Lichterman, E.: Critical infrastructure interdependencies. Int. J. Inf. Secur. **24**(1), 18 (2025)
9. Welburn, J.W., Strong, A.M.: Systemic cyber risk and aggregate impacts. Risk Anal. **42**(8), 1606–1622 (2022)
10. Yan, P., Zhang, F., Zhang, F., Geng, L.: Cascading disasters in urban infrastructure. Buildings **15**(7), 1011 (2025)
11. Renn, O., et al.: Systemic risks. Risk Anal. **42**(9), 1902 (2022)
12. Case, D.U.: Analysis of the cyber attack on the Ukrainian power grid. E-ISAC **388**(1–29), 3 (2016)
13. Geng, Y., et al.: LogicLocker ransomware attacks on industrial control systems. IEEE Internet Things J. **10**(10), 8331 (2022)
14. Rehak, D., et al.: Cascading impact assessment. Int. J. Crit. Infrastruct. Prot. **22**, 125–138 (2018)
15. Palleti, V.R., Adepu, S., Mishra, V.K., Mathur, A.: Cascading effects of cyber-attacks. Cybersecurity **4**, 8 (2021)
16. Wang, W., et al.: An approach for cascading effects. Phys. A **510**, 164–177 (2018)
17. Hempel, L., Kraff, B.D., Pelzer, R.: Dynamic robustness in the German electric power system. Int. J. Disast. Risk Reduc. **30**, 257–268 (2018)
18. Mottahedi, A., Sereshki, F., Ataei, M., Nouri Qarahasanlou, A., Barabadi, A.: Resilience by design. Energies **14**(6), 1571 (2021)
19. McDaniels, T., Chang, S., Cole, D., Mikawoz, J., Longstaff, H.: Critical infrastructure risk assessment. Glob. Environ. Chang. **18**(2), 310 (2008)

20. Hochrainer-Stigler, S., Colon, C., Boza, G., Poledna, S., Rovenskaya, E., Dieckmann, U.: Risk assessment and resilience. Int. J. Disast. Risk Reduc. **51**, 101868 (2020)
21. Adamos, K., Stergiopoulos, G., Karamousadakis, M., Gritzalis, D.: Risk management frameworks. Int. J. Inf. Secur. **23**(1), 187 (2024)
22. Roshanaei, M.: NIST cybersecurity framework. J. Comput. Commun. **9**(8), 80 (2021)
23. Magnusson, L., Iqbal, S., Elm, P., Dalipi, F.: Cybersecurity frameworks for critical infrastructure. Int. J. Inf. Secur. **24**(4), 177 (2025)
24. Rathnayaka, B., Siriwardana, C., Robert, D., Amaratunga, D., Setunge, S.: Resilience of critical infrastructure systems. Int. J. Disast. Risk Reduc. **78**, 103123 (2022)
25. Page, M.J., et al.: The PRISMA 2020 statement: an updated guideline for reporting systematic reviews. BMJ **372**, n71 (2021)
26. Cybersecurity: Critical infrastructure: framework for improving critical infrastructure cybersecurity (2018)
27. Papadopoulos, L., et al.: PRAETORIAN: a framework for the protection of critical infrastructures from advanced combined cyber and physical threats. In: Proceedings of the 18th International Conference on Availability, Reliability and Security (ARES), pp. 1–6 (2023)
28. Stelkens-Kobsch, T.H., Boumann, H., Piekert, F., Schaper, M., Carstengerdes, N.: A concept-based validation approach to validate security systems for protection of interconnected critical infrastructures. In: Proceedings of the 18th International Conference on Availability, Reliability and Security (ARES), Association for Computing Machinery, Article. 110 (2023)
29. Almoghathawi, Y., González, A.D., Barker, K.: Exploring recovery strategies for optimal interdependent infrastructure network resilience. Netw. Spat. Econ. **21**, 229–260 (2021)
30. Almoghathawi, Y., Barker, K., Albert, L.A.: Resilience-driven restoration model for interdependent infrastructure networks. Reliab. Eng. Syst. Saf. **185**, 12–23 (2019)
31. Rios, E., et al.: The DYNABIC approach to resilience of critical infrastructures. In: Proceedings of the 18th International Conference on Availability, Reliability and Security (ARES), Article 136. Association for Computing Machinery (2023)
32. Kong, J., Zhang, C., Simonovic, S.P.: A two-stage restoration resource allocation model for enhancing the resilience of interdependent infrastructure systems. Sustainability **11**(19), 5143 (2019)
33. He, X., Cha, E.J.: Post-disaster recovery planning of interdependent infrastructure systems: a game theory-based approach. In: 13th International Conference on Applications of Statistics and Probability in Civil Engineering (ICASP13), Seoul, South Korea, May 26–30 (2019)
34. Mouratidis, H., Giorgini, P.: Secure Tropos: a security-oriented extension of the Tropos methodology. Int. J. Software Eng. Knowl. Eng. **17**(2), 285–309 (2007)
35. Faraji, F., Javadpour, A., Sangaiah, A.K., Zavieh, H.: A solution for resource allocation through complex systems in fog computing for the internet of things. Computing **106**(7), 2107 (2024)

DIAPROD: A Framework for Formal Consent Modelling, Dynamic Storage Limitation, Data Minimisation, and LLM-Assisted Compliance for Data Protection by Design

Samuel Opoku Daniels(✉)

Institute for Analytics and Data Science (IADS), School of Computer Science and Electronic Engineering, University of Essex, Colchester, UK
s.daniels@essex.ac.uk

Abstract. Modern information systems process personal data across complex ecosystems where consent states change, purposes complete, and regulations evolve—all after deployment. Yet existing privacy engineering approaches remain static, with no formal mechanisms for dynamic runtime enforcement. An analysis of 29 named approaches from 73 primary studies confirms this gap. This research presents DIAPROD (DynamIc dAta PROtection by Design), a framework reconceptualising Data Protection by Design as a runtime capability through: (i) ConsentML, a formal modelling language with denotational and operational semantics for lifecycle-aware consent; (ii) a Certifying Kernel with an extensible constraint registry for consent, storage limitation, and data minimisation enforcement; and (iii) an LLM-assisted architecture where the LLM proposes and the kernel decides. ConsentML tool is implemented as a Secure Tropos add-on, verified via the MMQEF framework, and validated through a United Kingdom open banking case study. The remaining work extends the kernel to address storage limitations and data minimisation, with evaluation across healthcare and public sector domains.

Keywords: Data Protection by Design · GDPR · Consent Modelling · Storage Limitation · Data Minimisation · Domain-Specific Language · Formal Methods · LLM-Assisted Compliance

1 Introduction

The General Data Protection Regulation (GDPR) mandates that data controllers implement data protection by design and by default (DPbD)—Article 25, necessitating the integration of privacy safeguards into the system from the outset [1,22]. Article 25, does not limit this obligation to any single principle [3]; it encompasses the full spectrum of data protection requirements, including

A. Bernasconi et al. (Eds.): RCIS 2026, LNBIP 597, pp. 235–246, 2027.
https://doi.org/10.1007/978-3-032-35043-5_20

lawful consent management (Art. 6–7), purpose limitation (Art. 5(1)(b)), data minimisation (Art. 5(1)(c)), storage limitation (Art. 5(1)(e)), and cross-border transfer safeguards (Art. 44–49). In practice, however, existing DPbD approaches treat these principles as static design-time properties rather than dynamic run-time constraints that must adapt as consent states change, processing purposes are completed, retention periods expire, and regulations are revised.

A related work analysis of 29 named approaches drawn from 73 primary studies, assessed against four dynamicity dimensions (D1: Formal Modelling Language, D2: Consent Lifecycle Management, D3: Regulatory Evolution, D4: Runtime Adaptation & Enforcement), identified six converging gaps. **Gap 1:** No formal consent-modelling language exists—of 29 methodologies, none achieves full D1 support; the 7 with partial support repurpose existing formalisms (state machines, policy languages, ontologies, theorem provers) that lack consent-specific semantics. **Gap 2:** Few existing approaches comprehensively represent the dynamic consent lifecycle—none achieves full D2 support; the 15 with partial support address fragments (e.g., consent states, platform-level changes) without integrated state transitions, temporal enforcement, and downstream effect propagation. **Gap 3:** No automated regulatory evolution mechanism exists—none achieves full D3 support; the 8 with partial support require manual intervention with no systematic change detection or propagation. **Gap 4:** Storage limitation and data minimisation lack formal enforcement mechanisms linked to consent state—only 2 of 29 achieve full D4 support, and both enforce general privacy policies rather than formally defined consent states. **Gap 5:** From our review, no approach addresses all four dimensions even partially, forcing practitioners to combine incompatible approaches with no architectural coherence. **Gap 6:** Based on our analysis of the related work, no approach establishes formal safety boundaries for LLM assistance in privacy-critical contexts—the single LLM-integrating approach in the corpus employs the LLM as an unconstrained assistant without formal verification boundaries.

1.1 Research Questions

Research Questions. Four research questions drive the doctoral research. Each is derived from the gaps identified in the related work analysis and justified by the specific limitation it addresses:

- **RQ1:** *How can a formal modelling language be designed to represent dynamic consent with sufficient semantic rigour for machine-verifiable compliance?* (Gaps 1–2) This question is necessary because the absence of a formal consent language (Gap 1) is the most foundational gap identified in the analysis—it structurally prevents lifecycle management (Gap 2), automated validation, and computational reasoning over consent. Without a purpose-built language with denotational and operational semantics, consent remains an informal concept that cannot be systematically verified.
- **RQ2:** *How can LLM-enhanced intelligence be integrated into a formal consent management architecture while preserving deterministic compliance guarantees?* (Gaps 1, 6) This question is motivated by two observations: formal

methods, while rigorous, present an accessibility barrier for privacy officers and legal practitioners who lack formal training, and the single LLM-integrating approach in the corpus (Gap 6) employs the LLM without formal verification boundaries, risking non-deterministic compliance decisions. The challenge is architectural—how to harness LLM capability for interpretation an explanation while confining all authoritative decisions to deterministic, formally verified components.

RQ3: *How can a dynamic data protection framework enforce data minimisation and storage limitation at runtime through formally grounded, consent-derived mechanisms?* (Gaps 2, 4) This question addresses the design-time/runtime disconnect: only 2 of 29 methodologies achieve full runtime enforcement (Gap 4), and both enforce general policies rather than formally defined consent states. Data minimisation and storage limitation are inherently dynamic—the permissible data scope and retention period change as consent is withdrawn, purposes complete, and processing activities evolve—yet to the best of our knowledge, no existing approach formalises this dependency.

RQ4: *How can such a framework adapt to evolving privacy regulations without architectural redesign?* (Gaps 3, 5) This question is driven by the finding that none of the approaches we analysed provides automated regulatory change detection or propagation (Gap 3), and none of the analysed approaches integrates all four dynamicity dimensions (Gap 5). Given that privacy regulations are actively evolving across jurisdictions—the EU AI Act, national GDPR implementations, and sector-specific regulations—a framework that requires redesign in response to regulatory changes cannot sustain compliance in practice.

1.2 Research Objective

Existing privacy engineering approaches treat Data Protection by Design (DPbD) as a static, design-time analysis activity: privacy properties are specified once during requirements engineering and encoded into fixed system configurations. This static treatment is fundamentally misaligned with the reality of personal data processing, where consent states change, processing purposes complete or narrow, retention periods expire, regulations are amended, and new processing activities are introduced—all after deployment. We use the term *dynamic* to denote four specific capabilities that a DPbD framework must exhibit: (D1) formal representation of consent that can be computationally constructed, validated, and reasoned about; (D2) lifecycle management where consent transitions (grant, withdraw, modify, limit, object, override, expire) are formally specified with cascading effects on dependent processing; (D3) regulatory evolution where changes in legal requirements propagate into enforcement rules without architectural redesign; and (D4) runtime enforcement where compliance is continuously monitored and enforced against the formally defined consent state, not merely asserted at design time. The overarching objective of this research is therefore to design, formalise, implement, and evaluate a dynamic data protection framework that closes the gap between the regulatory mandate for continuous, adaptive privacy enforcement and the static, design-time nature of existing approaches—reconceptualising DPbD as a runtime enforcement capability that addresses all four dynamicity dimensions within a single coherent architecture.

1.3 Expected Contributions

This PhD research addresses these gaps through a layered framework spanning four dynamicity dimensions through three major contributions:

C1: ConsentML (Gaps 1–2)—a formal modelling language with denotational semantics defining a domain $\mathcal{C}$ of consent tuples and operational semantics for seven lifecycle transitions with temporal validity, expiry propagation, and purpose-specific granularity (D1, D2). *Novelty:* ConsentML, to our knowledge, is among the first modelling languages to formalise consent as a lifecycle entity with its own denotational semantics and certifying mechanism. Existing approaches either repurpose general-purpose formalisms (Event-B, type systems, theorem provers) that lack consent-specific semantic domains, or provide ontological descriptions (GConsent, DPV) without formal specification or validation capability. ConsentML introduces domain-specific abstractions, typed relationships encoding GDPR accountability structures, and well-formedness predicates integrated into the language definition rather than expressed as external constraints.

C2: A Certifying Kernel (Gaps 4–5)—a deterministic validation engine operating over ConsentML's formal consent state, with an extensible constraint registry supporting storage-limitation and data-minimisation rules (D4). *Novelty:* The two existing approaches achieving full runtime enforcement operate over general privacy policies; the kernel is among the first to enforce compliance directly over formally defined, dynamically evolving consent states. Its extensible registry architecture enables integration across all four dynamicity dimensions within a single validation interface—addressing the cross-dimensional integration gap (Gap 5) that forces practitioners to combine incompatible methodologies.

C3: An LLM-assisted architecture (Gaps 3, 6)—a YAML-based regulatory rule engine with a retrieval-augmented generation (RAG) pipeline enabling configuration-level regulatory adaptation without redeployment (D3). The LLM *proposes*; the kernel *decides* (Gap 6). *Novelty:* This is among the first architectures to establish formal methods as a safety boundary for LLM assistance in privacy-critical contexts. The single LLM-integrating approach in the literature employs the LLM as an unconstrained assistant; DIAPROD confines all authoritative compliance decisions to the deterministic kernel while leveraging LLM capability for interpretation, explanation, and regulatory corpus management. Table 1 clarifies the current status of each contribution.

The remainder of this paper presents the state of the art (Sect. 2), methodology (Sect. 3), solution and results (Sect. 4), remaining work and open science conclusion (Sect. 5).

2 State of the Art

A systematic literature review (SLR) was conducted in accordance with the PRISMA 2020 guidelines [18] of 73 primary studies (2014–2026). The SLR analysed 29 named approaches from the 73 primary studies against four dynamicity dimensions. No methodology achieved full support on D1, D2, or D3; only two achieved full support on D4; and no methodology addressed all four dimensions, even partially.

Table 1. Contribution maturity levels.

Contribution	Status	Evidence
ConsentML metamodel & semantics	Completed	Completed & Accepted at RCIS 2026
Lifecycle transitions & predicates	Completed	Formally verified
SecTro ConsentView tooling	Completed	Case study validated
Data minimisation extension	In progress	Metamodel designed
Storage limitation extension	In progress	Requirements defined
Certifying Kernel	Partially designed	Architecture specified
LLM-assisted architecture	Conceptual	Design principle defined
Regulatory evolution	Future work	Scoped, not yet started

Design-Time Approaches. Privacy requirements methods (Secure Tropos [16], PriS [13], PRIPARE [17], Gharib et al. [9], Islam et al. [12]), threat modelling (LINDDUN [7,28], PILLAR [15]), de Carvalho et al. [4] and privacy patterns (Hoepman [5]) provide systematic frameworks for early-stage analysis, threat identification, and architectural guidance. However, all treat privacy as a design-time concern—none offer formal consent constructs, lifecycle support, or runtime enforcement. PILLAR is the only LLM-integrating approach in the corpus, but it employs it without formal verification boundaries.

Model-driven and Ontological Approaches. TILT/TIRA [11] achieves partial coverage across all four dimensions—-one of only two methodologies to do so—but without formal semantic grounding for consent dynamics. GConsent [20] offers GDPR-aligned semantic representations (partial D1–D2) but cannot verify well-formedness predicates. COPri [8] and DEFeND [23] formalise privacy concepts through ontological and platform architectures. The common limitation is that consent can be documented but not formally validated or reasoned about.

Dynamic Consent and Runtime Enforcement. Peyrone-Wichadakul [21] provide the strongest partial D2, defining consent as tuples with state machine semantics, but within existing frameworks (Event-B) rather than a purpose-built language. Purpose-Based Access Control [19] and the Runtime Enforcement Survey [25] are the only two achieving full D4 but enforce general policies rather than formally defined consent states. DPL [2]—the second methodology with partial coverage across all four dimensions—and Tokas & Owe [26], Mechanised PbD [2], and Isabelle/HOL [14] repurpose general-purpose formalisms that cannot model consent as a first-class lifecycle entity.

The structural dependency is the central finding: the absence of a formal consent language (D1) renders complete lifecycle management (D2) structurally impossible, thereby preventing consent-specific runtime enforcement (D4).

3 Research Methodology

The research follows the Design Science Research (DSR) methodology as defined by Johannesson and Perjons, structured into five phases.

Problem Explication. The problem was identified through a related work analysis of existing privacy engineering approaches (Sect. 2), revealing three compounding limitations: absence of formal consent constructs, lack of dynamic enforcement for storage limitation and data minimisation, and no unified architecture spanning design time and runtime.

Design and Development. The framework is being developed iteratively. The first iteration produced ConsentML with its metamodel, formal semantics, and Certifying Kernel. The second iteration (in progress) extends the kernel with storage-limitation and data-minimisation constraints. The LLM-assisted architecture spans both iterations. Figure 1 illustrates the four-layer DIAPROD architecture. Layer 1 (Ingestion) accepts natural-language privacy policies through the LLM-assisted interface, producing candidate consent tuples. Layer 2 (Formal Modelling) provides the ConsentML metamodel-based environment where models are created or refined. Layer 3 (Certification) validates all models against GDPR-grounded well-formedness predicates through the deterministic Certifying Kernel—the sole authority boundary. Layer 4 (Enforcement & Evolution) drives runtime enforcement from certified consent states: withdrawal cascades to processing cessation, purpose narrowing contracts, data scope, and regulatory changes propagate through the YAML/RAG pipeline. The architectural invariant is that no compliance decision bypasses Layer 3.

Demonstration. ConsentML was demonstrated through a case study in the UK open banking domain (the ZipFin FinTech scenario), modelling two consent instances—one granted, one withdrawn—across six actors, which includes competency questions. Planned case studies in healthcare, financial services, and the public sector will demonstrate the storage-limitation and data-minimisation extensions.

Why DSR. The primary contribution is a design artefact—a modelling language, validation kernel, and tool—whose value is assessed through utility rather than empirical observation, making DSR more appropriate than action research, grounded theory, or case study research.

Response to Doctoral Consortium Panel Comments

Critical and Compelling Evaluation Criteria. The evaluation employs three complementary frameworks, each addressing a distinct concern. The *VVE taxonomy* [24], distinguishes verification (structural correctness via MMQEF across six quality viewpoints), validation (outcome correctness via competency questions answered from the model), and evaluation (practical utility via domain case studies). The *FEDS framework* [27], positions the ZipFin FinTech case study as a formative, ex ante evaluation (completed), and the three planned domain case studies (healthcare, financial services, and public sector) as summative, ex post, naturalistic evaluations. The most critical criterion is *regulatory fidelity*: every well-formedness predicate traces to a specific GDPR article, and the Certifying Kernel deterministically validates models against these predicates. The most compelling criterion is *dynamic expressiveness*: whether The framework can detect and enforce consent-derived scope contraction, purpose-linked erasure, and legal hold suspension across domains with conflicting regulatory requirements.

LLM Evaluation in a Privacy-Critical Context. LLM integration is justified by the accessibility gap: formal methods require specialist expertise that privacy officers and legal practitioners typically lack. However, the architectural principle is that *no compliance decision is made by the LLM.* The LLM proposes candidate consent models from natural-language privacy policies; the Certifying Kernel deterministically accepts or rejects them. Evaluation of LLM usage will measure: (i) *proposal accuracy*—proportion of LLM-generated candidates accepted by the kernel on first attempt; (ii) *bounded refinement convergence*—number of feedback-adapt cycles before kernel acceptance (capped at three); and (iii) *explanation fidelity*—whether RAG-generated natural-language justifications correctly cite the GDPR articles referenced in kernel validation reports, assessed through expert review. The LLM is not required for the framework to function—all operations can be performed manually through the SecTro tool—but it lowers the barrier to adoption. If LLM outputs are unreliable in a given deployment, the kernel's deterministic validation remains the sole authority.

Runtime Consent Decisions and Decision Changes. This is the core contribution. ConsentML's operational semantics define seven lifecycle transitions (grant, withdraw, modify, limit, object, override, expire) with formally specified preconditions, postconditions, and cascading effects. When a data subject withdraws consent at runtime, the ConsentWithdrawalContraction rule (data minimisation extension) contracts the permissible data scope to the empty set for all dependent processing activities, and the WithdrawalErasure rule (storage limitation extension) reassesses retention obligations. When a purpose narrows, the PurposeNarrowingContraction rule enforces monotonicity: the data scope can only reduce, never expand. When a retention period expires, the RetentionExpiry rule transitions the retention state and identifies erasure candidates. Legal holds suspend erasure without destroying the underlying policy, and release triggers reassessment. All transitions produce audit records citing the applicable GDPR article (Art. 5(2)). The Certifying Kernel validates the model state after every transition, ensuring that data owner decisions are immediately and deterministically enforced.

Cross-Domain Effectiveness Assessment. Effectiveness will be assessed through three domain-specific case studies, each testing different regulatory tensions: healthcare (special category data under Art. 9, NHS 30-year retention, multi-controller accountability), financial services (AML mandatory retention vs. right to erasure under Art. 17(3)(b), legal hold mechanisms), and public sector (non-consent legal bases under Art. 6(1)(e)(f), regulatory evolution). Each case study will evaluate: metamodel expressiveness (can all domain constructs be represented?), predicate coverage (do C-MIN and C-SL predicates detect domain-specific violations?), and regulatory fidelity (do enforcement outcomes match legal requirements?). Cross-case comparative analysis will assess whether the framework handles domain variation without architectural modification, confirming the regulation-agnostic design claim.

Consent Management and Consent Evolution. Consent management is operationalised through ConsentML's formal lifecycle. "Consent evolution" encompasses two dimensions: *consent-initiated change* (data subject withdraws, modifies, or limits consent) and *context-initiated change* (purpose narrows, completes, or regulatory interpretation shifts). Both are handled through the operational semantics: consent-initiated changes trigger lifecycle transitions directly; context-initiated changes trigger scope contraction events that cascade to dependent processing activities and retention states. Regulatory evolution (RQ4) is addressed through the YAML-based regulatory rule engine and RAG pipeline: when regulations change, updated rules are loaded as configuration without code changes, and the RAG corpus is refreshed with new regulatory text. The system is *not* designed for full autonomy—it enforces constraints deterministically but requires human decisions for consent granting, purpose definition, and regulatory interpretation. LLM-assisted interpretation *proposes*; the kernel *decides*; the human *authorises.*

Evaluation Scope Extension. The panel's suggestion to extend the evaluation to usability and acceptance is noted and will be incorporated. The early publication of the ConsentML tool is planned through a GitHub release upon thesis completion, with a Zenodo deposit of the tool source code and configuration files. A machine-readable case study instance is already publicly available at [6].

4 Proposed Solution and Results Achieved

4.1 Results Achieved

RQ1 (Formal Consent Modelling Language—Answered). The answer is ConsentML: a 24-class metamodel with a 14-tuple consent definition, denotational and operational semantics, seven lifecycle transitions, and GDPR-grounded well-formedness predicates. The Certifying Kernel implements deterministic validation against the full predicate catalogue. Tool support is provided as an add-on to the Secure Tropos Sectro tool with an integrated constraint validator. ConsentML has been verified via the MMQEF framework [10] (structural completeness across all six quality viewpoints) and validated through an expert-driven case study in the UK open banking domain (the ZipFin FinTech scenario), where six competency questions (CQ1–CQ6) were answered directly from the model. Verification and validation apply to the *consent modelling language only*; the remaining framework components have not yet been evaluated.

RQ2 (LLM Integration—Partially Answered). The architectural design is complete: the "LLM proposes, kernel decides" separation principle, the YAML-based regulatory rule engine, and the RAG pipeline for dynamic corpus refresh have been designed. Full end-to-end integration with the Certifying Kernel and empirical evaluation of explanation quality remain in progress.

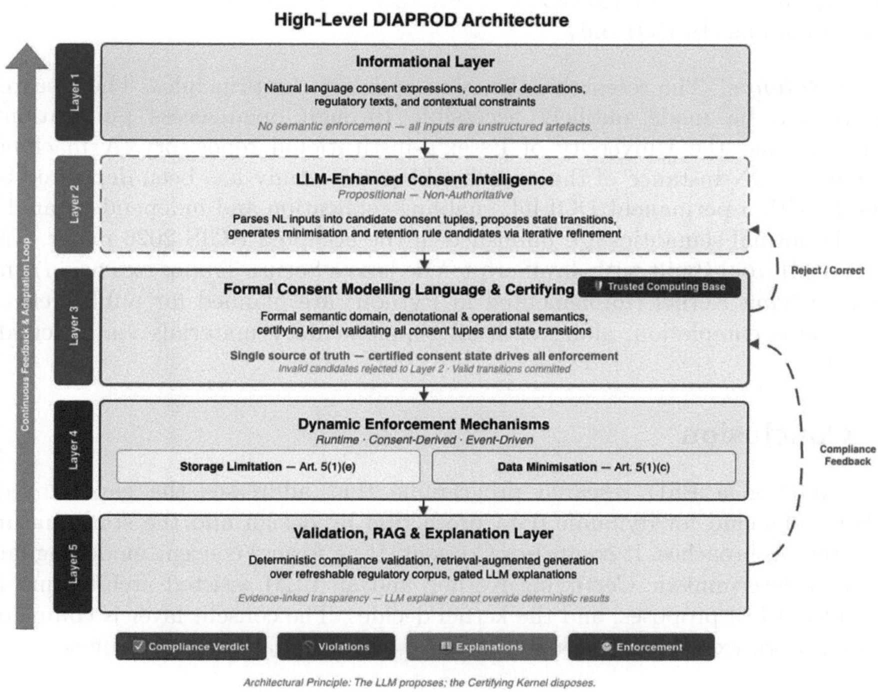

Fig. 1. High-Level DIAPROD Architecture.

4.2 Planned Work

RQ3 (Data Minimisation and Storage Limitation—Not Yet Answered). Dynamic storage limitation will be formalised as a kernel extension (RetentionPolicy and ErasureTrigger constructs with purpose-completion detection). Dynamic data minimisation will be formalised through purpose–data linkage models specifying minimum, optional, and contextual data sets with adaptive scope contraction. Neither has been designed or implemented.

RQ4 (Regulatory Evolution—Partially Answered). The YAML-based rule engine and RAG pipeline are architecturally designed as part of RQ2. Automated regulatory change detection and cross-dimensional propagation across the formal model, enforcement engines, and validation layer remain planned.

Summative Evaluation (RQ1–RQ4). Three domain-specific case studies are planned (healthcare, financial services, public sector) to evaluate the *complete framework* following the FEDS strategy [27]. MMQEF verification [10] will be rerun once the storage limitation and data minimisation extensions are designed. The current assessment (MMQEF + FinTech case study) is formative and ex ante and applies to RQ1 only.

Open Science. The research adheres to open science principles. The research papers will be made publicly accessible through open-access publications, preprints, and the University of Essex's institutional repository. A machine-readable JSON instance of the open banking case study has been deposited on Zenodo with a permanent DOI [6], enabling replication and independent analysis. The formal semantics are published in the accepted RCIS 2026 paper. The ConsentML tool (built with JavaScript/Vue.js as a Secure Tropos extension) and the Certifying Kernel (implemented in Python) are planned for public release upon thesis completion, along with all supplementary materials via a Zenodo deposit.

5 Conclusion

DIAPROD is a PhD research programme that addresses the gap between GDPR's demand for dynamic data protection by design and the static nature of current approaches. It comprises ConsentML, a formal consent modelling language; a deterministic Certifying Kernel; and an LLM-assisted architecture in which the LLM proposes, and the kernel decides. The consent layer is complete; ongoing work extends the kernel to storage limitation and data minimisation.

Doctoral Consortium History. This work is being submitted to a doctoral consortium for the first time and has not been presented at any previous doctoral consortium at other conferences.

Supervisor: Prof. Haralambos Mouratidis, School of Computer Science and Electronic Engineering, University of Essex, United Kingdom. **Prior Doctoral Consortium presentations:** None.

References

1. Alkubaisy, D., Piras, L., Al-Obeidallah, M., Cox, K., Mouratidis, H.: Confis: a tool for privacy and security analysis and conflict resolution for supporting GDPR compliance through privacy-by-design. In: Proceedings of the 16th International Conference on Evaluation of Novel Approaches to Software Engineering, pp. 80–91. SCITEPRESS - Science and Technology Publications (2021). https://doi.org/10.5220/0010406100800091
2. Basin, D., Hublet, F., Krstić, S., Nguyen, H.: Mechanizing privacy by design. In: Proceedings of the 2025 ACM SIGSAC Conference on Computer and Communications Security (2025). https://doi.org/10.1145/3719027.3748271
3. Bygrave, L.A.: Data protection by design and by default: deciphering the Eu's legislative requirements. Oslo Law Rev. **4**, 105–120 (2017). https://doi.org/10.18261/issn.2387-3299-2017-02-03
4. de Carvalho, R.M., et al.: Protecting citizens' personal data and privacy: joint effort from GDPR EU cluster research projects. SN Comput. Sci. **1**, 217 (2020). https://doi.org/10.1007/s42979-020-00218-8
5. Colesky, M., Ghanavati, S.: Privacy shielding by design — a strategies case for near-compliance. In: 2016 IEEE 24th International Requirements Engineering Conference Workshops (REW), pp. 271–275. IEEE (2016). https://doi.org/10.1109/REW.2016.051
6. Daniels, S.O.: CONSENTML case study instance: open banking consent model (supplementary material) (2026). https://doi.org/10.5281/zenodo.19141712. Accessed 21 Mar 2026
7. Deng, M., Wuyts, K., Scandariato, R., Preneel, B., Joosen, W.: A privacy threat analysis framework: supporting the elicitation and fulfillment of privacy requirements. Requirements Eng. **16**, 3–32 (2011). https://doi.org/10.1007/s00766-010-0115-7
8. Gharib, M., Giorgini, P., Mylopoulos, J.: COPri vol 2 – a core ontology for privacy requirements. Data Knowl. Eng. **135**, 101921 (2021). https://doi.org/10.1016/j.datak.2021.101921
9. Gharib, M., et al.: Privacy requirements: findings and lessons learned in developing a privacy platform. In: 2016 IEEE 24th International Requirements Engineering Conference (RE), pp. 256–265. IEEE (2016). https://doi.org/10.1109/RE.2016.13
10. Giraldo, F.D., España, S., Giraldo, W.J., Óscar Pastor: Evaluating the quality of a set of modelling languages used in combination: a method and a tool. Inf. Syst. **77**, 48–70 (2018). https://doi.org/10.1016/j.is.2018.06.002
11. Grünewald, E., Pallas, F.: TILT: A GDPR-aligned transparency information language and toolkit for practical privacy engineering. In: Proceedings of the 2021 ACM Conference on Fairness, Accountability, and Transparency (FAccT), pp. 636–646. ACM (2021). https://doi.org/10.1145/3442188.3445925
12. Islam, S., Mouratidis, H., Kalloniatis, C., Hudic, A., Zechner, L.: Model based process to support security and privacy requirements engineering. Int. J. Secure Softw. Eng. **3**, 1–22 (2012). https://doi.org/10.4018/jsse.2012070101
13. Kalloniatis, C., Kavakli, E., Gritzalis, S.: Addressing privacy requirements in system design: the PriS method. Requirements Eng. **13**(3), 241–255 (2008). https://doi.org/10.1007/s00766-008-0067-3
14. Kammüller, F.: Formal modeling and analysis of data protection for GDPR compliance of IoT healthcare systems. In: 2018 IEEE International Conference on Trust, Security and Privacy in Computing and Communications (TrustCom), pp. 1066–1071. IEEE (2018). https://doi.org/10.1109/TrustCom/BigDataSE.2018.00148

15. Mollaeefar, M., Bissoli, A., Landuyt, D.V., Ranise, S.: Pillar: LINDDUN privacy threat modeling using LLMS. In: 2025 IEEE European Symposium on Security and Privacy Workshops (EuroS&PW), pp. 278–286. IEEE (2025). https://doi.org/10.1109/EuroSPW67616.2025.00038
16. Mouratidis, H., Giorgini, P.: Secure tropos: a security-oriented extension of the tropos methodology. Int. J. Software Eng. Knowl. Eng. **17**(02), 285–309 (2007). https://doi.org/10.1142/S0218194007003240
17. Notario, N., et al.: PRIPARE: integrating privacy best practices into a privacy engineering methodology. In: 2015 IEEE Security and Privacy Workshops (2015). https://doi.org/10.1109/SPW.2015.22
18. Page, M.J., et al.: The Prisma 2020 statement: an updated guideline for reporting systematic reviews. PLoS Med. **18**(3), e1003583 (2021). https://doi.org/10.1371/journal.pmed.1003583
19. Pallas, F., et al.: Karl: towards application-layer purpose-based access control. In: Proceedings of the 35th Annual ACM Symposium on Applied Computing (2020). https://doi.org/10.1145/3341105.3375764
20. Pandit, H.J., Debruyne, C., O'Sullivan, D., Lewis, D.: Gconsent – a consent ontology based on the GDPR. In: The Semantic Web. ESWC 2019, LNCS, vol. 11503, pp. 270–282. Springer, Cham (2019). https://doi.org/10.1007/978-3-030-21348-0_18
21. Peyrone, N., Wichadakul, D.: Formal models for consent-based privacy. J. Log. Algebraic Methods Programm. **129**, 100826 (2022). https://doi.org/10.1016/j.jlamp.2022.100826
22. Piras, L., et al.: Defend architecture: a privacy by design platform for GDPR compliance. In: Privacy and Identity Management. The Smart Revolution, pp. 78–93. Springer (2019). https://doi.org/10.1007/978-3-030-27813-7_6
23. Piras, L., Al-Obeidallah, M.G., Mouratidis, H., Pavlidis, M.: DEFeND architecture: a privacy by design platform for GDPR compliance. In: Trust, Privacy and Security in Digital Business (TrustBus 2019). LNCS, vol. 11711, pp. 78–93. Springer (2019). https://doi.org/10.1007/978-3-030-27813-7_6
24. Ralyté, J., Koutsopoulos, G., Stirna, J.: Verification, validation, and evaluation of modeling methods: experiences and recommendations. Softw. Syst. Model. (2025). https://doi.org/10.1007/s10270-025-01304-2
25. Sion, L., Landuyt, D.V., Joosen, W.: An overview of runtime data protection enforcement approaches. In: 2021 IEEE European Symposium on Security and Privacy Workshops (EuroS & PW) (2021). https://doi.org/10.1109/EuroSPW54576.2021.00044
26. Tokas, S., Owe, O., Baramashetru, C.: Static checking of GDPR-related privacy compliance for object-oriented distributed systems. J. Log. Algebraic Methods Programm. **125**, 100737 (2022). https://doi.org/10.1016/j.jlamp.2021.100737
27. Venable, J., Pries-Heje, J., Baskerville, R.: Feds: a framework for evaluation in design science research. Eur. J. Inf. Syst. **25**, 77–89 (2016). https://doi.org/10.1057/ejis.2014.36
28. Wuyts, K., Sion, L., Joosen, W.: LindDun go: a lightweight approach to privacy threat modeling. In: 2020 IEEE European Symposium on Security and Privacy Workshops (EuroS&PW), pp. 302–309. IEEE (2020). https://doi.org/10.1109/EuroSPW51379.2020.00047

Combining Large Language Models with Named Entity Recognition to Automate Market Research Interview Transcription and Customer Journey Mapping – Empirical Feedback by Business Professionals

Tarmo Koppel(✉)

Tallinn University of Technology, Ehitajate tee 5, 19086 Tallinn, Estonia
correspondence@koppel.ee

Abstract. This paper presents an automated pipeline combining large language model inference with named entity recognition to extract structured customer journey insights from interview transcripts. The system, built on Langextract library, identifies pain points, journey stages, emotions, competitors, and stage transitions with structured attributes. A human expert feedback study with 66 business professionals, each serving as both interviewee and assessor, evaluated outputs across seven quality dimensions using three LLM backends. Overall quality was rated good to very good (M = 4.2/5). Gemini 2.5 Flash significantly outperformed alternatives on pain point characterization (g = 1.14). Thematic analysis identified input quality dependency and oversimplification as principal limitations. Results suggest that LLM-based extraction can meaningfully accelerate qualitative market research while requiring human validation. Beyond the pipeline itself, the study positions LLMs as a structured analytic instrument for qualitative market research and contributes a reusable seven-dimension evaluation rubric for assessing LLM-extracted customer journey artifacts.

Keywords: large language models · named entity recognition · market research · customer journey

1 Introduction

Customer journey mapping is a foundational practice in market research and service design, yet it remains predominantly manual: analysts must read interview transcripts, identify touchpoints, pain points, emotions, and stage transitions, and code them into structured representations. This process is time-consuming, subjective, and difficult to scale across large interview corpora. There is limited empirical evidence on whether LLM-based information extraction can produce customer journey mappings that business professionals judge as accurate, complete, and actionable.

Despite growing interest in leveraging artificial intelligence for customer experience analysis, relatively few computational frameworks have been developed specifically

A. Bernasconi et al. (Eds.): RCIS 2026, LNBIP 597, pp. 247–260, 2027.
https://doi.org/10.1007/978-3-032-35043-5_21

for automated customer journey mapping. Pati et al. (2025) introduced a multimodal framework that classifies user-generated content text, images, and videos-into journey stages (Awareness, Consideration, Decision, Post-Purchase) using large language models such as Meta-LLaMA and GPT-Neo, achieving up to 92.9% classification accuracy [1]. Lu and Kannan (2026) proposed a transformer-based architecture that models sequential touchpoint data to predict conversion outcomes and enable individualized marketing attribution [2]. At a conceptual level, Soni (2025) outlined how LLMs can be integrated across lifecycle stages for lead identification, personalized engagement, and churn prediction [3]. Torrens (2022) mapped customer journey constructs onto agent-based modeling components for retail high-street simulation [4]. Ferber et al. (2025) systematically reviewed NLP-based technologies along the customer journey, finding that research concentrates on pre-purchase chatbot interactions with limited coverage of post-purchase stages and advanced generative models [5]. These efforts remain largely stage-classification or prediction-oriented, leaving a gap in structured extraction of journey elements like pain points, emotions, transitions - from qualitative interview data, which the present work addresses.

The automation of qualitative text analysis has undergone a fundamental transformation with the emergence of natural language processing (NLP) techniques and, more recently, large language models (LLMs). Central to many extraction pipelines is named entity recognition (NER), the task of identifying and classifying meaningful elements, such as persons, organizations, locations, or domain-specific concepts within unstructured text. A consistent finding across domains is that domain-specific pre-training substantially improves NER performance. In materials science, MatBERT-pre-trained on domain literature, outperformed general BERT and SciBERT models by 1–4% F1, with advantages most pronounced in low-data regimes [6]. Analogous gains have been demonstrated in archaeology, where ArcheoBERTje significantly outperformed generic models and rendered explicit thesaurus knowledge unnecessary [7], and in social science, where SsciBERT improved text classification and entity recognition over general-purpose baselines [8].

The advent of generative LLMs has introduced prompt-based paradigms that reduce or eliminate the need for annotated training data. GPT-based models, using zero-shot and few-shot prompting, have achieved competitive performance on NER tasks in materials science, matching or surpassing fine-tuned BERT models in certain settings [9]. In clinical NER, prompt engineering with GPT-4 yielded relaxed F1 scores approaching those of supervised BioClinicalBERT, though exact-match performance remained lower [10]. Fine-tuned LLMs such as GPT-3 have also demonstrated strong joint NER and relation extraction capabilities in scientific text, outperforming specialized baselines with minimal annotation effort [11].

Extracting structured information from unstructured text remains a persistent bottleneck across domains, driven by heterogeneous document formats, domain-specific terminology, and the sheer volume of data that precludes manual processing. In municipal governance, heterogeneous writing styles and formats impede automated metadata extraction from meeting minutes, necessitating specialized pipelines that combine segmentation with entity recognition [12, 13]. Similarly, in cybersecurity, the overwhelming volume of unstructured threat intelligence renders manual extraction impractical, demanding robust NER systems [14]. LangExtract addresses these bottlenecks by providing a library that combines LLM inference with structured extraction capabilities, enabling researchers to systematically extract psychologically meaningful constructs from user-generated text [15], support persistent memory architectures through structured fact management [16], and facilitate text-mining workflows for thematic analysis [17]. By bridging LLM-based reasoning with named entity recognition in a unified pipeline, LangExtract reduces the engineering overhead of building domain-specific extraction systems while maintaining structured, interpretable outputs.

This study empirically evaluates an LLM-based customer journey extraction pipeline, deployed across three commercial backends (Gemini 2.5 Flash, Gemini 3 Flash Preview, and GPT-5.4 Mini), through structured feedback from business professionals. RQ1: to what extent can an LLM + NER pipeline produce customer journey extractions that domain-expert practitioners rate as accurate, complete, and actionable? RQ2: does the choice of commercial LLM backend systematically affect extraction quality across the seven quality dimensions?

2 Method

A within-subjects design with self-assessed interviews was chosen over third-party annotation because it ensures that each evaluator possesses ground-truth knowledge of the customer experience being analyzed, thereby eliminating the need for a separate gold-standard corpus and enabling direct validation of extraction accuracy by domain experts. The primary research objective is to assess perceived output quality and practical utility from the perspective of business professionals, which requires capturing subjective expert judgment. Each transcript was the verbatim record of a human expert describing their own lived experience with a product or service they personally use. A general purpose LLM acted as a structured-elicitation interviewer that asked the questions; the substantive content originated from the human, and the same human subsequently verified that the transcription faithfully captured what they had said before the transcript entered the extraction pipeline. The study is positioned as a proof of concept that an LLM + NER pipeline can produce a customer journey map at the level of a single, fully-validated interview. The principal implication is that scores reflect perceived accuracy against subjective recall rather than an objective ground truth; this inflates ecological validity but limits precision, and future work should complement self-assessment with independent annotation by a second domain expert per transcript.

2.1 System Architecture

The built system comprises a multi-stage pipeline that combines general-purpose large language models (LLMs) with a specialized named entity recognition (NER) extraction library to automate customer journey analysis from interview transcripts. Prior to entering this pipeline, users prepare a diarized interview transcript, which serves as the input to the system. The system was implemented as a cloud-hosted Python notebook application using the Langextract library, with commercial LLM APIs (Google Gemini and OpenAI) as the inference backends. A retrieval-augmented generation layer was not introduced at this stage, because the ground truth for each customer journey is contained in the single interview transcript itself rather than in external knowledge bases. The architecture comprises the following components (see Fig. 1):

- **Extraction prompt specification.** A domain-specific extraction prompt defines five entity classes relevant to customer journey analysis (Table 1). Each class includes structured attribute schemas specifying the metadata to extract entity span.
- **NER engine.** Langextract library orchestrates extraction by sending the transcript and the extraction prompt to the LLM API. It supports multi-pass extraction and parallel processing via configurable threads. It returns structured extraction objects containing the entity class, the exact extracted text span, and associated attributes.
- **Journey evolution tracker.** A second extraction pass uses a dedicated evolution-tracking prompt that captures stage transitions, shifts in customer expectations, stakeholder alignment changes, and key turning points.
- **Post-processing and aggregation module.** Extracted entities are aggregated by class, grouped by journey stage, and filtered. Summary statistics are computed, including counts per entity class and per journey stage.
- **Visualization generator.** Langextract visualization module produces an interactive HTML page where extracted entities are highlighted in labeled color-coded text.
- **Export module.** Results are exported in three formats: JSONL (full extraction data), CSV for pain points, and CSV for journey evolution tracking.

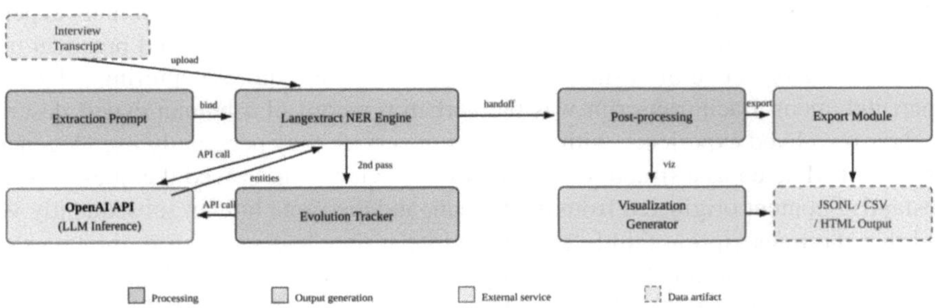

Fig. 1. System architecture of the customer journey analysis pipeline. The diarized interview transcript, produced from a human expert interview elicited externally via a general-purpose LLM and verified by the expert, serves as input.

The user uploads a transcript file to the cloud notebook, where the extraction prompt is bound to the Langextract NER engine. The engine sends the chunked transcript together with the prompt and few-shot examples to the OpenAI API, which returns structured extraction objects. These raw extractions are then passed to the post-processing module for aggregation and filtering. In parallel, the same transcript undergoes a second extraction pass through the evolution tracker, which issues its own API call to capture journey stage transitions. Finally, the post-processing module feeds aggregated results to the visualization generator, which produces an interactive HTML page, and to the export module, which writes JSONL and CSV output files that are automatically downloaded to the user's machine.

2.2 Data Flow

Input. The system accepts a single text or PDF file containing a diarized interview transcript. The interview covers a specific product or service that the interviewee (the customer) has personal experience with. Transcripts were produced using a general-purpose LLM guided by a structured interview plan that elicits information across the full customer journey: from initial awareness and triggers, through evaluation and purchase, to post-purchase experience and loyalty.

Processing. The transcript undergoes two extraction passes. In the first pass, the Langextract engine sends the text along with the customer journey extraction prompt and few-shot examples to Google or OpenAI API. The LLM identifies and returns named entities across five classes, each with structured attributes as defined in Table 1. Extraction is performed with 10 parallel workers for efficiency. In the second pass, a journey evolution prompt is applied to the same transcript, capturing stage-to-stage transitions, turning points, and shifts in customer priorities. After extraction, post-processing aggregates entities by class and journey stage, computes summary statistics, and applies optional filters (e.g., isolating pain points by severity).

Output. The system produces four output artifacts: (1) a JSONL file containing all extracted entities with their class labels, text spans, and attribute dictionaries; (2) an interactive HTML visualization with color-coded entity highlights, hover details, filtering controls, and summary statistics; (3) a CSV file of pain points with columns for stage, trigger, severity, and impact; and (4) a CSV file tracking journey evolution with stage-before, stage-after, trigger, and outcome fields. All output files are automatically downloaded to the user's local machine for review and assessment.

Table 1. Named entity classes and their attributes defined in the extraction prompt

Entity Class	Description	Extracted Attributes
product_service_solution	Mentioned offerings, bundles, or features	brand/provider, status (using, evaluating, former), context
competitor_alternative	Competitors, substitutes, DIY workarounds	comparison_basis, sentiment, decision_outcome
journey_stage_touchpoint	Journey stages (awareness → loyalty) or specific touchpoints	stage, channel, actor, timestamp/moment, success_metric
pain_point_friction_failure	Blocking issues, unmet needs, failure modes	stage, severity, trigger, workaround, impact
emotion_expectation_outcome	Expressed feelings, expectations, goals, actual outcomes	sentiment (pos/neg/mixed), intensity, target, outcome_status

2.3 Experimental Setup

The evaluation followed a within-subjects design in which each participant acted as the interview subject and the quality assessor of the tool output. This dual role was methodologically advantageous: since participants described their own experiences with a product or service they personally use, they possessed first-hand knowledge necessary to judge whether the system's extractions were accurate and complete.

Step 1 - Participants selected a product or service they personally use. They covered 12 product and service categories: apparel (N = 6), blue-collar services (N = 2), business services (N = 4), digital and subscription services (N = 13), food and beverages (N = 2), food delivery services (N = 7), health and beauty products (N = 5), home improvement products (N = 4), hospitality and leisure services (N = 8), infrastructure products (N = 3), technology products (N = 4), and transport and mobility services (N = 6). Each participant then conducted a customer journey interview in which they were the interviewee describing their own real experience with the chosen product or service. A general-purpose LLM (e.g., ChatGPT) acted as the interviewer, asking the participant the questions specified in a structured interview plan that was supplied as a system prompt. The participant reviewed and corrected the resulting diarized transcript before saving it as a text file with speaker labels. Interviews were conducted either in English (N = 38) or Estonian (N = 28). Average interview length was 1676 words, with min 249 to max 5601 words. Divided by median (1270 words) interviews were classified either as 'short' or 'long'.

Step 2 - Participants opened the provided app and uploaded interview transcript. Then executed the main extraction, which processed the full transcript through the NER engine using selected LLM provider's API. Test users per model: Gemini 2.5 Flash (N = 11), Gemini 3 Flash Preview (N = 28) and GPT-5.4 Mini (N = 27). The backends were chosen to span two major commercial providers (Google and OpenAI) at the low-cost end of the frontier, and to include an older/newer Gemini generation pair, enabling both within-provider and cross-provider comparison. Configuration was held fixed across runs: Gemini 2.5 Flash and Gemini 3 Flash Preview were called with temperature $\approx$ 0.5 and the Langextract default values for the remaining sampling parameters; GPT-5.4 Mini was called with reasoning effort = "medium" and verbosity = "medium". The same extraction prompt, few-shot examples, and chunking/parallelization settings were used for every model. The three cells contain unequal numbers of participants (11/28/27) because recruitment took place over an extended period across several professional training cohorts; participants were assigned to whichever model was active during their cohort, rather than self-selecting a model. This rules out the self-selection bias, but the unequal cell sizes do reduce statistical power for the smallest cell (Gemini 2.5), and the comparative results should be read accordingly.

Step 3 - Participants initiated JSONL, HTML visualization, journey insight summaries, journey evolution tracking, evolution CSV export, and pain point CSV export.

Test users (N = 66) were recruited from professional training courses and Master's-level programs at the authors' university business school. Participants thus represented the target user profile for the tool: business professionals and advanced students with domain knowledge of products and services but without specialist expertise in natural language processing or information extraction. No personally identifiable information was collected or stored, and all submissions were treated anonymously. Participants provided informed consent prior to taking part in the evaluation. Participants in this study described their own consumer experiences with commercially available products and services; no sensitive personal, clinical, or organizational data was processed, limiting the privacy risk inherent to the experimental setup. Each expert was the source of their own interview content and the assessor of the pipeline's output for that same interview, judging the extractions against their own first-hand knowledge of the experience rather than against the transcript. Each transcript is unique to one expert. This design eliminates the need for a separate gold-standard corpus and grounds every accuracy judgment in domain expertise.

2.4 Human Feedback Instrument

Expert feedback was collected using a structured questionnaire comprising 16 Likert-scale items and 4 open-ended questions. The Likert items used a five-point scale ranging from 1 (very poorly) to 5 (very well), with an additional NA option (see Table 2). This instrument design ensured that each output artifact was evaluated on multiple quality dimensions by assessors who had direct domain knowledge of the subject matter, thereby providing ecologically valid expert feedback on the system's performance.

The seven dimensions were not adopted wholesale from a single published instrument; rather constructed by combining established constructs from two complementary literatures and adapting them to the evaluation of an automated extraction-and-visualization tool. The five content-oriented dimensions (Pain Point Identification, Characterization, Actionability, Journey Evolution, Trigger–Outcome Logic) draw on the customer-experience and pain-point literature, which establishes that customer experience is a dynamic process in which customer cognitions, emotions, and behaviors evolve across the pre-purchase, purchase, and post-purchase stages [18]; the TCQ nomenclature, which formalizes touchpoints, context, and qualities as the three building blocks of CX [19]; and the experiential pain-point classification, which identifies pain point themes and their emotional consequences across journey stages [20]. The two visualization-oriented dimensions (Stage Representation, Visual Encoding Clarity) draw on the service-design CJM-modeling tradition [21, 22]. These foundations were extended with two adaptations specific to automated extraction: 1) an explicit Actionability dimension to capture whether extracted data is sufficiently grounded to inform CX improvement, and 2) the Trigger–Outcome Logic dimension to capture the causal coherence of LLM-extracted relationships, which has no direct precedent in manual CJM but is essential for evaluating generative tools.

2.5 Data Analysis

Quantitative analysis of the Likert-scale responses proceeded in two stages. First, each respondent's individual item scores were averaged into a single composite variable per output artifact (i.e., pain point extraction, journey evolution tracking, and visualization), yielding one summary score per participant per processing step. Also, for each test user, all individual items' ($N = 16$) scores were averaged to obtain overall means. To compare

Table 2. Human feedback questionnaire variables

Variable	Items (N)	Description
Pain Point Identification Accuracy	2	The extent to which the tool correctly identified friction points and assigned them to the appropriate journey stage
Pain Point Characterization Quality	3	The specificity and correctness of the triggers, severity ratings, and impact descriptions assigned to each pain point
Pain Point Actionability	1	The degree to which the extracted pain point data is specific enough to inform concrete customer experience improvements
Journey Evolution Accuracy	3	How accurately the tool captured shifts in customer mindset, goals, and priorities across journey stages
Trigger–Outcome Logic	3	The clarity and logical grounding of identified turning points, causal trigger–outcome connections, and relevant stakeholder roles
Journey Stage Representation	2	How well the visualization conveys the overall journey structure and highlights critical moments
Visual Encoding Clarity	2	The effectiveness of color-coding, labeling, and placement in distinguishing extraction classes, channels, and touchpoints

performance across subgroups, independent and paired-samples t-tests were conducted, with Hedges' g or Cohen's d computed as an effect size measure.

Pearson correlation coefficients were computed across all seven quality variables and interview word count, with Bonferroni correction applied to control for familywise error across 28 pairwise comparisons (adjusted $\alpha = .002$). Pairwise sample sizes ranged from $N = 55$ to $N = 66$ due to missing values. Internal consistency of the multi-item constructs (var1, var2, var4, var5, var6, var7) was assessed with Cronbach's α computed on listwise-complete cases; var3 (pain point actionability) is a single-item measure and is therefore reported without a reliability coefficient. For the three two-item constructs, the Spearman–Brown coefficient is reported alongside α as a more appropriate reliability estimate for short scales.

Qualitative analysis was applied to participants' open-ended responses regarding the tool's strengths and shortcomings. Responses were thematically coded through iterative reading of the free-text answers, identifying recurring patterns in perceived strengths and weaknesses of the AI-generated outputs. Codes were subsequently aggregated into higher-level themes, with approximate frequency counts assigned to each theme to indicate prevalence across the respondent pool.

3 Results

3.1 Descriptive Statistics

Instrument Reliability: Internal consistency of the six multi-item constructs was acceptable to good. Cronbach's α was .735 for pain point identification accuracy (k = 2, N = 56), .727 for pain point characterization quality (k = 3, N = 57), .693 for journey evolution accuracy (k = 3, N = 56), .733 for trigger–outcome logic (k = 3, N = 56), .778 for journey stage representation (k = 2, N = 57), and .426 for visual encoding clarity (k = 2, N = 54). The corresponding Spearman–Brown coefficients for the three two-item constructs were .735, .786, and .439, respectively. Pain point actionability (var3) is a single-item measure and is reported without a reliability coefficient. With the exception of visual encoding clarity, all multi-item constructs reached or approached the conventional $\alpha \geq .70$ threshold, supporting their use as composite scores in subsequent analyses. The lower reliability of visual encoding clarity (mean inter-item r = .28) suggests that its two items: the color-coding/labeling of extraction classes and the placement of channels and touchpoints within the journey tap somewhat distinct aspects of visualization quality and should be interpreted with caution; results involving this construct are reported but treated as exploratory.

Interview Language: An independent samples t-test revealed no significant difference in overall scores between English-language ($M = 4.31$, $SD = .43$) and Estonian-language ($M = 4.17$, $SD = .54$) interviews ($t(50.50) = 1.15$, $p = .255$, $g = .29$).

Interview Length: No significant difference in overall scores was found between short ($M = 4.31$, $SD = .51$) and long ($M = 4.19$, $SD = .46$) interviews, split at the median ($t(61.92) = 1.00$, $p = .321$).

Product or Service: No significant difference was found in overall scores between product-related ($M = 4.30$, $SD = .47$) and service-related ($M = 4.24$, $SD = .48$) customer journey analyses ($t(43.42) = .49$, $p = .629$, $g = .13$).

Best Model: Across all evaluation items, Gemini 2.5 Flash (Gem2.5f) achieved the highest average score (M = 4.50, SD = .41), followed by Gemini 3 Flash Preview (Gem3fp) (M = 4.25, SD = .46) and GPT-5.4 Mini (GPT5.4m) (M = 4.14, SD = .51). Independent samples t-tests showed that Gemini 2.5 Flash scored significantly higher than GPT-5.4 Mini (t(22.83) = 2.23, p = .036, g = .71), while differences between Gemini 2.5 Flash and Gemini 3 Flash Preview and between Gemini 3 Flash Preview and GPT-5.4 Mini were not significant.

Customer Journey Variables: Largest divergence - pain point characterization quality (var2) differed significantly across the three LLM backends. Gemini-2.5-flash achieved the highest mean score, followed by Gemini-3-flash-preview and GPT-5.4-mini. Welch's t-tests revealed that Gemini-2.5-flash scored significantly higher than both Gemini-3-flash-preview (t(19.05) = 2.66, p = .015, Hedges' g = 0.91) and GPT-5.4-mini (t(20.62) = 3.40, p = .003, Hedges' g = 1.14), with both comparisons yielding large effect sizes. The difference between Gemini-3-flash-preview and GPT-5.4-mini was not significant. These results suggest that Gemini-2.5-flash produced substantially more accurate pain point extractions, including friction identification, stage assignment, trigger specificity, severity ratings, and impact descriptions. The two newer models performed comparably to each other at moderate-to-good level (Table 3). A plausible interpretation is that Gemini 2.5 Flash, despite being the oldest of the three models tested, had undergone more extensive instruction-tuning and structured-output optimization at the time of its release, making it better suited to the precise attribute-extraction task. The two newer models may trade some extraction granularity for broader capability gains. Causal attribution is not possible from this design.

Table 3. Customer journey variables by product/ service and models, score M(SD)

Variable	ALL	Product	Service	Gem2.5f	Gem3fp	GPT5.4m
1. Pain point identification accuracy	4.44(.63)	4.48(.72)	4.41(.59)	4.82(.41)	4.48(.50)	4.24(.75)
2. Pain point characterization quality	4.13(.74)	4.21(.79)	4.12(.70)	4.73(.65)	4.11(.67)	3.91(.72)
3. Pain point actionability	4.24(.82)	4.39(.90)	4.14(.77)	4.23(.41)	4.39(.63)	4.07(1.07)
4. Journey evolution accuracy	4.16(.63)	4.33(.57)	4.12(.59)	4.72(.38)	4.19(.60)	4.06(.64)
5. Trigger outcome Logic	4.06(.73)	4.08(.76)	4.09(.68)	4.23(.82)	4.05(.82)	4.00(.60)
6. Journey stage representation	4.54(.61)	4.50(.66)	4.55(.60)	4.82(.34)	4.41(.67)	4.56(.61)
7. Visual encoding clarity	4.36(.60)	4.25(.61)	4.43(.58)	4.27(.75)	4.36(.58)	4.39(.58)

3.2 Correlation Analysis

Interview word count showed no significant correlation with any quality variable, indicating that transcript length did not systematically influence the perceived quality of the tool's outputs. Among the quality variables, the strongest correlations emerged within two interpretable clusters. The first cluster linked the pain point extraction variables: pain point identification accuracy correlated strongly with both characterization quality ($r = .62$, $p < .001$) and actionability ($r = .68$, $p < .001$), while characterization quality and actionability were also strongly associated ($r = .55$, $p < .001$). This suggests that when the tool correctly identified friction points, it also tended to characterize them well and produce actionable output. See Table 4.

The second cluster linked the journey evolution variables, where journey evolution accuracy and trigger-outcome logic were strongly correlated ($r = .72$, $p < .001$), indicating that accurate stage-transition tracking co-occurred with logically grounded trigger-outcome mappings. Bridging the two clusters, pain point characterization quality correlated significantly with both journey evolution accuracy ($r = .56$, $p < .001$) and trigger–outcome logic ($r = .62$, $p < .001$), serving as a connecting variable between the pain point and evolution dimensions. The visualization variables formed a separate pair: journey stage representation and visual encoding clarity were significantly correlated ($r = .51$, $p < .001$), but showed weaker and mostly non-significant associations with the extraction quality variables, suggesting that visualization quality was partially independent of the underlying extraction performance. The one exception was the moderate correlation between journey stage representation and pain point identification accuracy ($r = .45$, $p = .005$), which indicates that correct identification of pain points contributed to a clearer visual depiction of the journey.

Table 4. Correlation analysis of customer journey variables

Variable	Words	1	2	3	4	5	6	7
Interview length (Words)	1							
1. Pain point identification accuracy	−0.11	1						
2. Pain point characterization quality	−0.03	0.62***	1					
3. Pain point actionability	−0.10	0.68***	0.55***	1				
4. Journey evolution accuracy	−0.20	0.27	0.56***	0.37	1			
5. Trigger outcome Logic	−0.11	0.38	0.62***	0.40*	0.72***	1		
6. Journey stage representation	0.07	0.45**	0.39*	0.23	0.09	0.40*	1	
7. Visual encoding clarity	−0.09	0.23	0.35	0.18	0.32	0.31	0.51***	1

3.3 Thematic Analysis

Useful Features: Two outputs emerged as the most valued: pain point extraction ($N = 37$) and the interactive HTML visualization ($N = 36$). Participants highlighted that the pain point CSV provided an immediately actionable summary of friction points with severity ratings and triggers, while the visualization enabled rapid comprehension of the full journey structure through color-coded highlights and filtering. Structured output formats more broadly, including CSV and JSONL files, were valued for enabling downstream analysis ($N = 20$). Journey evolution tracking was cited by 13 respondents as particularly useful for revealing how customer priorities and mindsets shifted across stages. The tool's speed and efficiency in reducing manual analysis time was noted by 15, with several emphasizing that the automated extraction would otherwise require hours of manual coding.

Inaccuracies: The most frequently reported inaccuracy was overgeneralization, where extracted entities or triggers were too broad or lacked sufficient contextual specificity (N = 21). Closely related, 16 respondents noted misclassification of journey stages, where entities were assigned to incorrect stages or stage boundaries were drawn too coarsely. Triggers were perceived as too vague or insufficiently grounded in the interview text by 11 respondents. Severity misratings, where the tool's assigned severity (low/moderate/high) did not match the evaluator's own judgment, were reported by 9 respondents. Misclassification of sentiment polarity, including cases where positive statements were erroneously tagged as pain points, was noted by 8 respondents. Fragmented extractions, where the tool isolated single words or incomplete phrases rather than meaningful spans, were flagged by 7 respondents. Missing or empty attribute fields (e.g., null triggers, blank severity columns) were reported by 6 respondents. Eight respondents reported no inaccuracies or stated they could not identify any.

Risks and Limitations: The most prevalent concern was dependency on input quality (N = 30), as respondents recognized that the depth and structure of the interview transcript directly determined the quality of the tool's output, and that shallow or poorly conducted interviews would yield superficial results. Over-reliance on automated outputs without human validation was raised by 22 respondents, who cautioned that structured outputs may appear more authoritative than warranted, potentially leading to false confidence in flawed extractions. Privacy and data protection risks were cited by 20 respondents, particularly regarding the transmission of sensitive customer interview data to external LLM APIs and the associated GDPR implications. An equal number (N = 20) expressed concern about the risk of hallucination or false associations, where the tool might generate interpretations not grounded in the source text. Oversimplification - the loss of emotional nuance, context, or nonverbal cues inherent to qualitative interviews, was noted by 14 respondents. Massive queries related API costs and rate limits were mentioned by 11 as a practical barrier to organizational-scale deployment.

4 Conclusions

This study demonstrates that combining LLM inference with structured named entity recognition can automate customer journey analysis from interview transcripts at a quality level that business professionals rate as good to very good (overall M = 4.2/ 5). Unlike existing approaches that rely either on generic text summarization or on manually defined coding schemes, the proposed pipeline produces structured, multi-attribute entity extractions, including severity ratings, stage assignments, and trigger–outcome mappings. This is fetched directly from raw transcripts without requiring prior annotation or domain-specific model training. The comparative evaluation across three LLM backends revealed that model choice matters: Gemini 2.5 Flash significantly outperformed both alternatives on pain point characterization, with large effect sizes. Correlation analysis confirmed that extraction quality dimensions cluster coherently. Thematic analysis identified input quality dependency, over-reliance without human validation, and data privacy as the principal deployment concerns. Several issues remain unresolved. Construct validity is constrained by reliance on participant self-assessment, which may conflict accurate extraction with subjective recall of experience. Internal validity may be affected by common-method bias, as both interview and extraction stages use generative models. External validity is limited by single-interview units, a university-based sample, and restriction to English and Estonian. Conclusion validity is influenced by unequal cell sizes and uncorrected comparisons.

The study contributes a primary artefact as the LangExtract-based extraction pipeline bundled with a seven-dimension evaluation rubric. This forms a reusable, empirically validated system for automated customer journey analysis. The pipeline is domain-agnostic and generalizable beyond customer journey mapping to other qualitative interview contexts such as usability studies or employee experience analysis. Compared to prior NLP-based approaches (e.g., rule-based NER, fine-tuned BERT classifiers), it requires no annotated training corpus, trading recall precision for zero-shot deployability. The tradeoff is empirically supported by the 'good' to 'very good' ratings observed here. The rapid advancement of LLMs strengthens rather than undermines the contribution, as improved model performance validates the architectural direction, and both the rubric and pipeline design remain backend-agnostic. The differences between LLMs are significant on pain point characterization (Gemini 2.5 Flash vs GPT-5.4 Mini) but non-significant on five of the seven dimensions, suggesting backend choice matters primarily for attribute-level extraction granularity.

Future work should incorporate inter-rater reliability metrics alongside self-assessment, benchmark the pipeline against non-LLM NLP baselines to quantify the added value of LLM inference, and investigate fine-tuned or domain-adapted models to reduce the overgeneralization and stage misclassification errors most frequently reported. Scaling to multi-interview corpora per product and extending the architecture to adjacent qualitative research domains such as employee experience analysis represent natural generalizations of the approach.

References

1. Pati, S., Pasam, V.K., Toxtli Hernandez, C.: Customer journey mapping with multimodal large language models. In: Companion Proceedings of the ACM on Web Conference 2025, pp. 2744–2748 (2025)
2. Lu, Z., Kannan, P.K.: AI for customer journeys: a transformer approach. J. Mark. Res. **63**, 1–26 (2026)
3. Soni, V.: Large language models for enhancing customer lifecycle management. J. Empirical Soc. Sci. Stud. **7**, 67–89 (2023)
4. Torrens, P.M.: Agent models of customer journeys on retail high streets. J. Econ. Interact. Coord. **18**, 87 (2022)
5. Ferber, T., Vaziri, D., Boden, A.: Natural language processing-based technologies along the customer journey—a systematic review and co-occurrence analysis. Hum. Behav. Emerg. Technol. **2025**, 1205909 (2025)
6. Trewartha, A., et al.: Quantifying the advantage of domain-specific pre-training on named entity recognition tasks in materials science. Patterns. **3** (2022)

7. Brandsen, A., Verberne, S., Lambers, K., Wansleeben, M.: Can BERT dig it? Named entity recognition for information retrieval in the archaeology domain. J. Comput. Cult. Heritage (JOCCH). **15**, 1–18 (2022)
8. Shen, S., et al.: SsciBERT: a pre-trained language model for social science texts. Scientometrics **128**, 1241–1263 (2023)
9. Choi, J., Lee, B.: Accelerating materials language processing with large language models. Commun. Mater. **5**, 13 (2024)
10. Hu, Y., et al.: Improving large language models for clinical named entity recognition via prompt engineering. J. Am. Med. Inform. Assoc. **31**, 1812–1820 (2024)
11. Dunn, A., et al.: Structured information extraction from complex scientific text with fine-tuned large language models. arXiv preprint arXiv:2212.05238 (2022)
12. Batista, R., et al.: MiNER: a two-stage pipeline for metadata extraction from municipal meeting minutes. arXiv preprint arXiv:2602.00316 (2026)
13. Campos, R., et al.: Citilink-minutes: a multilayer annotated dataset of municipal meeting minutes. arXiv preprint arXiv:2602.12137 (2026)
14. Ranzinger, M.: A systematic comparison of named entitiy recognition approaches for cyber threat intelligence (2026)
15. Harry, T., Ngong, I., Nweke, C., Feng, Y., Near, J.: Beyond fixed psychological personas: state beats trait, but language models are state-blind. arXiv preprint arXiv:2601.15395 (2026)
16. Kende, R.: K-lines: a cognitively-grounded four-memory architecture for persistent conversational AI. Available at SSRN 6234218 (2026)
17. Ural, G., Balkul, H.İ: AI literacy as a core competence in the age of human-computer interaction: the case of translation studies. Karamanoğlu Mehmetbey Üniversitesi Edebiyat Fakültesi Dergisi. **8**, 387–396 (2025)
18. Lemon, K.N., Verhoef, P.C.: Understanding customer experience throughout the customer journey. J. Mark. **80**, 69–96 (2016). https://doi.org/10.1509/jm.15.0420
19. De Keyser, A., Verleye, K., Lemon, K.N., Keiningham, T.L., Klaus, P.: Moving the customer experience field forward: introducing the touchpoints, context, qualities (TCQ) nomenclature. J. Serv. Res. **23**, 433–455 (2020)
20. Holz, H.F., Becker, M., Blut, M., Paluch, S.: Eliminating customer experience pain points in complex customer journeys through smart service solutions. Psychol. Mark. **41**, 592–609 (2024)
21. Halvorsrud, R., Kvale, K., Følstad, A.: Improving service quality through customer journey analysis. J. Serv. Theory Pract. **26**, 840–867 (2016)
22. Rosenbaum, M.S., Otalora, M.L., Ramírez, G.C.: How to create a realistic customer journey map. Bus. Horiz. **60**, 143–150 (2017). https://doi.org/10.1016/J.BUSHOR.2016.09.010

Towards a Knowledge Graph-Based and LLM-Assisted Support System for University Curriculum Engineering

Aritha Kumarasinghe(✉)

Institute of Applied Computer Systems, Riga Technical University, 6A Kipsalas Street, Riga 1048, Latvia
balasuriyage-aritha-dewnith.kumarasinghe@rtu.lv

Abstract. The adaptability of systems to changes in both internal and external conditions remains an open research issue. This doctoral thesis seeks to address this issue by adopting a fractal approach to the development of the architecture of a knowledge graph-based and LLM-assisted information system to support continuous curriculum development and maintenance (curriculum engineering). This task requires consideration of multiple rapidly changing influential factors while respecting the hierarchical and distributed nature of curriculum development. The design science research approach is applied to develop the architecture of a KG-based and LLM-assisted information system to support continuous curriculum engineering by the approach that addresses fractal features and external challenges of curriculum engineering.

Keywords: Fractal Approach · LLMs · Knowledge Graphs · Curriculum Engineering

1 Introduction

Recent developments in Knowledge Graphs (KGs) and Large Language Models (LLMs) open new opportunities for Information Systems (ISs) in semantically rich domains such as education [1]. One such use case is curriculum engineering (CE), which is a systematic process that consists of designing and redesigning curriculum across different composition levels (e.g., department, study program, and study course [2]) within a Higher Education Institution (HEI), while addressing several external factors/drivers (e.g., knowledge/skill requirements and regulatory reasons) [3]. This process can be supported using the knowledge representation capabilities of KGs to represent a curriculum and information relating to external factors (e.g., competencies for employment [4], external current mathematical knowledge [5]), and LLMs can be used for the extraction, organization, and reasoning on contents of KGs. However, solutions that combine KGs and LLMs still face several issues due to the static nature of knowledge representation, and low cross-task adaptability [1]. These issues combined with the fact that CE within HEIs is a collaborative process [6] that requires a balance between autonomy and governance [7] and requires the consideration of several external factors [3], highlight the

A. Bernasconi et al. (Eds.): RCIS 2026, LNBIP 597, pp. 261–272, 2027.
https://doi.org/10.1007/978-3-032-35043-5_22

need for an IS that supports flexibility and adaptability (across different factors) for both the curriculum representation and its engineering activities.

Considering the similarities in the curriculum representation structure, and the similarity in activities of CE at different organizational levels corresponding to the curriculum composition levels, a potential solution to this issue is a fractal approach to ISs. 'Fractal' is a geometric based concept characterized by patterns/structures that recur across different scales (hierarchical levels) [8]. When applied from a conceptual perspective during system design, it results in the system being self-similar, goal-oriented, self-organized, and having dynamism [9]. Previous research has identified how a fractal approach can be applied for developing ISs that could be capable of adapting to a changing external environment [10] and has demonstrated how an IS developed using a fractal approach can be used to improve collaboration across decentralized organizational units [11]. However, currently, to the best of the authors' knowledge, no prior research has considered the application of a fractal approach to ISs that combine KGs and LLMs within the application domain of CE. Therefore, this thesis seeks to design a system architecture for a KG-based, LLM-assisted IS employing a fractal approach to support CE across different composition levels.

The remainder of this paper is organized as follows: Sect. 2 introduces the problem domain of the thesis. Section 3 presents the solution domain and the key concepts underlying this work, including KGs, LLMs, and a fractal approach to systems development. Section 4 sets out the research questions, the methodology, and the expected outcomes. Section 5 summarizes the preliminary progress to date. Section 6 concludes the paper.

2 Problem Domain

In this thesis, curriculum engineering (CE) refers to a systematic designing and redesigning of curriculum within an HEI [12]. This process is organized hierarchically, given that a curriculum itself has a hierarchical structure. For instance, a department/faculty/institute within a HEI defines several study programs, while each study program consists of several study courses [13]. This results in decisions made at higher curriculum organization levels constraining and guiding the decisions made at lower levels, with each curriculum element (i.e., study course) being specified at a higher level, being further developed by curriculum developers at lower levels according to their disciplinary and pedagogical expertise [14].

This results in each curriculum organization level consisting of several relatively autonomous curriculum developers who should conform to the governance provided by higher levels. Therefore, CE within a HEI requires a balance between governance and autonomy [7]. The decentralized and hierarchical structure of HEIs can enable autonomy and agility but can also result in data use variations and data fragmentation, which make the implementation of data-driven solutions challenging [15]. In practice, this can result in curriculum developers working in silos, with limited visibility of related curriculum elements. Additionally, curriculum developers responsible for higher-level curriculum elements (e.g., study programs) must coordinate lower-level curriculum elements, a task that becomes more difficult as the number of curricula grows. The first challenge that will be addressed in this thesis is therefore to support CE as a hierarchical and distributed process with an IS.

A second challenge that must be addressed arises from curriculum developers having to consider not only internal governance factors but also external factors such as the educational needs of society [3]. A recent scoping review identified 77 different drivers of curriculum redesign/renewal (within health professional education) organized into 10 categories (Knowledge/skill issues, Exceptional events, Regulatory reasons, etc.) [3]. This, in essence, means that a curriculum developer must be capable of assessing several different factors and then be able to renew their curriculum accordingly. The assessment itself will entail the consideration of one or more information sources that relate to the driver, e.g., changes due to regulatory reasons would entail the assessment of educational standards. From an IS perspective, designing a separate system/subsystem for each driver would increase the complexity associated with design, development, and maintenance. The second challenge is therefore adaptability in terms of representing, analyzing, and adjusting curriculum based on information relating to external factors, with the resulting requirement for an IS being the ability to support different drivers without requiring a separate, rigid solution for each driver. Additionally, given the decentralized structure of CE, different organizational units can consider different drivers as relevant based on the domain to which the course relates; for instance, a course on 'Business Law' would focus on factors relating to regulatory knowledge, whereas the course 'Artificial Intelligence' would focus on factors relating to changing technological and scientific knowledge.

A high-level abstraction of the problem domain is represented in Fig. 1, with the dotted arrows representing information flow. Overall, the problem domain suggests two core challenges for the proposed IS architecture. First, it must support CE as a hierarchical and distributed process, enabling autonomous work while maintaining coordination across curriculum levels. Second, it must adapt to heterogeneous drivers of curriculum renewal, each of which may involve different information sources, assessment methods, and implications for curriculum change.

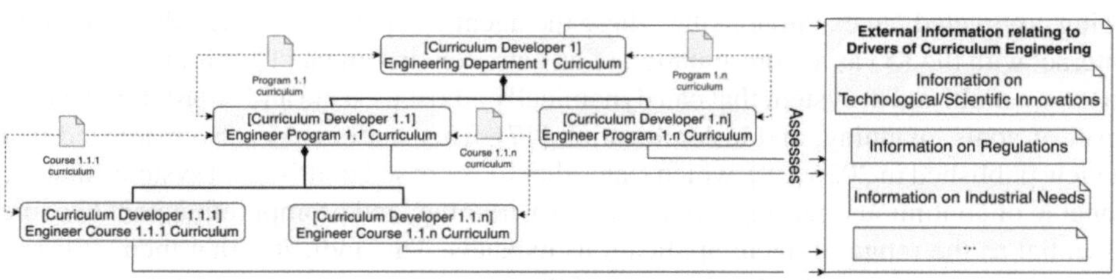

Fig. 1. Model representing how CE is organized within a university.

3 State of the Art

This section introduces the concepts that are related to the solution that is proposed in this thesis, with the first subsection discussing KGs and LLMs, and the second subsection discussing a fractal approach to systems.

3.1 Knowledge Graphs and Large Language Models

A KG is a formal representation of knowledge in the form of a graph that consists of nodes that represent entities and edges that represent the relationship between these entities. KGs are organized according to an underlying ontology, which specifies the types of entities, their attributes, and the relationships that thereby represent the domain in which the KG is used [1]. Additionally, since KGs represent information using a graph-based data model, they can offer higher schema flexibility than other data models (e.g. relational data models). This makes it easier to adapt the data model based on changing informational needs [16].

However, the main drawback that is associated with KGs is their costs of construction and maintenance, which require domain expertise and systems that are capable of updating knowledge [1]. The construction cost of a KG can be elevated by utilizing LLMs, which can be applied for the extraction, organization, and reasoning about knowledge, thereby reducing KG construction and maintenance costs. KGs can also be utilized to support the natural language processing tasks that are carried out by LLMs, by providing a factual representation of knowledge that can be used during the generation and verification of responses [1].

The related research has combined these technologies to support some CE tasks by representing required competencies based on job market analysis [4] and supporting the structuring of curriculum knowledge [5]. However, within the context of the two challenges of CE that were discussed in the previous section (decentralized and hierarchical organization structure and the need to consider dynamic factors/drivers), LLM and KG solutions have some limitations [1]. For instance, when using LLMs to support the functions related to KGs (e.g., KG construction), one of the drawbacks relates to prompting methods that require the manual definition and maintenance of prompts, which limit the adaptability to changing tasks. Conversely, when using KGs as an external knowledge source to support the functions of an LLM, similar to the previous case, the LLMs have limited adaptability across tasks due to the static nature of the integration mechanism between the two technologies [1]. In terms of a more synergistic integration of KGs and LLMs, LLMs can be used to dynamically update the KG based on the tasks that are being completed, an example being where the agentic capabilities of LLMs are used to interact with the KG as a representation of an active environment [1]. It is also suggested to have a multi-agent system that can dynamically adjust its scale and coordination at the level of goals, planning, and task allocation [17]. This sentiment is also resonated in a review published in 2026 [18], which states that LLM-based multi-agent systems that are capable of continuous learning, coevolutionary training, and prompt refinement become essential as the range of agent applications expands. This indicates that there still is a research gap relating to the development of ISs that are composed of LLM-based agents that are capable of evolving collectively and collaboratively, which must be addressed if these technologies are to be used in support of CE.

To conclude, although previous research has explored the application of KGs and LLMs (LLM-based agents) in education and demonstrated its application for specific CE tasks [4, 5], there remains a gap in creating KG-based and LLM-assisted ISs that are capable of adapting across different tasks [1] which in the context of CE relates to supporting curriculum engineers in creating and maintaining curricula based on evolving internal and external factors [3].

3.2 Fractal Approach to Systems Development

As a concept, "fractal" originates in geometry and refers to patterns that recur across different scales, resulting in self-similarity [8]. A fractal system demonstrates how a complex system can consist of relatively simple self-similar systems in terms of system structure [8] or system behavior [9]. In this thesis, both uses are relevant as there can be similarity in curriculum representation and the CE task across different levels. To leverage these features, a fractal approach will be applied to exploit self-similar structures (physical or functional) in systems analysis, design, or maintenance.

To apply a fractal approach, four design principles must be considered [9, 11]. The first being *self-similarity*, which results in the whole of the system being similar (in terms of organizational structure, system results, information flows, and/or functionality) to its parts [9]. This principle enables similarity across different system levels, with the system being composed of fractal units having similar structures referred to as Basic Fractal Unit (BFU), where each fractal unit then evolves to achieve a certain goal [9]. This ability to evolve the BFU to achieve a goal enables the next two principles, namely, self-organization and goal-orientation. *Self-organization* imposes that a fractal unit is free to change its organizational structure, information flows, and/or functionality, in order to adapt to internal and external conditions [9]. To enable self-organization, the next principle of *goal-orientation* is needed, which entails the definition of a goal at a system level with each fractal unit carrying out a sub-goal to achieve this main goal [9]. The implementation of goal-orientation enables real-time feedback between operation and environment, as well as cooperation within and outside each fractal unit [9]. All the aforementioned principles enable the final principle of *dynamism*, which results in a system that is capable of adapting to both internal and external changes.

The applicability of these principles to the context of CE is shown in Fig. 2, where self-organizing units (curriculum developers) carry out the task of CE. This results in the definition of curricula (self-similar in terms of the unit's function and the data produced) based on the units' domain knowledge and the organizational goals (resulting in the units being goal-oriented). Additionally, each unit can adapt its curricula, due to its awareness of the environment (based on external factors), with the changes made in terms of information sources used can then be adopted globally thereby allowing the constituent systems to evolve collectively (dynamisms).

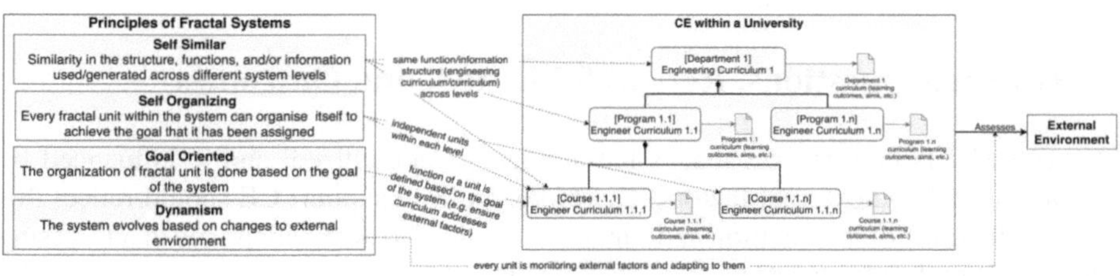

Fig. 2. Fractal principles and how they relate to CE.

Focusing on the application of a fractal approach to ISs, several research articles were identified through a literature search using the keywords 'fractal approach' and "information system" in Scopus, Web of Science, and IEEExplore. The first of the identified works [10] presents a theoretical argument for the applicability of a multifractal (system having multiple fractal patterns) approach for developing ISs that address the need for businesses to continuously adapt to external conditions. A subsequent work by the same author presents a flexible information architecture [19] that can be used by a fractal IS using HEI as an example organization for which a fractal IS can be implemented. Enabling change propagation is a key feature of this information architecture, which allows changes made by a single fractal units to be adopted by other fractal units [19].

The proposed flexible information architecture [19] for fractal ISs is implemented in [20], with study program creation within a higher education institution being considered as a use case. This architecture is created for defining the change and knowledge propagation method [20]. This knowledge propagation method allows for the sharing of information across organizational units, thereby addressing the decentralized organizational structure of universities. In terms of applying a fractal approach to an IS that is tested in a real-world use case, a doctoral thesis published in 2013 [11] details the design and implementation of a fractal-based model of a distributed healthcare IS to improve cooperation among physicians. The results from questionnaires and semi-structured interviews conducted before and after implementing the fractal healthcare IS revealed that there was an improvement in the efficiency in providing productive information across healthcare workers [11].

These previous works demonstrate the applicability of a fractal approach to ISs and how it can support the propagation of knowledge across organizational units and can support the adaptability of the real-world system [11, 20]. They used an IS to propagate the manually generated internal knowledge of a unit across the organization [11, 20]; however, in the context of KG-based systems, the underlying shared ontology (referred to as core ontology in [19]) of the sub-KG relating to a specific unit could be expanded based on activities of that unit with these expansions then being propagated to other units. Additionally, the changes to the KG and the underlying ontology can be made based on prompts to LLMs, which can then be propagated in terms of the changes made to the ontology, as well as the LLM-related knowledge (prompts, tools, information sources, models, etc.) to other units. This can help address the task-specific nature of KG-based and LLM-assisted systems through, for instance, the sharing of prompts used and changes to the underlying ontology of the KG across fractal units.

4 Research Questions, Research Method, and Outcomes

CE across HEIs is a process that is hierarchical, decentralized, and is influenced by several internal and external factors. Therefore, an IS to support CE should reflect this structure in terms of information and functions, while also being adaptable to addressing selected internal and external factors. The selection of technologies for the IS is driven by CE requiring both the representation of information relating to curricula and the factors that influence it, and the ability to interpret heterogeneous data (in terms of the factors to which they relate) to provide recommendations on how curricula can be adapted to address the changes in internal and external factors. KGs are suitable for representing the information relating to external factors in a machine-readable format, and LLMs can be used to interpret this knowledge to support the maintenance of a curriculum to address these factors [1, 4, 5].

Although existing KG-based and LLM-assisted solutions can support some CE tasks, they typically rely on static KGs, manually defined prompts, and limited update mechanisms, which limit KG adaptability across tasks, organizational levels, and changing factors [1, 17, 18]. This reduces the applicability of solutions that combine KGs and LLMs in the context of CE, where the system must be capable of addressing several internal and external factors and adapting to these factors, and where different organizational units can consider differing factors at a given time.

A possible solution to this problem is the adoption of a fractal approach to ISs by applying fractal design principles when developing a KG-based and LLM-assisted system, such that it addresses the organizational structure of CE while supporting the maintenance of curriculum based on changing internal and external factors. The fractal approach is applicable within the context of CE given the self-similar characteristics of curriculum engineering itself (Fig. 2) with the expected result being an IS that is capable of supporting CE across university organizational units and is able to collect, represent, analyze, and provide recommendations based on data relating to multiple curriculum related internal and external factors. To the best of the author's knowledge, no previous works have applied fractal design principles to KG-based and LLM-assisted ISs.

Therefore, **the goal of this thesis is to design (develop an architecture of) a KG-based and LLM-assisted IS, enriched with fractal system features, to support CE.** The expected artifact is an architecture of a KG-based, LLM-assisted IS that supports CE. The expected theoretical contribution is to deepen scientific evidence on (1) forms of representation of curricula, (2) combined use of LLMs and KGs in supporting CE, and (3) the applicability of a fractal approach to ISs and how it can be used to improve the integration of KGs and LLMs, especially for dynamic, collaborative, and multi-organizational unit settings. The currently defined research questions of this thesis are:

RQ: How can a KG-based and LLM-assisted IS for university CE be designed to respect CE's fractal features and multifactor dependence on changing internal and external influencing factors?

- RQ1: *How can CE knowledge be modelled and dynamically updated to support CE within universities using KGs?* This question will be answered by collecting existing knowledge from the scientific literature relating to KGs that support CE within universities (on program and course level) and by designing and implementing a KG that not only represents curricula across levels but also can accumulate and maintain knowledge relating to external factors.
- *RQ2: How can LLM and KGs be combined to support CE tasks within a university*? To answer this question, research relating to the combination of KGs and LLMs applicable to support CE, respecting its dependence on ever evolving factors, will be reviewed and a system architecture will be designed, implemented, and evaluated. By answering this question, the thesis seeks to provide novel insight into how KGs and LLMs can be combined such that is not task specific, as is the case with previous research [1, 4, 5].

- *RQ3: How can fractal principles be applied when designing a KG-based and LLM-assisted IS to support CE?* As discussed in Sect. 3.2, although previous research has considered a fractal approach to ISs, the combination of KGs, LLMs, and fractal principles represents a novel research direction, with the potential of fractal principles to support the adaptability of KG-based and LLM-assisted ISs. By answering this research question, the thesis seeks to provide insight into how the fractal features of CE (and other systems that have fractal features) can be leveraged and respected in ISs that utilize KGs and LLMs.

To reach the stated goal of designing the proposed artifact (architecture of the IS), the design science research methodology [21] is adopted. This methodology is applicable because the goal of the thesis is to design an artifact that addresses real-world problems, namely the need for an IS that can address the essence and challenges of CE. The research will be conducted according to the following iterative phases.

The *first phase* formalizes the problem to be addressed. This is done by analyzing CE as an application domain (Sect. 2) and exploring the current body of scientific knowledge by considering previous research (through a structured literature review) that considers combining KGs and LLMs (Sect. 3.1) and a fractal approach to systems development (Sect. 3.2). Insights from this analysis will be used to define the requirements that the artifact being developed must satisfy.

In the *second phase*, the IS architecture will be designed through iterative design cycles, exploring how a fractal approach can be combined with the KGs and LLMs. This entails creating models of IS components, such as a KG, to represent a curriculum and knowledge on its external factors, and an LLM capable of interacting with said KG; and gradually combining these models into the architecture of an IS. Each iteration has a forward link to the next phase and an optional return to the first phase.

In the *third phase*, through the demonstration of how the proposed architecture of the IS or its components (based on models developed in the previous phase) can be used to support CE stakeholders in real-world use cases, this thesis seeks to ensure the efficacy of the proposed IS components and architecture. Each iteration of this phase has a forward link to the next phase and an optional return to the previous phase.

In the *fourth phase*, the evaluation of the performance (potential effectiveness and efficiency of use) of the architecture or components of the IS will be done using quantitative and qualitative approaches. For instance, the performance of the developed system for supporting a specific CE task can be experimentally evaluated by comparing the IS's outputs to a human-defined benchmark [22], through evaluation frameworks such as RAGAs, which include metrics (e.g., faithfulness) that can be used to evaluate the responses generated by an LLM [23]. The evaluation results will be made publicly available via GitHub repositories to ensure transparency and replicability. Each iteration in this phase can return to previous phases.

Figure 3 represents an overview of the research methodology. The methodology prescribes iterative artifact development. The green elements represent activities where preliminary results have already been achieved, the yellow elements represent ongoing activities, and the red elements represent planned activities.

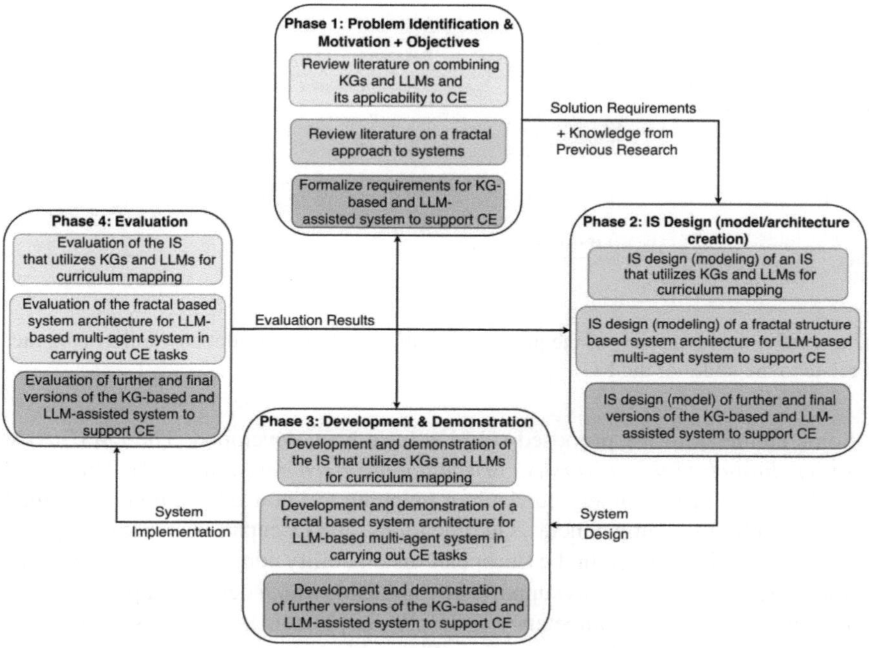

Fig. 3. An Overview of the research methodology.

5 Current Progress

This section discusses the current progress that has been made in this doctoral thesis (represented by green elements in Fig. 3).

A *preliminary model of CE* as a domain that is to be supported by the IS has been developed and is presented in Fig. 1. Based on this model and by applying a fractal approach to how a curriculum and CE is organized, the *preliminary overview of a proposed solution* has been designed (Fig. 4). This envisions an IS where the hierarchical nature of the curriculum is represented through a KG, which can then be altered by LLM agents based on the natural language queries defined by the user with the agents being organized according to the structure of the KG. Additionally, an LLM agent's ability to use tools [18] can be leveraged to automate the access of information relating to the factors influencing the CE, thereby supporting the amalgamation of related information, which can then be represented using nodes and edges in a KG.

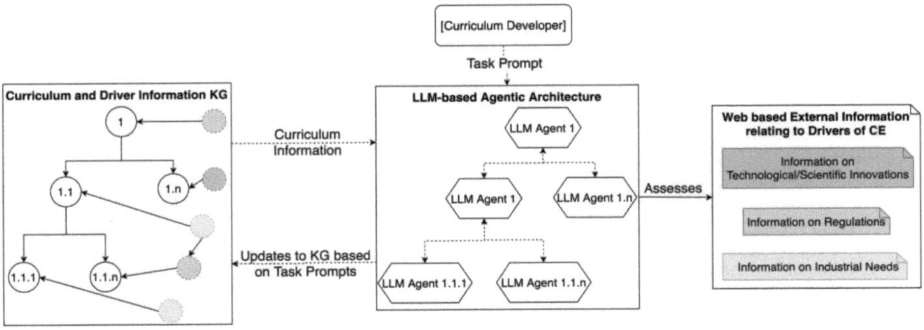

Fig. 4. Preliminary overview of the proposed solution (colors represent how influencing factors can relate to the nodes of the KG).

Some elements of the proposed solution have been developed. The *LMM-assisted KG of one branch of the university curriculum has been developed* [24]. This KG was used for two CE tasks, namely curriculum mapping to knowledge elements defined in the European Skills, Competencies, Qualifications, and Occupations taxonomy [25] and green competencies defined in the European sustainability competence framework [26]. The mapping process itself was supported by an LLM, which defines mappings between curriculum elements and competency nodes. An implementation of this system design is currently being prepared for evaluation (Phase 4 in Fig. 3).

The *preliminary implementation of an LLM-based multi-agent system* performing some simple CE-related tasks has been made using a fractal approach [27]. An agent system architecture organizes fractal units of LLM instances that enable profile definition, planning, and action execution as described in [17]. These fractal units mimic the fractal structure of CE (Fig. 1), which is represented by the aforementioned KG, with each fractal unit being responsible for updating existing nodes that represent a curriculum element or adding new nodes based on the task that is assigned via prompting. A change propagation mechanism [20] enables the knowledge gained by one fractal unit to be implemented in other fractal units, with preliminary (LLM-based) evaluation with two separate CE-related tasks demonstrating an approximate 20% increase in the performance of the system (represented by the mean evaluation results across all fractal units) after change propagation. These initial results indicate how organizing LLM-based agents according to the fractal structure of curriculum can be used to support more than one CE task, and how KG can be used as a formal representation of curricula and information related to drivers of CE (e.g., Green Competencies).

6 Conclusion and Future Works

This thesis seeks to develop, for supporting CE, an IS architecture that utilizes the potential of KGs, LLMs, and a fractal approach. The purpose of this combination is to achieve effective representation and processing of CE information to address structural and functional features of CE and to help keep up with the increasing speed of changes in the CE environment.

The current progress of the thesis has addressed the fractal nature of CE and has created an initial model to represent this. Additionally, from a technological perspective, a fractal-based system architecture that utilizes LLM-based agents that are organized in a fractal structure and are used to alter a curriculum KG based on user instructions has been implemented, with the initial results demonstrating an improved performance in CE-based tasks. Planned further research will expand on the fractal perspective of CE to arrive at models and IS architecture that effectively utilize the capabilities of KGs and LLMs in ISs operating in the context of CE.

Acknowledgments. This work was developed under the supervision of Prof. Marite Kirikova of Riga Technical University (RTU), Latvia, and supported by the doctoral studies grant of RTU.

References

1. Cai, L., Yu, C., Kang, Y., Fu, Y., Zhang, H., Zhao, Y.: Practices, opportunities and challenges in the fusion of knowledge graphs and large language models. Front. Comput. Sci. **7**, 1590632 (2025)
2. European Commission: ECTS Users' Guide 2015. Publications Office of the European Union, Luxembourg (2015)
3. Kok, D.L., Xu, M., Trumble, S., Matthews, K., Wright, C.: Categorising drivers of curriculum renewal in entry-to-practice health professional education: a scoping review. Med. Educ. **59**(8), 812–822 (2025)
4. Tang, Z., Wei, W., Yang, Y., Zhang, S., Lam, C.K.: Building AI competency knowledge graphs with LLMs: from job market analysis to educational guidance. In: 2025 IEEE International Conference on Teaching, Assessment, and Learning for Engineering (2025)
5. Lobo-Santos, A., Borrego-Díaz, J.: Enhancing mathematical knowledge graphs with large language models. Modelling **6**(3), 53 (2025)
6. Zeivots, S., Hopwood, N., Wardak, D., Cram, A.: Co-design practice in higher education: practice theory insights into collaborative curriculum development. High. Educ. Res. Dev. **44**(3), 769–783 (2025)
7. Velthuis, F., Dekker, H., Coppoolse, R., Helmich, E., Jaarsma, D.: Educators' experiences with governance in curriculum change processes; a qualitative study using rich pictures. Adv. Health Sci. Educ. **26**(3), 1027–1043 (2021)
8. Frischer, R., Singh, M.P., Porcaro, C., Namazi, H.: Fractal theory: a comprehensive review of applications across engineering disciplines. Fractals **33**(7), 1–18 (2025)
9. Peralta, M.E., Soltero, V.M.: Analysis of fractal manufacturing systems framework towards industry 4.0. J. Manuf. Syst. **57**, 46–60 (2020)
10. Kirikova, M.: Towards multifractal approach in IS development. In: Information Systems Development, pp. 295–306. Springer, Boston, MA (2009). https://doi.org/10.1007/978-0-387-68772-8_23
11. Ahmed, N.S.: A fractal-based model to improve cooperation among physicians in distributed healthcare information systems. PhD thesis, University of Malaya (2013)
12. Ersoy, M.: An IDEA for design pedagogy: devising instructional design in higher education 4.0. Des. Technol. Educ. **26**(1), 118–136 (2021)
13. Hoare, A., Dishke Hondzel, C., Wagner, S., Church, S.: A course-based approach to conducting program review. Discov. Educ. **3**, 5 (2024)
14. Annala, J., Lindén, J., Mäkinen, M., Henriksson, J.: Understanding academic agency in curriculum change in higher education. Teach. High. Educ. **28**(6), 1310–1327 (2023)

15. Trinidad, J.E., San Andres, A.N., Garnace, P.L., Guevarra, S.: Effective subunits in ineffective systems: simultaneous tight and loose coupling in higher education data use. Acta Sociologica **67**(1), 50–65 (2024)
16. Tuteja, S., Kumar, R.: A unification of heterogeneous data sources into a graph model in e-commerce. Data Sci. Eng. **7**(1), 57–70 (2022)
17. Dong, X., Zhang, X., Bu, W., Zhang, D., Cao, F.: A survey of LLM-based agents: theories, technologies, applications and suggestions. In: 2024 3rd International Conference on Artificial Intelligence, Internet of Things and Cloud Computing Technology (AIoTC), pp. 407–413. IEEE (2024)
18. Chowa, S.S., et al.: From language to action: a review of large language models as autonomous agents and tool users. Artif. Intell. Rev. **59**(2), 71 (2026)
19. Kirikova, M.: Towards flexible information architecture for fractal information systems. In: Proceedings of the International Conference on Information, Process, and Knowledge Management (eKNOW 2009), pp. 135–140. IEEE Computer Society (2009)
20. Grabis, J., Kirikova, M., Vanags, J.: Information architecture of fractal information systems. In: Proceedings of the 11th International Conference on Enterprise Information Systems - Information Systems Analysis and Specification, pp. 150–155. SciTePress (2009)
21. Peffers, K., et al.: The design science research process: a model for producing and presenting information systems research. In: DESRIST International Conference on Design Science Research in Information Systems and Technology, pp. 83–106 (2006)
22. Chang, Y., et al.: A survey on evaluation of large language models. arXiv preprint arXiv:2307.03109 (2023)
23. Es, S., James, J., Anke, L.E., Schockaert, S.: Ragas: automated evaluation of retrieval augmented generation. In: Proceedings of the 18th Conference of the European Chapter of the Association for Computational Linguistics: System Demonstrations, pp. 150–158 (2024)
24. Kumarasinghe, A.: Multi level curriculum mapping to knowledge taxonomies using graph retrieval augmented generation. GitHub. https://github.com/ArithaRTU/Multi-Level-Curriculum-Mapping-to-Knowledge-Taxonomies-using-Graph-Retrieval-Augmented-Generation. Accessed 2 Apr 2026
25. European commission: What is ESCO? ESCO portal, https://esco.ec.europa.eu/en/about-esco/what-esco. Accessed 2 Apr 2026
26. Bianchi, G., Pisiotis, U., Cabrera Giraldez, M.: GreenComp the European sustainability competence framework. Eur **30955**, 1–40 (2022)
27. Kumarasinghe, A.: FSA-LMAS: GitHub repository. https://github.com/ArithaRTU/FSA-LMAS. Accessed 2 Apr 2026

Author Index

A
Andreswari, Rachmadita 153
Arduin, Pierre-Emmanuel 29

B
Baâzaoui, Abir 90
Baâzaoui, Malek 90
Barhoumi, Walid 90
Bellara, Ikram 90
Bouchekif, Djalal 213
Bunea, Andrei-Alexandru 115
Burnay, Corentin 127

D
Danet, Brieuc 191
Daniels, Samuel Opoku 235
De Coster, Elisa 66
Deneckère, Rébecca 102
Dias, Mariana 165

F
Fahrenkrog-Petersen, Stephan A. 153
Falconi, Matteo 41
Faraji, Fatimah 223
Faulkner, Alexandre 127
Fortina, Valeria Maria 41

G
Giunta, Benito 127

H
Hakala, Ismo 16
Harjumaa, Lasse 16
Hemmati, Yasaman 202

J
Jyrkkä, Petri 16

K
Kivelä, Ilkka 16
Koppel, Tarmo 247
Krauze, Beāte 78
Kruglova, Laura 78
Kumara, Indika 41
Kumarasinghe, Aritha 261

L
Liu, Hao Yue 102
Liu, Haoyue 179
Lundberg, Jan 140

M
Mendling, Jan 153
Merad, Myriam 29
Micu-Hontan, Valentin 115
Milune, Megija Krista 54

N
Nikiforova, Oksana 54

O
Olbricht, Adrian 140

P
Plebani, Pierluigi 3, 41
Popovici, Dan-Matei 115

A. Bernasconi et al. (Eds.): RCIS 2026, LNBIP 597, pp. 273–274, 2027.
https://doi.org/10.1007/978-3-032-35043-5

S

Simonofski, Anthony 66
Skopac, Jennifer 140
Snoeck, Monique 66
Stirna, Janis 140

T

Tantaroudas, Nikolaos Dimitrios 115
Teixeira Lopes, Carla 165
Tonnarelli, Marco 41

W

Wojciechowska-Dzięcielak, Paulina 29

Y

Yang, Shudan 3

Z

Zdravkovic, Jelena 140

MIX
Papier aus verantwortungsvollen Quellen
Paper from responsible sources
FSC® C105338

If you have any concerns about our products,
you can contact us on
ProductSafety@springernature.com

In case Publisher is established outside the EU,
the EU authorized representative is:
Springer Nature Customer Service Center GmbH
Europaplatz 3, 69115 Heidelberg, Germany

Printed by Libri Plureos GmbH
in Hamburg, Germany

SpringerBriefs in Business

More information about this series at http://www.springer.com/series/8860

Hanadi Mubarak Al-Mubaraki
Ali Husain Muhammad · Michael Busler

Innovation and Entrepreneurship

Powerful Tools for a Modern Knowledge-Based Economy

Springer

Hanadi Mubarak Al-Mubaraki
College of Engineering
Kuwait University
Kuwait
Kuwait

Michael Busler
Richard Stockton College
Galloway, NJ
USA

Ali Husain Muhammad
College of Business Administration
Kuwait University
Kuwait
Kuwait

Translated by Arch Tait

ISSN 2191-5482 ISSN 2191-5490 (electronic)
SpringerBriefs in Business
ISBN 978-3-319-13682-0 ISBN 978-3-319-13683-7 (eBook)
DOI 10.1007/978-3-319-13683-7

Library of Congress Control Number: 2014955794

Springer Cham Heidelberg New York Dordrecht London

Printed on acid-free paper

Springer International Publishing AG Switzerland is part of Springer Science+Business Media (www.springer.com)

Preface

Today there is a broad recognition that entrepreneurial, knowledge-based enterprises are prime creators of economic growth and that such ventures need unique business development services. Innovation is a powerful engine for expansion and for addressing societal and global challenges. Innovation drives economic growth and job creation and is important not only for high-tech sectors but for all economic sectors. While economists and policymakers worldwide have recognized the importance of technological innovation for growth, the relationship between innovation and entrepreneurship has been appreciated only recently, specifically with regard to the important role entrepreneurs play in fomenting innovation.

The objective of this research is threefold: (1) to identify the impact of innovation, incubators, and entrepreneurship in economic development from different perspectives; (2) to identify the key performance indicators of innovation, incubators, and entrepreneurship including economics, policy, industry, and culture; and (3) to establish a roadmap for shaping economic development as a modern economy based on knowledge.

The research methodology to be used is a mixed-methods approach with quantitative (survey) and qualitative (multi-case study and interview) components to examine the innovation and incubator best practices worldwide.

Although the research analysis focuses on ten case studies from developed countries, all of the cases presented here underscore the value of innovation and business incubators. Each case study analysis includes the ratio of performance indicators over the years a particular program has been in operation since the early 1980s. It is evident that some innovation centers and incubation programs are performing better than others. Austria presents the highest rate of job creation and graduate companies, at 25 and 12.63, respectively, per year. Also, the UK presents the highest rate of client companies, at 5.53, and Austria indicates the oldest operating program, at 32 years.

The results of ten international interviews conducted by the authors in 2013–2014 of incubator programs located in the US, UK, and GCC focused on the four categories, include cultural, economic, policy, and industry using 16 indicators to measure all the categories. In 2013–2014, the authors studied ten incubator programs in the US, UK, and GCC member states. Sixteen key performance indicators (KPI) in the four categories of economic development, policy, cultural change, and industry change were used to analyze the success of the incubators. All the interview respondents presented positive impacts from different contexts, including economic development, innovation, entrepreneurship, technology transfer, and commercialization.

Analysis of the survey, case studies, radar charts (international interview), and literature review resulted in the following key findings:

1. The case studies of innovation centers and incubation programs indicated that programs tend to become more successful with experience and maturity, making years in operation a very important metric.
2. Innovation centers and incubation programs that construct strategic relationships with an international organization such as the European Business and Innovation Centre Network (EBN) and governmental bodies are more successful in technology transfer.
3. Innovation centers and incubator programs that provide continuing tangible and intangible services are able to add value to the companies they support. This leads to a high number of tenants in incubator programs and a high number of start-up companies.
4. The most common outcome from innovation is economic growth.
5. Innovation systems in developed and developing countries lead to a platform for policy decisions and high technology derivatives.

Seventy-four incubators and innovation centers from both developed and developing countries participated in a survey. A descriptive analysis of the survey responses revealed the following:

1. Two-thirds (67 %) of the services offered by incubators were strong tangible and specialized services.
2. Three-fourths (73 %) of incubators' goals focused on the entrepreneurial climate.
3. The main sponsors of incubators were non-profit economic development (48.0 %).
4. Most incubators (72.0 %) have a staff of one to five persons.
5. The majority of incubators (59.0 %) reported creating more than 50 jobs.
6. A high percentage (43.0 %) reported having graduated between six and twenty-five companies.
7. A majority of respondents (55.0 %) indicated a low increment of incubators worldwide.
8. Half of the incubators surveyed (50.0 %) indicated having tenants in the range of six to twenty-five.
9. Half of the incubators and innovation centers (49.0 %) indicated that the survival rate of their tenants ranged between 81 and 90 %.
10. Nearly half (45.0 %) of the incubators and innovation centers indicated a poor role of science parks.
11. One-third (34.0 %) of the programs indicated active cooperation with R&D.

12. A high percentage (72.0 %) of programs indicated fostering entrepreneurship at the market rate.
13. A significant percentage of programs (40.0 %) indicated a very active role for innovation.
14. Nearly half (48.0 %) of incubators and innovation centers indicated technology transfer at a modest level.
15. Most incubators and innovation centers present low rate of patents (48.0 %).
16. Low rate of licensed intellectual property (IP) in most of the innovation and incubators (47.0 %).

In addition, some of the leading rankings revealed in the study were as follows:

1. Among the ten programs located in the US, UK, and GCC that were interviewed based on selected best practices, Sussex Innovation Centre presented the highest outcomes (93 %) with respect to the other innovation centers.
2. An Austrian organization, Styrian Business Promotion Agency, SFG, was the oldest program studied, at 32 years.
3. The Austrian case study also presented the highest ratio of jobs creation, at 25.00 per year.
4. The UK case study demonstrated the highest ranking based on the ratio of client companies, at 6 per year.

In summary, the study recommends guidelines for practitioners such as government, policymakers, funding organizations, and academic institutions. Implementation of the insights from this study can be expected to result in: (1) enhanced economic development through job creation, (2) a stronger entrepreneurship climate, (3) technology commercialization and transfer for graduated companies, (4) sustainability of graduated companies in the market with high rate of survival, (5) innovation acceleration with smart product and services, and (6) diversification of the economy from companies' outcomes such as innovation and technology.

Acknowledgments

The project team would like to express their genuine appreciation to the Kuwait Foundation for the Advancement of Sciences (KFAS) for the financial support provided for the project (2012-1103-01). A special acknowledgment and appreciation is due to Prof. Rashed Al-Ajmei, Dean of the College of Business Administration, Kuwait University and Chairman of Center of Excellence in Management (CEM) for all managerial advice and support required for the project. The team also provides deep thanks to the National Business Incubation Association (NBIA) and the United Kingdom Business Incubation (UKBI) for providing successful international programs to be interviewed.

Contents

1 **Introduction** 1
  1.1 Problem Statement 3
  1.2 Project Objectives 4
  1.3 Research Question 4
  1.4 Research Methodology 4

2 **Literature Review** 7

3 **Data Collection** 13
  3.1 Interview Design 14
  3.2 International Case Studies 15
    3.2.1 US Case Study 15
    3.2.2 UK Case Study 17
    3.2.3 France Case Study 17
    3.2.4 Germany Case Study 18
    3.2.5 Spain Case Study 19
    3.2.6 Italy Case Study 19
    3.2.7 Austria Case Study 20
    3.2.8 Netherlands Case Study 21
    3.2.9 Luxemburg Case Study 21
    3.2.10 Belgium Case Study 22
  3.3 International Interviews 23
  3.4 Survey 24

4 **Data Analysis** 27
  4.1 International Cases Studies 27
  4.2 International Interview 30
    4.2.1 Interview 1: High Tech Rochester Inc. 31
    4.2.2 Interview 2: Blue Valley Schools—Center for Advanced Professional Studies—"CAPS Incubator" . . . 31

4.2.3 Interview 3: South Side Innovation Center—"Inclusive Entrepreneurship" . . . 33
4.2.4 Interview 4: Qatar Science and Technology Park, Doha, Qatar . . . 33
4.2.5 Interview 5: Institute of Knowledge Transfer . . . 35
4.2.6 Interview 6: Sussex Innovation Center, University of Sussex, Brighton, UK . . . 37
4.2.7 Interview 7: St. John's Innovation Centre . . . 38
4.2.8 Interview 8: The Petchey Center of Entrepreneurship, University of East London . . . 40
4.2.9 Interview 9: University of Birmingham, UK . . . 44
4.2.10 Interview 10: Southampton Management School, University of Southampton . . . 48
4.3 Survey Results . . . 50
4.3.1 Services of Incubator . . . 50
4.3.2 Goals of Incubator . . . 51
4.3.3 Sponsors of Incubator . . . 52
4.3.4 Staff of Incubator . . . 52
4.3.5 No. of Jobs Creation from the Incubator . . . 53
4.3.6 No. of Graduate Companies from Incubator . . . 53
4.3.7 No. of Incubators Worldwide . . . 54
4.3.8 No. of Tenants Inside the Incubator . . . 55
4.3.9 Survival Rate of Tenants . . . 55
4.3.10 Incubators Focus . . . 56
4.3.11 The Role of Science Park . . . 57
4.3.12 The Role of Cooperation of R&D . . . 58
4.3.13 The Rate of Fostering Entrepreneurship . . . 59
4.3.14 The Role of Innovation . . . 59
4.3.15 The Technology Transfer of Incubator . . . 60
4.3.16 The Rate of Patents . . . 61
4.3.17 The Rate of Licensed Intellectual Property (IP) . . . 61
4.4 Summary . . . 62
4.4.1 Summary of Surveys . . . 62
4.4.2 Summary of Interviews . . . 63
4.4.3 Summary of Case Studies . . . 63

5 Conclusion and Recommendation . . . 67
5.1 Overview . . . 67
5.2 Key Findings . . . 68
5.3 Policy Recommendations . . . 69
5.4 Kuwait Implications Policy . . . 70
5.5 Proposed Model . . . 71
5.6 Future Studies . . . 71

References . . . 73

About the Authors

Dr. Hanadi Mubarak Al-Mubaraki is an Assistant Professor in Kuwait University. She teaches project management in civil engineering for undergraduate and graduate courses as well as management course in business schools. She has published scientific articles in different academic journals and a book and has presented her research papers in many countries. Dr. Al-Mubaraki is the recipient of several international awards and medals for contribution to International Scientific Research in the Who's Who in the World 2009, 2010, 2011 and 2012 and Deputy Director General of Asia—IBC, Life Fellowship—IBA, International Peace Prize—United Cultural Conventions—UN, IBC Illuminated Diploma of honors of Professional Education twenty-first century award for Achievement, International Educators of the year 2004, Medal 2005, International Who's Who of Professional Educators 2003, 2004, 2005, 2006, 2007, 2012, Madison Who's Who professionals Life Fellowship, Marquis Who's Who, Master Degree Honor Medal 1996—Kuwait University from HH Sheikh Jaber Al-Ahmed Al-Sabah, the Amir of Kuwait. Dr. Al-Mubaraki serves on the Editorial Board of several international journals: Business, Economic Development, Management and International Business Entrepreneurship. She has substantial experience in research innovation and Entrepreneurship in developing countries, Economic Development, Green Economy, application of business incubators Worldwide, in-depth case studies, lessons learnt with best practical Incubators implications for strategy worldwide, relationships between innovation and technological transfer, sustainable development approaches and methods.

Dr. Ali Husain Muhammad is an Associate Professor of Management and head of the Management and Marketing Department in Kuwait University. He teaches management courses in the business school for undergraduate and graduate students. He has published scientific articles in different academic journals. Dr. Muhammad worked as consultant in developing work systems and organizational structures for a number of public and private organizations in Kuwait. His research interests include organizational development, organizational design, and employees' attitudes and behaviour.

Professor Michael Busler is an Associate Professor of Finance, Finance Track Coordinator and a Fellow at the William J. Hughes Center for Public Policy at Richard Stockton College. He teaches undergraduate courses in Finance and Game Theory as well as Managerial Economics and Corporate Finance in the MBA Program. He has published in eight different academic journals and has presented his research in ten countries. In addition, he has worked as a Financial Analyst for Ford Motor Company and FMC Corporation and has been an entrepreneur having owned several businesses mostly in the Real Estate development field. He earned his Doctorate at Drexel University.

Abstract

Innovation is the process of making change, difference, and novelty in products and services, to create economic and social benefits such as entrepreneurship. Innovation will create new jobs and catalyze broadly shared economic growth. The aim of this paper is to investigate and identify key factors of innovation and entrepreneurship that facilitate the outcomes of successful implementation of business incubation programs. To achieve the aim, the research uses a mixed-method approach consisting of a review of the literature, survey questionnaires, case studies, and international interviews. The research findings suggest a roadmap of the best practices for innovation and entrepreneurship worldwide. The adoption of such elements will add value to the economic development in Kuwait and other GCC member states. The research adds to the body of the current literature on sustainability of innovation and entrepreneurship as powerful tools for a modern economy. Policy makers, governments, and practitioners will benefit from the knowledge of experiences of worldwide successful implementation of innovation and incubator programs.

Keywords Innovation · Entrepreneurship · Incubators · Modern economy · Economic development · Technology transfer · Gulf cooperation council states

Chapter 1
Introduction

Abstract Business incubators (BIs) have been established around the world to stimulate new business creation. This book can be considered as the classical reading on business incubation and it provides a clear understanding of this novel and modern tool—business incubation to promote entrepreneurship. It will give both practitioners and academics a thorough overview on ins-outs of business incubation. This chapter provides the general introduction; it explains the research problem, the importance, contribution, objectives, question, method and structure of the research. The aim of this book is to investigate and identify key factors of innovation and entrepreneurship that facilitate the outcomes of successful implementation of business incubation programs. Its 5 chapters, featuring the advice and insight of United States (US), United Kingdom (UK) and Gulf Cooperation Council (GCC) states incubation professionals and interviews, are divided into five main sections.

Keywords Innovation · Entrepreneurship · Incubators

Innovation is the process of making change, difference, and novelty in the products, services, add value, and business practices to create economic and social benefit [42]. The Organisation for Economic Cooperation and Development [66] defines innovation as the implementation of a new or significantly improved product, service, process, a new marketing method, or a new organizational method in business practices, workplace organization, or external relations. The definition also considers the importance of innovation within the economic cycles and entrepreneurship. Innovation deals with: (1) new products, (2) new production methods, (3) new markets, and (4) new forms of organization. Therefore, the basic concepts of entrepreneurship, innovation, and incubation and the associated terminology must be commonly accepted and shared when putting into practice actions towards the creation of new innovation-based incubators [42].

The European Business and Innovation Centre Network [34] has reported more than 25 years of practices and policies, with lessons learned on the concept of innovation-based incubation and innovation-based incubators. Today, it is possible

H.M. Al-Mubaraki et al., *Innovation and Entrepreneurship*,
SpringerBriefs in Business, DOI 10.1007/978-3-319-13683-7_1

to understand the key elements of their success, and the tools needed to strengthen and carry out policies for innovation, small and medium-sized enterprises (SME) support, and entrepreneurship [42].

In 2010, the Strategy for American Innovation [75] shaped the ideas and technologies suitable for building the 21st century. Innovation will create new jobs and catalyze broadly shared economic growth. The strategy consists of three parts: (1) to invest in the building blocks of American innovation and to ensure the economic tools for successful innovation, from research and development to the transfer of those innovations, (2) to promote competitive markets that spur productive entrepreneurship to allow companies to be internationally competitive in innovation, and (3) to catalyze breakthroughs for national priorities.

Innovation is the driver of our future growth [34, 42, 43, 49, 75]. This requires improving the quality of our education, strengthening our research performance, promoting innovation and knowledge transfer, making full use of information and communication technologies, and ensuring that innovative ideas can turn into new products and services that create growth, quality jobs and help address global societal challenges. However, to succeed, this must be combined with entrepreneurship, finance, and a focus on user needs and market opportunities [42].

Innovation-based incubators are local economic development tools [3, 22, 26, 43, 49], which favor the conditions for creation and growth of novel business activities and contribute actively to the development of the regions where they operate. In addition, innovation-based incubators support innovative business projects, which could be either technologically-oriented or non-technologically oriented. Technology therefore is not the only unit by which to measure the degree of innovation of a business idea [42].

The National Business Incubation Association [62] defines incubation as a process that tends to be activated whenever a need to support entrepreneurs in developing their own business. The process or parts of it put in place whenever a need for nurturing would-be entrepreneurs to think over and further develop their business idea and transform it into a viable and sustainable activity. Incubators provide new high-tech venture creation, technological entrepreneurship, commercialization, and transfer of technology [1, 6–8, 26, 56, 57, 67].

According to Monkman [61], for 50 years, incubators have been helping entrepreneurs turn their ideas into viable businesses, promoting innovation, and creating jobs by providing emerging companies with business support services and resources tailored to young firms to increase their chances of success. Business incubators (BI) nurture the development of entrepreneurial companies, helping them survive and grow during the start-up period, when they are most vulnerable. These programs provide their client companies with business support services and resources tailored to young firms. The most common goals of incubation programs are creating jobs in a community, enhancing a community's entrepreneurial climate, retaining businesses in a community, building or accelerating growth in a local industry, and diversifying local economies [13–16, 29].

According to the European Business and Innovation Centre (BIC) observatory report [36], BICs are playing a leading role in Europe and have supported more than 67,700 business plans. BICs have directly assisted the creation of more than 31,700 new jobs and have identified more than 66,900 enterprise-creation projects. More than 17,400 new innovative enterprises were created with the assistance of these centres. More than 15,300 tenants (enterprises) are located in the BIC incubators, supporting more than 92,200 jobs. The average survival rate of the enterprises supported by BICs is 89 % [36].

Many reports demonstrate [36, 66] that the EBN is instrumental in shaping the knowledge-based economy of the future, anchored in value-added jobs, robust economic models, technology and innovation-based enterprises, academic spin-offs, dynamic clusters, and efficient accelerators of the "Growth and Jobs" challenge. BICs are organizations that promote innovation and entrepreneurship. They help enterprises to innovate; they drive the creation of start-ups (support to innovation, incubation, and internationalization) and they promote economic development through job and enterprise creation and development.

Today, according to various reports [36, 66] the innovation research and development (R&D) spending in Europe is below 2 %, compared to 2.6 % in the US and 3.4 % in Japan, mainly due to lower levels of private investment. The aim of this is to re-focus R&D and innovation policy on the challenges facing our society. Every link needs to strengthen in the innovation chain, from *blue sky* research to commercialization. The essential benefits of economic growth should reach the public because the direct relationship between the innovation, entrepreneurship and job creation basis on incubators [36, 66].

1.1 Problem Statement

While innovation and entrepreneurship have been recognised to offer a range of advantages to the modern economy based on knowledge, their actual impact on economic development has generally been measured in terms of basic quantitative measures such as number of jobs created, firm graduation rates and tax receipts [55].

The theoretical arguments and the majority of research in this area has been conducted in United States and Europe; however, there are significant gaps in knowledge regarding the impact of innovation and entrepreneurship in promoting economic development in developing countries, such as Kuwait and others Gulf Cooperation Council (GCC) states. The findings of this research will add value to both academia and practitioners such as policy makers and governments, based on qualitative and quantitative approaches.

1.2 Project Objectives

Following are the objectives of current project:

1. To study the impact of innovation programs and business incubation programs in economic development from different perspectives.
2. To identify and measure key performance indicators of innovation centers and incubators in the four categories of culture, policy, industry, and economy.
3. To establish a roadmap for shaping modern knowledge-based economic development.

1.3 Research Question

The current project has following research question:

1. What are the impacts of the four categories—culture, policy, industry, and economy—on innovation programs and incubators?

1.4 Research Methodology

The proposed research will employ a mixed-methods approach in which both quantitative (survey questionnaire) and qualitative (literature review and case studies) data will be gathered (see Fig. 1.1). Specifically, the steps to be followed will consist of the following:

1. **Identification**: The impact of innovation and entrepreneurship on economic development.
2. **Analysis**: Analysis of the criteria of innovation and entrepreneurship as powerful tools for a modern economy based on knowledge, with each criterion to be analyzed through literature review, case studies, and an international questionnaire survey.
3. **Roadmap**: Roadmaps will elaborate as a modern tool for shaping economic development and as an aid to government to implement this program. This roadmap focuses on the analysis of best practices for innovation and entrepreneurship worldwide.

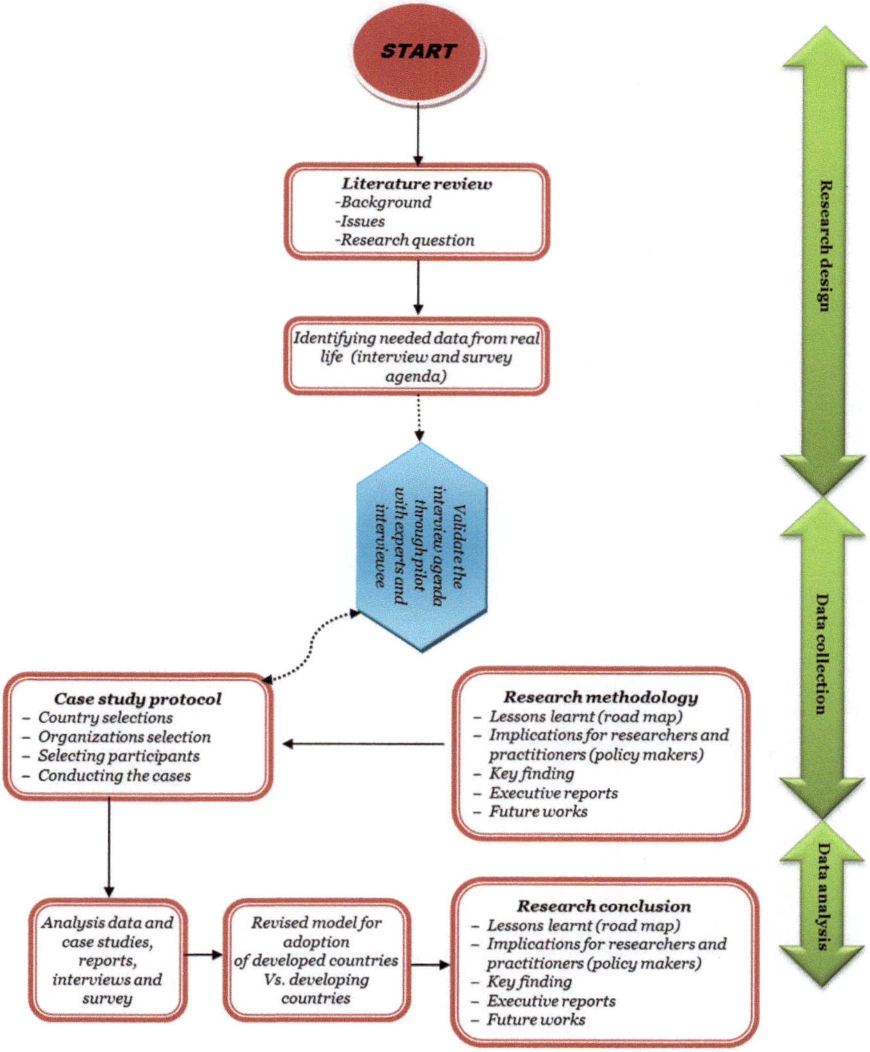

Fig. 1.1 The process of developing a research methodology

Chapter 2
Literature Review

Abstract This chapter systematically reviews the literature on business incubators and business incubation. Focusing on the primary research orientations—i.e. studies centering on incubator development, configurations, development, incubator-incubation impacts, and theorizing about incubators-incubation—problems with extant research are analyzed and opportunities for future research are identified. From our review, it is clear that research has just begun to scratch the surface of the incubator-incubation phenomenon. This chapter discusses the main results from the international cases studies and provides the conceptual framework and a thorough review of the international literature.

Keywords Innovation · Entrepreneurship · Incubators · Literature review · Conceptual framework

Innovation centers provide entrepreneurs with expertise, networks, and tools they need to make their ventures successful, Al-Mubaraki and Busler [22] discussed European models based on their adoption as case study examples: the United Kingdom, France, and Germany. They account for 83 % of all the incubators located throughout Europe today.

Innovation plays an important role in generating policy-related information about innovation processes, behaviour, and performance. Innovation is persistent in every phase of life, but its significance becomes clearer when observed in the perspective of economic development. Incubator programs in the developing and restructuring countries are naturally focused on technology ventures. The incubation process that was developed has included services for on-the-spot diagnosis and treatment of business problems, dramatically lowering the usual early stage failure rate.

Innovation is one of the main drivers of sustained economic growth from practitioner perspectives such as governments, policy maker and funded foundation and academic institution. In addition, innovation indicators consist of basic research, applied research, development, and commercialization which contributed

H.M. Al-Mubaraki et al., *Innovation and Entrepreneurship*,
SpringerBriefs in Business, DOI 10.1007/978-3-319-13683-7_2

to economic benefits includes productivity improvements and economic growth [30, 31, 46]. Furthermore, innovation system is an important tool for governments and commonly used performance indicators are R&D intensity and GDP per capita [50–52]. Another study used the combination of science, technology and innovation (STI) as an indicators and the used of science and technology can contributed positively on the social and economic which business competition is increasingly based on innovation [44, 45, 69]. Moreover, European countries used innovation indicators as methodology, for example, European Innovation Index have been published [38–41, 65]. Finally, Japan used S&T activities includes inputs in R&D, staff, output and number of scientific paper citations as ranking multiplicity of indicators [54, 63, 64].

Al-Mubaraki and Busler [14] indicated that innovation programs can help young firms to survive and grow during their start-up years, and can play a key role in the economic development of a community or region. In developing countries, business incubators are particularly valuable in helping to develop local economies, promote technology transfer, create new enterprises, and generate jobs. These outcomes are used to make recommendations for maximizing the success of incubators. Successful innovation systems merge economic, social, political, organizational, and institutional elements, while supporting the development and diffusion of new technologies in compliance with governmental initiatives.

The European Business and Innovation Network [37] indicates the percentages of the groups of innovation as divided in Europe into three groups: technological innovation 51.49 %, non-technological innovation 38.34 % and non-innovation 10.16 %. The main focus of business innovation centers (BICs) was to support start-ups firms. In fact, BICs supported 2,491 companies and requested 666 patents for companies and entrepreneurs, resulting in a 307 granted patents.

Al-Mubaraki and Busler [5] studied incubator programs in France, Spain, Netherlands, Luxemburg and Portugal. They utilized a SWOT analysis to reflect the strengths of each program according to their compliance with their missions and objectives, opportunities for future plans, and the performance of each program.

A business incubator (Business and Innovation Center) is a physical facility aimed at promoting economic development of its community by supporting start-up companies and their business development. Business incubator programs, often called "new entrepreneur creation projects" help extend new entrepreneurs and support them to start businesses and be better able to survive on a longer-term sustainable basis. Typical business incubators include small entrepreneurs that want to develop new businesses and those who would like to expand their talent and ideas or engage in the transfer of technology. Its measures indicate economic growth within the community rather than outside of it [1, 56, 57, 59, 67].

Al-Mubaraki and Busler [5] considered innovation centers as cost-effective economic development processes. Innovation could be an effectual driver for economic development at appreciably higher cost than originally anticipated unless a thorough and objective feasibility study is planned, performed, and best-practice is applied.

Business incubators can play an active role in local, regional, and national economic development efforts. However, business incubators cannot transform an economy and must be integrated into broader economic policy reform infrastructure [27]. Innovations that are novel to the world are predominantly found in the progressive economies and are based on research and development at the frontiers of global knowledge. In developing countries far removed from the international technological boundary, innovations tend to be new to the market or new to the firm. Innovations new to the market in developing countries refer to the international diffusion and absorption of technology.

Al-Mubaraki and Busler [6] found that business incubators are innovative tools in supporting start-up businesses. Their empirical results highlight implications for successfully developing and implementing best practices of business incubation programs that contribute to knowledge about the process of business incubation.

Al-Mubaraki and Busler [7] used a mixed-method approach to study business incubation as a tool for economic development based on economic indicators from incubation outcomes such as: (1) entrepreneurs, (2) companies created, (3) jobs created, and (4) incubator companies.

Al-Mubaraki and Schrödl [23] evaluated aspects of innovation centers or business incubation that can be used to measure the effectiveness of business incubation. Their study used a multi-method approach combining desk-research, interviews, and a multi-case study of five incubator organizations in the GCC member states. From these findings, a model for measuring the effectiveness of business incubation in a standardized way was developed. The developed model supports the work of incubator managers, policy makers, researchers, practitioners, stakeholders, and government parties for effective execution of business incubation enterprises. The researchers were able to determine the effectiveness of business incubators (or innovation centers) individually and as an industry using four dimensions for gauging the health of programs: (1) the number of businesses graduated over a period of time, (2) the number of businesses still in business over a period of time, jobs created by incubator clients, and (3) salaries paid by incubator clients.

Al-Mubaraki and Wong [25] studied innovation centers in Europe and developed countries. Incubators are used as much for spurring regional economic progress and establishing industry clusters as they are for revitalizing urban environments and industry. The study describes the important role and the impact of incubation on the economic development process of GCC countries.

Business innovation centre's goals are to produce successful businesses that will leave the program financially viable and freestanding. They create jobs, revitalize communities, commercialize new technologies and create wealth for local and national economies [8].

Al-Mubaraki and Busler [9] summarized 40 innovation centers in the Middle East according to seven categories: (1) type, (2) financial model, (3) services, (4) objectives, (5) number of clients, (6) number of graduates, and (7) jobs creation with a description of each incubator. In addition, chain incubators program created greater than 61 (20 %) jobs with a total number of graduated companies greater than 14 (40 %). Furthermore, the number of the client companies inside the incubators greater than 21 (45 %), the percentage of financial model of not-for profit incubators 80 % and the 98 % of incubators in Middle East have as their main objectives the support of entrepreneurship and profitable enterprises.

Al-Mubaraki and Busler [10] applied a mixed-method approach, which states that the business incubation is a tool for economic development with typical outcomes such as assisting entrepreneurs, starting new companies, and creating new jobs. It was proved in both the United States and developed countries, but is still taking shape in the developing countries such as the GCC member states.

Al-Mubaraki and Busler [11] compared the research grounded on a mixed-method style using both qualitative and quantitative methods. The exploration indicates the incubator or innovation program is a place where the incubation activities are carried out, and where the would-be entrepreneurs and the existing SMEs find a suitable place, in terms of facilities and expertise, to address their needs and develop their business ideas, and transforms them into sustainable realities.

Al-Mubaraki and Busler [12] present innovation programs as effective tools for technology transfer, innovation, new technology, and research incentive. The findings of the study included: (1) Technology commercialization from the university innovation programs leads to economic development based on the technology transfer license income and the high rate of total royalty income of $1.3 B; (2) Innovation incubators increased the research incentive fund to $4.5 M; (3) Innovation based on the university incubators produced over 8,000 inventions; and (4) The new technology adds value to the market with an accumulative total $878 M.

Al-Mubaraki and Wong [25] present four reasons for why some incubators or innovation program perform better than others: (1) incubators may differ in terms of their size, (2) the services provided by each incubator may not be the same, (3) incubators may differ in their focus of services, and (4) the source of sponsorship differs across different incubators.

Anderson and Al-Mubaraki [2] pinpointed the key components for developing a successful business incubator as an effective tool for economic development, based on the case study of an unsuccessful effort. Examination about the short life of a highly anticipated business incubator in an area of Southern California, Gateway Innovation Center was reported. It sought to foster new businesses and create jobs in a region of relatively high unemployment. They explored various missteps in the formation of the Gateway Innovation Center and provided a better understanding of key issues in developing successful incubators. The lack of success was attributed to five features: function following form; lack of planning; lack of expertise; lack of due diligence; and being placed in a market area that would not support a technology incubator.

Al-Mubaraki et al. [28] ranked business incubators according to four key indicators, such as financial data, mission, size, and obstacles, using mathematical techniques which can contribute to the acceleration of the economic development process of incubators within business incubation programs.

Al-Mubaraki and Busler [13] compared incubator programs in the US and Brazil. There are six key findings: (1) The strategic focus in both countries is economic development, technology transfer, and job creation. (2) Entrepreneurship is very active in both the U.S. and Brazil. (3) For incubator funding, the stakeholders are mainly the government, businesses and universities. (4) Incubator services in both countries include both tangible and intangible services. (5) The culture in the U.S. is more risk-taking whereas Brazil is more risk-averse. (6) Innovation is very active in both the U.S. and Brazil.

Al-Mubaraki and Busler [14] conducted a quantitative and qualitative study of success rates and the key indicators of incubators in several countries. They then used their insights to develop a model of best practices learned from case studies. The model suggests that an incubatee's progress to sustainable graduation depends on: (1) clear objectives, (2) the incubator's location, (3) access to services, (4) employment creation, and (5) economic development strategy. Moreover, when followed, the best practice model can lead to a 90 % survival rate of companies and replicates sustainability in the market.

Al-Mubaraki and Busler [15] specified innovation and entrepreneurship as critical drivers of social and economic development. They found evidence of increasing global awareness of the importance of incubators, especially in developing countries such as Asian countries, for the purpose of promoting innovation and entrepreneurship, which was reported by 96 % of those surveyed. Moreover, policymakers and other stakeholders increasingly view business incubation as an important tool to create sustainable jobs (60 %), with youth (90 %) and technology sectors (82 %) being common targets.

Al-Mubaraki and Busler [16] used a SWOT analysis to evaluate innovation systems models and their potential use worldwide. The study adds new and useful knowledge for both academics and practitioners who are interested in systems innovation. Five European case studies: France, Spain, Netherlands, Luxemburg, and Portugal were chosen based on their successful innovation centers throughout Europe. Five criteria used to evaluate the European case studies were: (1) the legal status, (2) target market, (3) stakeholders, (4) entrepreneurship, and (5) job creation. All the criteria are related to the economic development of European countries.

Al-Mubaraki and Busler [19] gave a solid base of evidence to support recommendations of forward-looking strategy measures to advance the rate of successful commercialization in Latin America through a more focused framework for business and technology incubation activities and organizations.

Al-Mubaraki and Busler [20] investigated the youth entrepreneurship dimension as an outcome from business incubation program. The empirical results highlight some implications for successful development and implementation of best practices for creating an entrepreneurial generation to support economic development. The work contributes to the knowledge about the youth entrepreneurship.

Al-Mubaraki and Hamad [21] concluded that the incubator organizations or innovation programs are model accelerator tools for the 21st century. This study has clearly stated that business incubation programs are designed to accelerate the successful development of entrepreneurial companies through an array of business support resources and services, developed by incubator management.

Al-Mubaraki and Schrödl [24] identified and assessed critical dimensions of business incubation with focus on the GCC. These are suitable to measure the effectiveness of business incubation to support startup and entrepreneurial businesses by providing a number of services and resources to the clients. Their proposed model is suitable in more highly developed economic regions and helps practitioners and government parties with future implementation of incubators program. Furthermore, it adds new knowledge for the academic literature on a commonly agreed model for effective measurement of business incubators.

Al-Mubaraki et al. [17] provide useful implementation guidelines to both academics and practitioners involved with incubators worldwide. They focused on six key success indicators and pointed out proposed incubators model in the years to come. The design methodology is based on survey and interviews of 100 selected incubators. The research findings indicate the similarities of incubator programs that are (1) creating jobs, (2) enhancing the community's entrepreneurial climate, and (3) providing tangible services. Differences noted were (1) Incubator's type, (2) Financial status, and (3) Incubator's age.

Al-Mubaraki and Busler [18] categorized the impact of innovation and entrepreneurship as a tool for a dynamic economic model based on the successful practice of innovation program case studies. They found that the adoption of innovation programs leads to: (1) a high rate of networking and outcomes, (2) high potential financing and strategic planning, (3) fostering entrepreneurship and innovation, research commercialization, and supporting technological entrepreneurship, (4) high number of jobs created, and (5) successful start-up companies with high survival rates. The empirical results highlight implications for practitioners such government and academic institution and makes a contribution to knowledge about the innovation and entrepreneurship in developing countries.

Chapter 3
Data Collection

Abstract Business incubators provide an important service network for new and fledgling Small and Medium Enterprises (SMEs). To ascertain the perception of the importance of business incubation process, perceptions of international interviews are compared. The survey endeavoured to cover all types of business-incubation programme in United States (US), United Kingdom (UK) and Gulf Cooperation Council (GCC). This chapter discusses the main findings from the international interviews and focuses on detailed issues associated to innovation, incubators and modern economy based on the knowledge; this implies that the use of mixed-method approach is most appropriate. The study has included a survey of a sample of seventy-four organizations, ten real interviews focused in the US, UK and GCC, and ten international multi-case studies that investigated, addressed and explained the successful adoption of innovation centres and business incubators worldwide. This study is based on field interviews conducted by the authors in the United States, the United Kingdom, and Qatar during 2013–2014. The interview instruments for the semi-structured in-depth interviews were developed after a thorough literature review. UK case study demonstrated the highest ranking based on the ratio of client companies, at 6 per year.

Keywords Business incubators · International interview · Modern economy · Mixed method approach

This study concentrates on a specific context, i.e. the innovation, incubators and modern economy based on the knowledge, making the mixed-method approach most appropriate. The investigation and analysis of literature is an accepted form of desk-based research that compares the works of different authors [47]. This approach uses quantitative data (survey questionnaire) of trends derived from a study of a population. Then, the author generalizes the results and tests the impact of outcomes, controlling for all other indicators that might influence that outcome. Furthermore, qualitative data from multi-case studies, literature review and interviews combined with the quantitative research [32] in order to a powerful mixed-methods study. This approach allows a broader assessment of a particular and real situation [77]. The notion of applying different methods of data

H.M. Al-Mubaraki et al., *Innovation and Entrepreneurship*,
SpringerBriefs in Business, DOI 10.1007/978-3-319-13683-7_3

collection focuses on valid and reliable case findings and reports [32, 78]. In a case study strategy, many sources of evidence can be used [78]. Such sources include documentation, archival records, interviews, observation, and physical artifacts. Ten interviews provided the main source of evidence used in the current study. The case study allows researchers to gain an in-depth understanding of the phenomenon under investigation [79]. Furthermore, it provides both an understanding of the research context and a rich insight into the issue being examined [33, 76].

3.1 Interview Design

Many studies in the literature discuss the importance of semi-structured interviews as an important tool for gathering qualitative data [48, 53, 70, 72]. Furthermore, semi-structured interviews are a good technique using open-ended questions in which the respondents are encouraged to answer in their own words and questions whose content and sequence are not fully specified in advance [48].

Interviews will conduct with ten innovation directors in the United States (US), United Kingdom (UK) and GCC states will be interviewed. The interview questions designed to better explain the findings of secondary data in the best way. The key questions of the interview are summarized in the radar chart with the chart divided into four categories: (1) economy, (2) culture, (3) policy, and (4) industry. Each category includes key indicators to be measured (see Fig. 3.1).

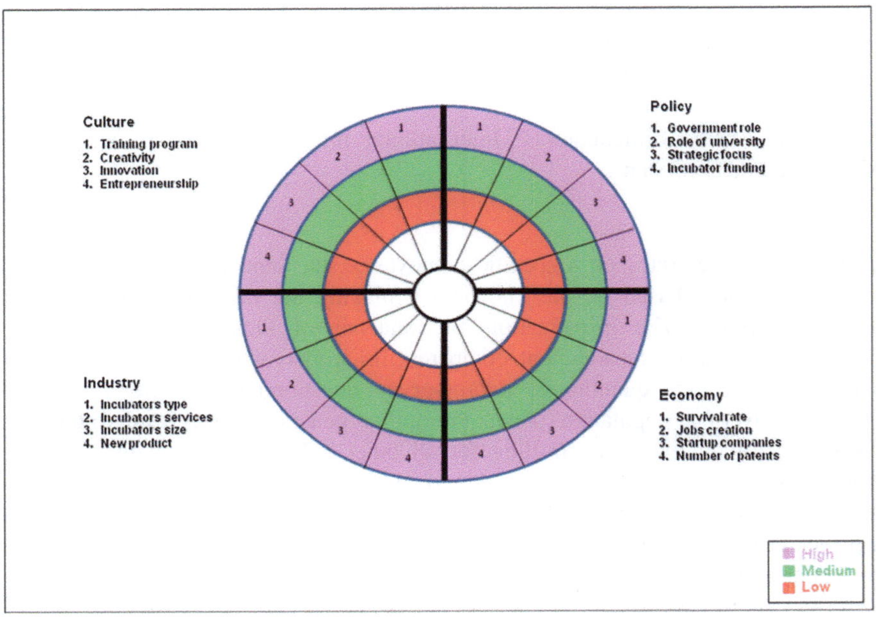

Fig. 3.1 Radar chart

3.2 International Case Studies

In case study research, the researcher can collect detailed information by using a variety of data collection approaches. Gordon and Langmaid indicated that the use of case study research leads to a deeper understanding of how things happen rather than just testing relationships between them. However, case study research focused on the understanding of areas of organizational functioning are not yet well documented.

The ten case studies to be examined are structured following the research design. The authors collected data from successful international case studies of innovation programs in selected developed countries. Table 3.1 shows each case study including the innovation program's profile such as region, companies in the region (in 2008), year founded, stakeholders, and catchment area for each country.

3.2.1 US Case Study

Technology continues to be the driving force for the new economy and is critical to Maryland's future. Maryland is richly endowed with technology assets, having the largest concentration of federal laboratories of any state in the nation and host to nationally recognized research universities, a federal contractor base with a specialty in science, security and communications, and an emergent biotechnology cluster. The Maryland Technology Development Corporation (TEDCO) was created by the state legislature in 1998. Its first objective was to assist in transferring to the private sector and commercializing the results and products of scientific research and development conducted by colleges and universities. Its second objective was to assist in the commercialization of technology developed in the private sector. Its third objective was to foster the commercialization of research and development and to create and sustain businesses throughout all regions of the state [71].

The mission of TEDCO [71] is to foster the development of a technology-driven economy that will create and sustain businesses throughout all regions of the State of Maryland. TEDCO's identified role is to be Maryland's leading source of funding for seed capital and entrepreneurial business assistance for the development, transfer, and commercialization of technology. The goals and objectives of business incubation are to encourage, promote, stimulate, and support Research and Development (R&D) activity using different investments, which leads to commercialization of new products and services by small businesses.

Research Triangle Institute (RTI) International [68] conducted a comprehensive study of TEDCO. Maryland currently has 19 technology incubators and seven proposed incubator projects. The key results and findings of RTI international are as follows; all figures are related to TEDCO in 2006. First, the gross state product increased by $1.2 billion. Second, the total annual employment impact of technology incubators was 14,044 jobs. Third, new jobs contributed $845 m

Table 3.1 Profile of case studies

Courtiers case studies

Profile	USA	UK	France	Germany	Spain	Italy	Austria	Netherlands	Luxemburg	Belgium
Region	Maryland	North East of England	Normandy	Bavaria	Navarra	Trentino province	Styria	Friesland	N/A	Province of antwerp
Companies in the region (2008)	N/A	–	13,200	>560,000	43,847	48,525 (registered at chamber of commerce)	51,000	35,983	28,800	114,617
Region population (2008)	N/A	2,515,442	220,000	12,520,000	620,377	470,000	1,200,000	642,230	483,800	1,700,000
Founded year	N/A	1994	1987	2005	1988	1987	1981	2002	1987	1987
Stakeholders	Government of Maryland state	Existing and potential tenants, general public of North East England, members of the regional business community, existing and potential partner organizations	Local government-enterprises	AZO Shareholders	45 companies and regional organizations; the government of navarra is the major stakeholders	Local government (provincia autonoma di trento), trento university, provincial research centres, industrial and artisan associations	Styrian government, banks	Regional innovation agency and private companies	Ministry of culture, higher education and research, the ministry of the economy and the European regional development fund	Governmental bodies Universities and Research centres regional development institutions bank and insurance companies industrial companies
Catchment area	N/A	North East England	1,450,000 citizens	ESNC: international/worldwide	10,391 m^2	Inconsistent with other examples	Styria	Friesland	Luxemburg	Province of antwerp and Kempen region

Source [43]

in annual salary. Fourth, the technology incubators in the state increased the state and local tax revenue by approximately $104 million per year. Fifth, gross state product contributions totaled $1.2 billion to increase state output by $2.7 billion per year. Sixth, TEDCO contributed $104 million in state and local taxes. Seventh, for every $1 of incubator assistance funding provided by TEDCO, tenant companies contributed $1,800 to Maryland's gross state product. Eighth, TEDCO made an average investment of $120 per incubator company job.

3.2.2 UK Case Study

The North East Business and Innovation Centre (BIC)'s story began in 1994, when the make-up of the North East region, and British industry itself, was undergoing a transformation. Traditional industries were being resigned to the history books, and with this followed a great deal of uncertainty. Eventually, however, their demise led to a new breed of businesses. Organizations like the BIC began to emerge, and helped to set the foundations for the development of an enterprising and innovative culture in the region. The BIC designed to nurture new businesses by providing access to a network of experts, accommodation and business support. The BIC opened in June 1994, providing a home for 22 businesses and support for many more region-wide. In 2010, the BIC site could accommodate over 160 businesses from a range of sectors, and covers a 14-acre site. Their business support team assists over 300 businesses in the startup process each year [34–36, 43].

The North East BIC has invested around £170 K in establishing four unique divisions: (1) an education arm for ideas generation, (2) youth enterprise facility, (3) a product development company, and (4) a commercial marketing firm and an IT service provider. The BIC created a new role, tenant business mentor [34–36, 43].

Key Performance Indicators for 2009 indicated that the number of tenants in incubators was 105, employing a total of 911 people. The number of tenants since the beginning of the incubator was 489. The number of start-ups created in 2009 was 1,114,502 jobs created. The number of jobs created in SMEs (client companies of BICs) was 85 with an Enterprise Survival Rate of 75 %. In 2009, 155 enterprise creation projects were developed and 167 existing SMEs were supported [34–36, 43].

3.2.3 France Case Study

During the last decades, institutions of higher education all over the world have experienced a transformation by broadening their traditional mission of teaching, research, and public service to include a more active participation in their region's economic development. This is the case of the University of Bourgogne, which in 1999 decided to create with seven other key regional stakeholders, *Pôle de Ressources et de Management de l'Innovation et de la Création d'Entreprises*

(PREMICE). PREMICE was founded in 2000 by the French Research Ministry, regional agencies, and economic entities and the University of Bourgogne [34, 43].

PREMICE's mission is to promote, encourage, and develop high innovative and technology firms based on university or R&D centre discoveries. For that reason, the centre works intensively with universities and research centres in the region, in order to identify new ideas that have commercial potential and help to bring them to the market. Moreover, PREMICE has extended its services to SMEs in the region that are involved in innovative activities [34, 43]. PREMICE's value-added services depend on financial perspectives. The business incubators offer or assist their clients in obtaining capital by acting as a broker to facilitate the interaction between the clients and the potential investors. PREMICE provides: (1) support and services to new knowledge-based ventures, (2) an integrated range of services from scientific and economic evaluations of innovative projects, and has (3) contributed to the creation of the START EST Network in five regions of the East of France [34, 43].

Results confirmed that incubator programs in France in 2007 assisted 15 entrepreneurs and since the start of PREMICE, 25. In 2007, 17 companies were formed creating 112 jobs; cumulatively 44 companies were created with PREMICE support, generating 155 jobs. The number of companies in the incubator in 2007 was 11, which created 44 jobs and graduated 75 companies [35, 36].

3.2.4 Germany Case Study

Anwendungszentrum GmbH Oberpfaffenhofen (AZO) incubation centre is located in Bavaria in Oberpfaffenhofen, near Munich. The region has a population of 12,520,000 and approximately 560,000 companies. Anwendungszentrum GmbH Oberpfaffenhofen–AZO has helped to found 38 companies and create 650 new jobs since 2002. Phase I ran as a German Aerospace Center (DLR) project between 2002 and 2005 and was one of the most successful business incubators in the European aviation industry. The AZO Incubation Centre accepts ten incubatees per year for a duration of four years until 2013. Their 2005–2009 funding was €1,842,000 [34–36, 43].

AZO's close ties with the community of Oberpfaffenhofen and the ambitions of DLR–particularly in positioning their common location to play a major role in the implementation of the Galileo satellite navigation system with the support of the Free State of Bavaria–have helped bring about a number of successes in recent years. The fourth ESA Business Incubation Centre (BIC) has been existed since August 2009 and has funding of €300,000 until 2013. ESA BIC Oberpfaffenhofen's mission is to expand AZO's current start-up portfolio to include all areas of the aviation industry, as well as the transfer of technology to other branches of industry [43].

Key Performance Indicators of 2009 indicated that the number of tenants in incubators was three, employing 16 people. The number of start-ups created in

2009 was six; however, figures for jobs created are not available. The number of jobs created in SMEs (client companies of BICs) was 16 with Enterprise Survival Rate of 100 %. Six enterprise creation projects were developed and 10 existing SMEs were supported during 2009 [35, 36].

3.2.5 Spain Case Study

BIC Berrilan is a non-profit-making Public Limited Company, founded in 1993 by a group of public and private bodies in accordance with the Business and Innovation Centre (BIC) model of the European Commission. BIC Berrilan's aim is to promote the self-generation of wealth and employment in Gipuzkoa through the creation of innovative businesses with capacity for growth and long-term stability and the incorporation of innovation in existing SMEs. The centre does this by means of active industrial promotion, the mobilisation of resources and the competitive added value contribution and the incorporation of innovation in existing SMEs [34, 43].

Within the creation of the Basque Network of Science, Technology and Innovation, BIC Berrilan is an intermediate agent, which facilitates and enables the Technology and Knowledge transfer processes to the market through the creation of innovative and technology based start-ups and the incorporation of innovation in SMEs. BIC Gipuzkoa Berrilan's main objectives are to be the preferred instrument for qualified entrepreneurs as a support for the creation and development of innovative, technology-based businesses and to be recognised for guaranteeing the satisfaction of customers, partners and collaborators, as well as its own workforce [34, 43].

ENTREPRENARI is a strategic alliance between the University of the Basque Country, the "Entrepreneurial Campus", whose mission is the generation and promotion of the entrepreneurial culture inside the University. This supports new entrepreneurial initiatives, which come from technology transfer, and innovation management processes inside the University [34, 43].

Results confirmed that incubator programs in Spain in 2007 assisted 33 entrepreneurs and since the start of the BIC, 163. In 2007, 23 companies were formed creating 99 jobs; cumulatively 83 companies were created with BIC support generating 640 jobs. The number of companies in the incubator in 2007 was 23 and 60 companies had graduated from the program [34–36, 43].

3.2.6 Italy Case Study

Business Innovation Centre Lazio is located in one of the most economically dynamic regions in Italy: the Region of Lazio. With a population of 5.3 million, Lazio is one of the most important regions in Italy, accounting for 10.1 % of the

GDP. Business Innovation Centre Lazio, founded in 1990 by the Latium Region. Lazio is a joint stock (PLC) corporation, part of a network of regional development offices, whose main aim is to promote and encourage new enterprises, update and modernize existing projects, introduce an enterprise culture throughout the region, and to offer managerial and technical knowledge to local public offices [34, 43].

BIC Lazio's catchment area is one with 5.3 million of habitants (55 % of the population concentrated in Rome), 5 provinces, and 17,208 km^2. In order to reach every single entrepreneur and SME in the region, not only is it necessary to have offices in different locations throughout the region but also to customize the services offered to the reality of the each area [34, 43].

Results confirmed that incubator programs in Italy in 2007 assisted 3,856 entrepreneurs. In 2007, 256 companies were formed creating 386 jobs; cumulatively 2,196 companies created with BIC support generating 3,294 jobs. The number of companies in the incubator in 2007 was 42 and 62 companies had graduated from the program [34–36, 43].

3.2.7 Austria Case Study

In 1981, Styrian government established the Styrian Business Promotion Agency and SFG. The Impulse Centres play an important role in Styria's business strategy laid down in the strategic guideline "Regions and Infrastructure". The thirty centres, distributed over the whole of Styria, offer space for more than 460 companies with approximately 4,000 employees, allowing them to implement ideas, initiatives, and innovations. It is SFG's goal to expand these platforms of entrepreneurial activity and to offer an improved general framework to innovative companies in their foundation, development, and growth phase. To make this possible, SFG maintains the already existing centres and promotes access to knowledge infrastructure for companies in the regions [34, 43].

Value-added services of SFG include: (1) company development; (2) strengthening of regions by focusing on different sectors of technology, new jobs through settlement, and growth of companies, and (3) visualization of new technologies [34, 43].

Key Performance Indicators of 2009 indicated the number of tenants in incubators was 170, employing 1,750 people. The number of start-ups created in 2009 was three; however, figures for jobs created were seven. The number of jobs created in SMEs (client companies of BICs) was 15 with the number of start-ups created since the beginning of the incubator at 404. The enterprise survival rate was 15 % [34–36, 43].

3.2.8 Netherlands Case Study

Business Development Friesland (BDF) is located in Friesland, the northern part of the Netherlands. The region tries to improve its economy by stimulating investments, export, and innovative entrepreneurship. The centre contributes to this objective by setting up and carrying out projects for start-ups and existing SMEs [34, 43].

BDF was founded in October 2002. The BIC is involved in the most important start-up programs and the BDF is now a specialist on start-ups in the Province of Friesland. Furthermore BDF strongly supports local SMEs with coaching, export development, and financing [34, 43].

The Office of Small Business Development Centres (SBDC) in the United States of America, a cooperative effort of the private sector, the educational community, and federal, state and local governments, provides management assistance to current and prospective small business owners. The SBDC model followed by BDF is based on long-term relationships with their clients. BDF knows very well that the ability to access global markets has become a strategic tool to (1) guarantee survival, (2) ensure the expansion of many companies, (3) assist entrepreneurs, (4) help companies to introduce their products and services into international markets, and (5) network with large international companies, intermediaries, local governments and knowledge institutes [34, 43].

Results confirmed that incubator programs in Netherlands in 2007 assisted 48 entrepreneurs and since the start of BIC, +100. In 2007, 23 companies were formed creating 97 jobs; cumulatively 63 companies were created with BIC support generating +300 jobs [34–36, 43].

3.2.9 Luxemburg Case Study

Henri Tudor Research Centre was created in 1987 with the main mission of strengthening the economic and social innovation network of the Grand-duchy of Luxembourg with a European open-minded way of thinking. "From research to innovation" is the leitmotiv that the Centre has kept in mind at each stage of its development. The centre is carrying out a large scope of services and activities from applied and experimental research, doctoral research, development of tools, methods, labels, certifications and standards, technological assistance, consulting and watch services, knowledge and competences transfer, training and high-level qualification as well as incubation of high-tech companies [34, 43].

The centre is divided into eight departments; one of them is the Technoport, which is a technology-oriented business incubator created in 1998 with the objective of supporting and promoting the creation and development of technological and innovative companies in Luxembourg. Technoport is an initiative of the Henri Tudor Researcher Centre, supported by the Ministry of Culture, Higher Education and Research, the Ministry of the Economy, and the European Regional Development Fund [34, 43].

The value-added services Technoport offers include typical incubation services, namely physical resources, business assistance, access to networks, and also a series of research centre related benefits, such as access to scientific and technological knowledge as well as laboratories and equipment in specific cases [34, 43].

University business incubators seem to provide a nurturing environment for the development of technology based firms through a combination of much needed university-related inputs and other typical incubator services [34, 43, 58].

Results confirmed that incubator programs in Luxemburg in 2007 assisted 55 entrepreneurs. In 2007, four companies were formed creating 26 jobs; cumulatively 56 companies were created with BIC support generating 364 jobs. The number of companies in the incubator in 2007 was 13, which created 62 jobs. The program graduated 12 companies with 260 being created in the market [34–36, 43].

3.2.10 Belgium Case Study

In 2005, the Walloon government put in place a new revival plan aimed at regenerating Wallonia's economy. This plan tries to promote the setting up and development of new companies and to stimulate economic activity through, among other initiatives, a straightforward and attractive system of business incentives and promoting the entrepreneurship spirit. La Maison de l'Entreprise (LME), established in 1996 by two Economic Development Agencies (IDEA-IDETA), is one of the companies working to revitalize the area. Located on four different sites in Wallonia: Mons, Binche, Tournai, and Enghien, and funded by the EU and the Economic Affairs Ministry of the Région Wallonne, the BIC provides an environment in which small companies (in creation or expanding) can grow through the provision of professional and expert business guidance [34, 43].

A special focus of value-added services of LME is fostering creativity among start-ups and SMEs. The Creativity service of LME drives innovation and is a key factor for the development of personal, occupational, entrepreneurial and social competences and the well-being of all individuals in society. Creativity support methodologies includes two programs: (1) *CreaZoom Programme* was set up to help entrepreneurs to define, with experts, a participative innovation procedure within their company. (2) *Creamons Club* is an innovation circle composed of entrepreneurs, researchers, and students who help each other to innovate [34, 43].

Results confirmed that incubator programs in Belgium in 2007 assisted 138 entrepreneurs and since the start of the BIC, 1,302. In 2007, 11 companies were formed creating 34 jobs; cumulatively 174 companies were created with BIC support generating 582 jobs. The number of companies in the incubator in 2007 was 46, which created 158 jobs, with 174 companies graduated and 582 created in the market [35, 36].

3.3 International Interviews

This study was based on field interviews that conducted by the authors in the United States, the United Kingdom, and Qatar during 2013–2014. The interview instruments for the semi-structured in-depth interviews were developed after a thorough literature review. In addition, the supplementary information provided by innovation managers during the author's visit, form the research information on which the interviews are based. Table 3.2 shows the international interviews of innovation programs.

Table 3.2 Details of international interviews

No.	Institute	Website	Contact detail
1	High Tech Rochester Inc., West Henrietta, NY, US	http://www.htr.org/	Doreen Kula, Venture Development Manager, 150 Lucius Gordon Drive, Suite 208, West Henrietta, NY 14586, doreen.kula@htr.org
2	Blue Valley Schools—Center For Advanced Professional Studies—"CAPS Incubator", Overland Park, Kansas, US	http://www.bvcaps.org/s/1403/start.aspx	Donna Deeds, Executive Director ddeeds@bluevalleyk12.org 913-239-5920 Chad Ralston, Director ccralston@bluevalleyk12.org 913-239-5915 Center for Advanced Professional Studies, Blue Valley Schools 7501 W. 149th Terrace Overland Park, Kansas 66223 913-239-5900 Fax: 913-239-5948
3	South Side Innovation Center—"Inclusive Entrepreneurship", New York, US	http://www.southsideinnovation.org/	South Side Innovation Center (SSIC), 2610 South Salina Street Syracuse, New York 13205 Tel: (315) 443-631 email mmbutler@syr.edu
4	Qatar Science and Technology Park, Doha, Qatar	http://www.qstp.org.qa/	Mr. Hamad Mohammed Al Kuwari Managing Director, Al Gharrafa Street, Al Rayyan, Education City, PO Box 210000, Doha, Qatar O +974 4454 7070 D +974 4454 7029 M +974 5557 2288 F +974 4454 7011 hamalkuwari@qstp.org.qa
5	Institute of Knowledge Transfer, London, UK		Russ Hepworth MInstKT FRSA, Business Development Manager, 76 Portland Place, London W1B 1NT Telephone 0207 127 8081

(continued)

Table 3.2 (continued)

No.	Institute	Website	Contact detail
6	Sussex Innovation Center, University of Sussex, Brighton, UK	www.sinc.co.uk	Mike Herd Science Park square, Falmer, Brighton, BN1 9SB. Telephone 01273 704 400 | laura@sinc.co.uk
7	St. John Innovation, Cambridge, UK	http://stjohns.co.uk/	David Gill Cambridge, UK Telephone: 01223 420252 info@stjohns.co.uk m.edwards@stjohns.co.uk
8	The Petchey Center of Entrepreneurship, University of East London, UK	http://www.uel.ac.uk/petcheycentre/	Professor Sonny Nwankwo Director, University Way, London E16 2RD, United Kingdom Email: mostert@uel.ac.uk Tel: 020 8223 6535 Fax: 020 8223 3327
9	Birmingham University, Birmingham, UK	http://www.birmingham.ac.uk	Prof. Francis Greene University of Birmingham Edgbaston, Birmingham B15 2TT, UK Tel +44 (0) 121 414 3277 Fax +44 (0) 121 414 2982 Email: f.greene@bham.ac.uk
10	Southampton Management School, University of Southampton, Southampton, UK	http://www.southampton.ac.uk	Dr. Franz Huber Highfield, Southampton SO17 1BJ, United Kingdom Room Number: 2/6009 Telephone: (023) 8059 2545 Email: F.Huber@soton.ac.uk

3.4 Survey

Bryman indicated the aim of survey is to determine the answers to questions like "what", "when", "where", and "how many". The survey questionnaire is a good technique and an appropriate tool for collecting quantitative data especially when the information sought is reasonably specific and familiar to the respondents and when the researcher has knowledge of particular problems and the range of responses likely to emerge.

The survey design for this study is based on reviewing several existing survey instruments related to innovation program best practices and outcomes. The survey consists of 17 questions; each question is built on these instruments by refining the relevant questions to meet the objectives of the study. See Table 3.3.

Table 3.3 Survey sample

1	Services of incubator Strong Tangible and specialize services ☐ Partially tangible services ☐ Intangible ☐
2	Goals of incubator Employment ☐ Entrepreneurial climate ☐ Diversifying local economies ☐ Commercializing technologies ☐ Innovation ☐
3	Sponsors of incubator University ☐ Hybrid ☐ Government ☐ Nonprofit economic development ☐ For profit entity ☐
4	Staff of incubator 1-5 ☐ 6-25 ☐ 26-50 ☐ >50 ☐
5	No. of jobs creation from the incubator 1-5 ☐ 6-25 ☐ 25-50 ☐ >50 ☐
6	No. of graduate companies from incubator 1-5 ☐ 6-25 ☐ >25 ☐
7	No. of incubators worldwide High increment ☐ Medium increment ☐ Low increment ☐
8	No. of tenants inside the incubator 1-5 ☐ 6-25 ☐ >25 ☐
9	Survival rate of tenants Less 80% ☐ 81-90% ☐ >90% ☐
10	Incubators focus Economic development ☐ Jobs creation ☐ Technology transfer ☐ High technology & Patents ☐ Entrepreneurship ☐ Create ventures ☐
11	The role of science park Very active ☐ Present ☐ Active ☐ Poor ☐
12	The role of cooperation of R&D Very active ☐ Present ☐ Active ☐ Poor ☐
13	The rate of fostering entrepreneurship Below market rate ☐ At market rate ☐ Low rate ☐
14	The role of Innovation Very active ☐ Present ☐ Active ☐ Poor ☐
15	The technology transfer of incubator Strong ☐ Modest ☐ Poor ☐
16	The rate of patents High rate ☐ Medium rate ☐ Low rate ☐
17	The rate of licensed intellectual property High rate ☐ Medium rate ☐ Low rate ☐

Chapter 4
Data Analysis

Abstract This Chapter discusses the main findings of the research using the data collected from the international case studies, the survey and interview. Ten interviews made up the main source of evidence used in the current study. Each interview began with questions about the program based on the radar chart, with in-depth discussion of the 16 indicators. All of the interviews were structured to best understand the situation and were recorded and transcribed for clarity and were then sent to the interviewees for confirmation of accuracy. The maximum time for the interviews was 1 h and 30 min. All of the data from the interviews and documents were linked together. The international interview design is based on two charts. First, the radar chart consists of four categories: culture; policy; industry; and economy. Second the average for each category is measured on a scale of 100 % and the average for each indicator is measured on a scale of 25 %. This chapter explain that our findings from the case studies of innovation centers and incubation programs provide several interesting implications. The results from the case studies of innovation centers and incubation programs indicated that programs tend to become more successful with experience and maturity, making years in operation a very important metric.

Keywords Innovation · Case study · Incubators · Radar chart · Interview

4.1 International Cases Studies

An analysis of the case studies of incubator programs in all the countries studied yielded a composite of their outcomes. The number of companies in the incubator was 547, which created 15,751 jobs and graduated 1,049 companies. The total ratio between companies in the incubator (KPI2) to graduation companies (KPI3) was 1:2, total ratio between companies graduated (KPI3) to jobs created (KPI1) was 1:15 and total ratio between companies in the incubator (KPI2) to jobs created (KPI1) was 1:28. Table 4.1 presents the total ratio of KPI. In summary, these

H.M. Al-Mubaraki et al., *Innovation and Entrepreneurship*,
SpringerBriefs in Business, DOI 10.1007/978-3-319-13683-7_4

Table 4.1 Total ratio of KPI[a]

Total ratio of KPI[a]	
KPI2:KPI3	1:2
KPI3:KPI1	1:15
KPI2:KPI1	1:28

Table 4.2 Key performance indicators (KPI) for case studies

Key performance indicators (KPI)						
Countries	KPI1		KPI2		KPI3	
	Jobs created		Companies in the incubators (client companies)		Companies graduated	
	No.	%[a]	No.	%[b]	No.	%[c]
USA	14,044	89.2	99	18.1	32	3.1
UK	115	0.73	105	19.2	111	10.6
France	112	0.71	11	2	75	7.2
Germany	16	0.1	10	1.8	6	0.6
Spain	121	0.77	39	7.13	110	10.5
Italy	386	2.5	42	8	62	6
Austria	800	5.1	170	31.1	404	38.5
N. Land	97	6.2	23	4.2	63	6
Luxemburg	26	0.17	13	2.4	12	1.2
Belgium	34	0.22	45	8.2	174	17
Total	15,751	100	557	100	1,049	100

[a]Total number of countries divided by number of jobs created
[b]Total number of countries divided by number of client companies
[c]Total number of countries divided by number of companies graduated

figures and the total ratios reflect the effectiveness of incubators and innovation programs in terms of some points. Such as (1) economic development strategy in job creation, (2) promotion of entrepreneurship in the companies involved, (3) commercialization and technology transfer for graduated companies, (4) sustainability of graduated companies in the market with high rates of survival, and (5) diversification of the economy through outcomes such as innovation and technology. Table 4.2 shows the Key Performance Indicators (KPI) for case studies in different countries.

Table 4.3 demonstrates the case studies' characteristics such as (1) location, (2) year of founding, and (3) economic outcomes, for example, jobs created, client companies, and graduate companies. In addition, the highest number of jobs created was in the US, at 14,044 and the highest number of client companies and graduate companies was in Austria, at 170 and 404, respectively. Overall, high rates of positive economic outcomes were noted.

Table 4.3 Main characteristics of case studies

Case study	Location	Foundation	Countries outcomes		
			Jobs created	Client companies	Graduate companies
USA	Maryland	N/A	14,044[a]	99	32
UK	North East of England	1994	115	105	111
France	Normandy	1987	112	11	75
Germany	Bavaria	2005	16	10	6
Spain	Navarra	1988	121	39	110
Italy	Trentino province	1987	386	42	62
Austria	Styria	1981	800	170[a]	404[a]
N. Land	Friesland	2002	97	23	63
Luxemburg	N/A	1987	26	13	12
Belgium	Province of Antwerp	1987	34	45	174

Source [43]
[a]Highest program

Table 4.4 Ratio of performance indicators for each incubator and innovation over the years

Case study	Foundation	No. of years till 2013	Ratio of performance indicators for each incubator and innovation over the years[a]		
			Jobs created	Client companies	Graduate companies
UK	1994	19	6.05	5.53[b]	5.84
France	1987	26	4.31	0.42	2.88
Germany	2005	8	2.00	1.25	0.75
Spain	1988	25	4.84	1.56	4.40
Italy	1987	26	14.85	1.62	2.38
Austria	1981	32[b]	25.00[b]	5.31	12.63[b]
N. Land	2002	11	8.82	2.09	5.73
Luxemburg	1987	26	1.00	0.50	0.46
Belgium	1987	26	1.31	1.73	6.69

[a]Ratio of performance indicators = the countries' outcomes divided over the number of year till 2013
[b]Highest program

Table 4.4 presents the ratio of performance indicators for each incubator and innovation program over the years a particular program has been in operation since the early 1980s. It is evident that some innovation centers and incubation programs are performing better than others. Austria presents the highest rate of jobs created and graduate companies, at 25 and 12.63, respectively, per year. Also, the UK presents the highest rate of client companies, at 5.53 and Austria indicated oldest operated program, at 32 years.

4.2 International Interview

Insights from the literature review were combined with analysis of the interviews to measure the performance of each program to provide an in-depth understanding of the research landscape and a rich insight into the research objectives [33, 76]. The international interview design is based on two charts. First, the radar chart consists of four categories: (1) culture; (2) policy; (3) industry; and (4) economy. In addition, each category is measured by four indicators and each indicator is rank-ordered as an independent variable [e.g., low (L), moderate (M), and high (H)] with total of 16 indicators. Second, the average for each category is measured on a scale of 100 % and the average for each indicator is measured on a scale of 25 %. There are three groups of outcomes, for example, a total percentage of categories between 80 and 100 % indicated a high outcome, a total percentage of categories between 60 and 79 % indicated medium outcomes, and a total percentage of categories less than 60 % indicated low outcomes.

Ten interviews made up the main source of evidence used in the current study. The interviewees included the director of innovation and incubation program officers such as the president and vice president. The ten interviewees were asked to provide various strategic documents to support the data-gathering effort. The interviews were conducted systematically. Each interview began with questions about the program based on the radar chart, with in-depth discussion of the 16 indicators. All of the interviews were structured to best understand the situation while also giving the interviewees sufficient direction to ensure that they would provide as much information as possible. All of the interviews were recorded and transcribed for clarity and were then sent to the interviewees for confirmation of accuracy. The maximum time for the interviews was 1 h and 30 min. All of the data from the interviews and documents were linked together.

The in-depth knowledge of the authors led to the selection of programs to interview, which were in the United States (three programs), the United Kingdom (six programs), and Qatar (one program). The US has the largest number of business incubator programs in the world, around 1,800 in 2010 [62]. In addition, there are around 900 business incubators in Europe and 300 programs in the UK [61, 62]. The estimated number of incubators worldwide is 7,000.

The radar charts in the following section show the average percentage of the four categories and the 16 indicators. The four categories consist of 16 indicators and each category includes four indicators. For example, the economy category includes the four indicators: survival rate, jobs creation, start-up companies, and number of patents. Each indicator is scaled into three ranges: High (81–100 %), Medium (41–80 %), and Low (40 % and below), and each range is given a value: High (25 %), Medium (20 %), and Low (10 %). For example, the average percentages of key performance indicators in the four categories of culture, policy, industry and economy for all the programs interviewed were, respectively, 95, 65, 90, and 90 %. The division of the scales and the weighting of the values were based on previous studies and the authors' professional experience.

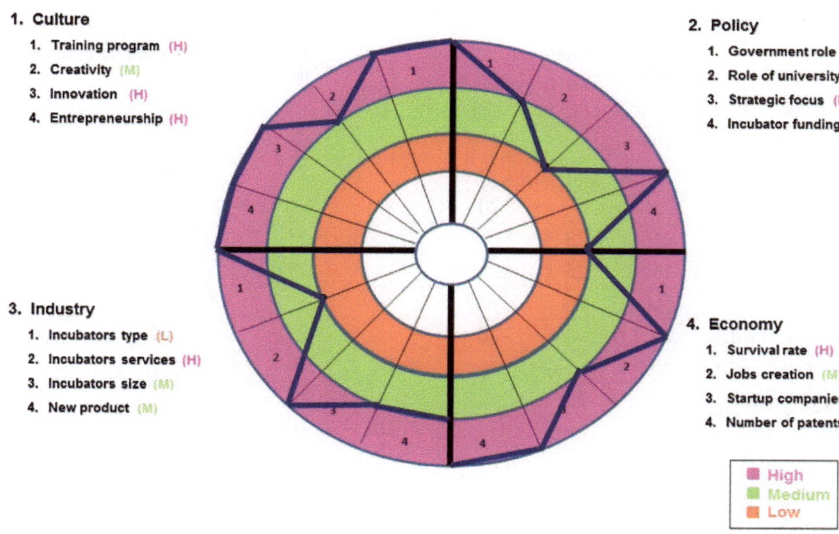

Chart 4.1 Radar chart of High Tech Rochester Inc., West Henrietta, NY, US

4.2.1 Interview 1: High Tech Rochester Inc.

Chart 4.1 shows the ratings for High Tech Rochester, Inc. The categories of culture and economy received high ratings, while the categories of policy and industry received medium ratings. Three indicators–role of university, incubator funding, and incubators type—received low ratings.

Table 4.5 presents the average of indicators as the result from the radar chart, at 80 %, which indicated an average outcome in the high range. The categories of culture and economy both received ratings of 95 %, while the categories of industry and policy were rated at 65 %. See Chart 4.2.

4.2.2 Interview 2: Blue Valley Schools—Center for Advanced Professional Studies—"CAPS Incubator"

Chart 4.3 shows the ratings for Blue Valley Schools-Center for Advanced Professional Studies–"CAPS Incubator." The culture and policy categories had high ratings, but the economy and industry categories received low ratings. Only two indicators received low ratings, incubators size and number of patents.

Table 4.6 presents the ratings of indicators included on the radar chart, with an average rating of 80 %. The percentage of each category from the radar chart indicated high outcomes in the two categories culture and policy, at 100 and 85 % respectively. However, the industry and economy categories received ratings in the medium range at 75 and 60 %, respectively. See Chart 4.4.

Table 4.5 Result of average indicators of High Tech Rochester Inc

	% 100	Scale			Indicators %	Total categories %
		High (25 %)	Medium (20 %)	Low (10 %)		
Culture	100					
Training program	25	25			25	95
Creativity	25		20		20	
Innovation	25	25			25	
Entrepreneurship	25	25			25	
Policy	100					
Government role	25		20		20	65
Role of university	25			10	10	
Strategic focus	25	25			25	
Incubator funding	25			10	10	
Industry	100					
Incubators type	25					65
Incubators services	25	25			25	
Incubators size	25		20		20	
New product	25		20		20	
Economy	100					
Survival rate	25	25			25	95
Jobs creation	25		20		20	
Startup companies	25	25			25	
Number of patents	25	25			25	
Total	400					320
Average	100					80

Chart 4.2 Percentage of total outcomes from radar chart

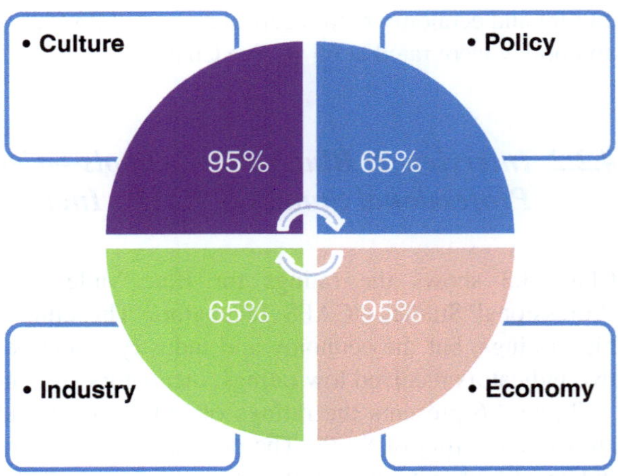

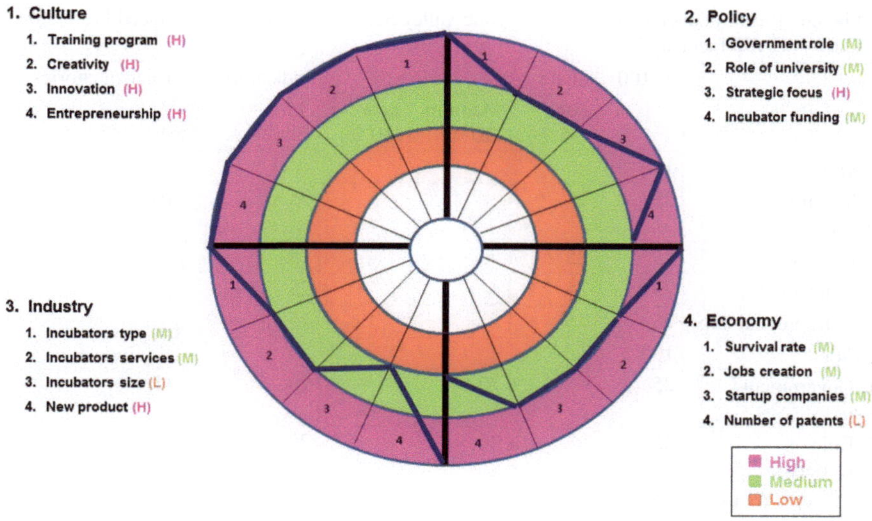

Chart 4.3 Radar chart of Blue Valley Schools—Center for Advanced Professional Studies—"CAPS Incubator", Kansas, US

4.2.3 Interview 3: South Side Innovation Center—"Inclusive Entrepreneurship"

Chart 4.5 shows the responses to the interview for South Side Innovation Center—"Inclusive Entrepreneurship." The culture and economy categories are rated as high, but the policy and industry categories are rated as medium. Low ratings were given to the indicators of government role, new product, and number of patents.

Table 4.7 presents the average of indicators as the result from the radar chart, with an average rating of 84 %. The percentage of each category from the radar chart indicated high outcomes in the two categories culture and economy, at 100 and 85 %, respectively. However, the policy and industry categories indicated medium outcomes at 75 %. See Chart 4.6.

4.2.4 Interview 4: Qatar Science and Technology Park, Doha, Qatar

Chart 4.7 shows responses for the interview for Qatar Science and Technology Park. All four categories received high ratings, with policy at the lowest point (80 %) and the other three-culture, economy, and industry-at significantly higher points within the overall range designated as "high." Only two indicators received low ratings: incubator funding and creativity.

Table 4.6 Result of average indicators of Blue Valley Schools—Center for Advanced Professional Studies—"CAPS Incubator"

	% 100	Scale			Indicators %	Total categories %
		High (25 %)	Medium (20 %)	Low (10 %)		
Culture	100					
1. Training program	25	25			25	100
2. Creativity	25	25			25	
3. Innovation	25	25			25	
4. Entrepreneurship	25	25			25	
Policy	100					
1. Government role	25		20		20	85
2. Role of university	25		20		20	
3. Strategic focus	25	25			25	
4. Incubator funding	25		20		20	
Industry	100					
1. Incubators type	25		20		20	75
2. Incubators services	25		20		20	
3. Incubators size	25			10	10	
4. New product	25	25			25	
Economy	100					
1. Survival rate	25		20		20	60
2. Jobs creation	25		20		20	
3. Startup companies	25		20		10	
4. Number of patents	25			10	10	
Total	400					320
Average	100 %					80 %

Table 4.8 presents the average of indicators as the result from radar chart, with an average rating of 90 %, which indicated an average outcome in the high range. The percentage of each category from the radar chart indicated high outcomes in all four categories: culture, economy, policy, and industry, which were rated at 85, 100, 80, and 95 %, respectively. See Chart 4.8.

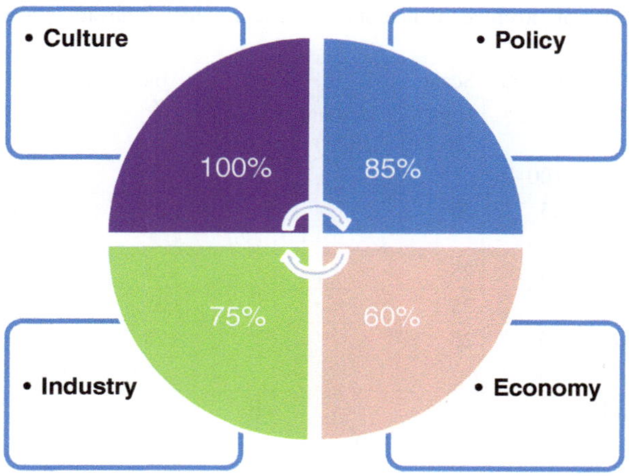

Chart 4.4 Percentage of total outcomes from radar chart

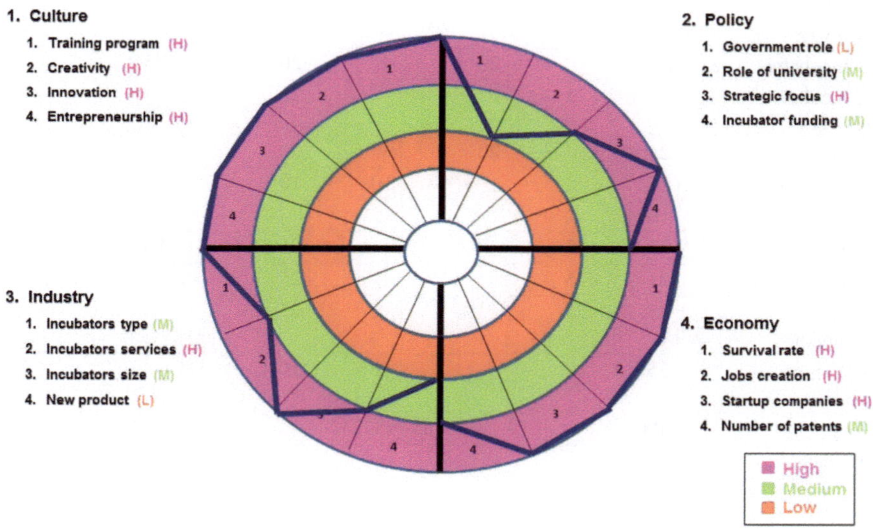

Chart 4.5 Radar chart of South Side Innovation Center—"Inclusive Entrepreneurship", New York, US

4.2.5 Interview 5: Institute of Knowledge Transfer

Chart 4.9 shows the ratings for the Institute of Knowledge Transfer in London. A single category-culture-received a high rating, with the three categories policy, industry and economy receiving medium ratings. Only incubator funding received a low rating among the indicators.

Table 4.7 Results of average indicators of South Side Innovation Center—"Inclusive Entrepreneurship"

	% 100	Scale			Indicators %	Total categories %
		High (25 %)	Medium (20 %)	Low (10 %)		
Culture	100					
1. Training program	25	25			25	100
2. Creativity	25	25			25	
3. Innovation	25	25			25	
4. Entrepreneurship	25	25			25	
Policy	100					
1. Government role	25			10	10	75
2. Role of university	25		20		20	
3. Strategic focus	25	25			25	
4. Incubator funding	25		20		20	
Industry	100					
1. Incubators type	25		20		20	75
2. Incubators services	25	25			25	
3. Incubators size	25		20		20	
4. New product	25			10	10	
Economy	100					
1. Survival rate	25	25			25	85
2. Jobs creation	25	25			25	
3. Startup companies	25	25			25	
4. Number of patents	25			10	10	
Total	400					335
Average	100 %					84 %

Table 4.9 presents the average ratings as the result from the radar chart, at 76 %, which indicated an average outcome in the medium range. The percentage of the culture category from the radar chart received high outcomes at 95 %. However, the policy, economy and industry categories received medium outcomes of 70 %. See Chart 4.10.

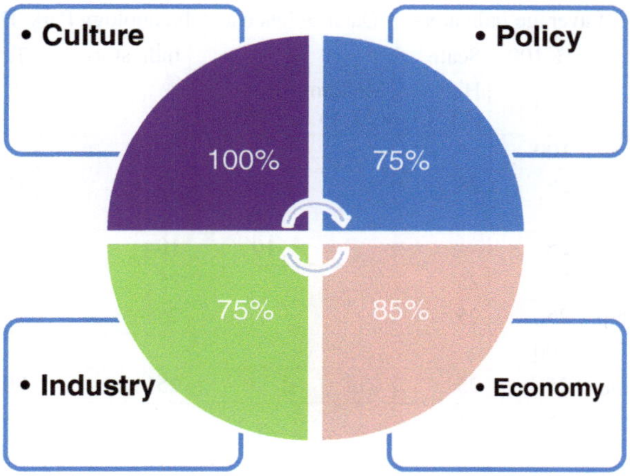

Chart 4.6 Percentage of total outcomes from radar chart

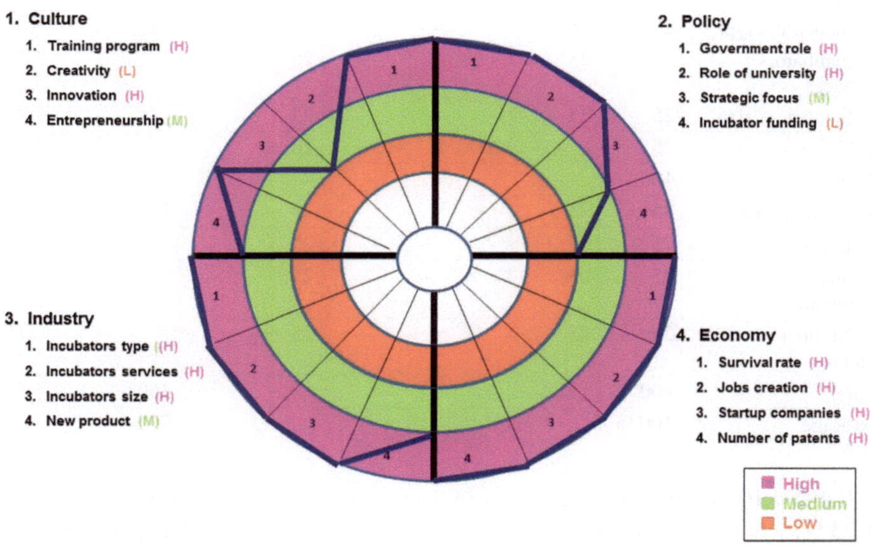

Chart 4.7 Radar chart of Qatar Science and Technology Park, Doha, Qatar

4.2.6 Interview 6: Sussex Innovation Center, University of Sussex, Brighton, UK

Chart 4.11 shows the ratings for Sussex Innovation Center, University of Sussex, Brighton, UK. All four categories-culture, policy, industry, and economy-received high and medium ratings.

Table 4.8 Result of average indicators of Qatar Science and Technology Park, Doha, Qatar

	% 100	Scale			Indicators %	Total categories %
		High (25 %)	Medium (20 %)	Low (10 %)		
Culture	100					
1. Training program	25	25			25	85
2. Creativity	25			10	10	
3. Innovation	25	25			25	
4. Entrepreneurship	25	25			25	
Policy	100					
1. Government role	25	25			25	80
2. Role of university	25	25			25	
3. Strategic focus	25		20		20	
4. Incubator funding	25			10	10	
Industry	100					
1. Incubators type	25	25			25	95
2. Incubators services	25	25			25	
3. Incubators size	25	25			25	
4. New product	25		20		20	
Economy	100					
1. Survival rate	25	25			25	100
2. Jobs creation	25	25			25	
3. Startup companies	25	25			25	
4. Number of patents	25	25			25	
Total	400					360
Average	100 %					90 %

Table 4.10 presents the average of indicators as the result from the radar chart, at 93 %, which indicated an average outcome in the high range. The percentage of each category from the radar chart showed high outcomes in all categories, including culture, policy, industry, and economy, which were rated at 90, 90, 100, and 90 %, respectively. See Chart 4.12.

4.2.7 Interview 7: St. John's Innovation Centre

Chart 4.13 shows the ratings for St. John's Innovation Centre. All four categories received ratings in the "High" range, with policy and industry at the lowest end of that range whereas culture and economy were ranked toward the upper end of the

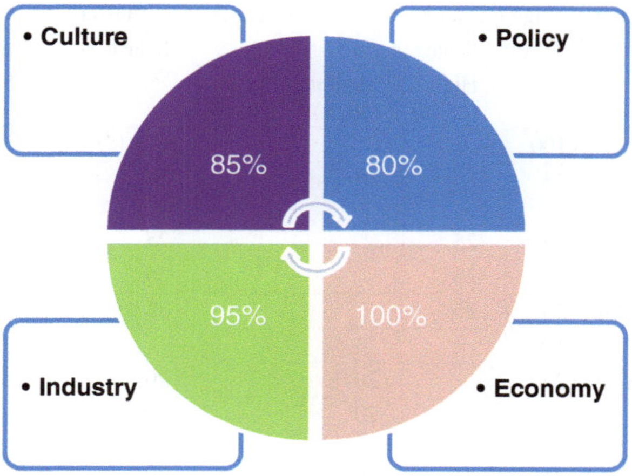

Chart 4.8 Percentage of total outcomes from radar chart

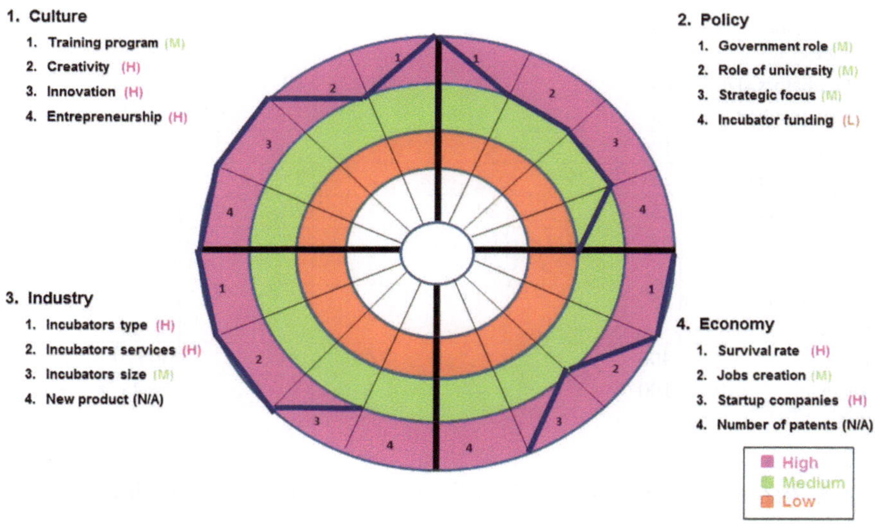

Chart 4.9 Radar chart of Institute of Knowledge Transfer

range. Moreover, no individual key performance indicators received low ratings at St. John's Innovation Centre.

Table 4.11 presents the average of ratings as the result from the radar chart, at 88 %, which indicated an average outcome in the high range. The percentage of each category from radar chart indicated high outcomes in all the categories, including culture, policy, industry and economy, which were rated at 95, 80, 80, and 95 %, respectively. See Chart 4.14.

Table 4.9 Result of average indicators of institute of knowledge transfer

	% 100	Scale			Indicators %	Total categories %
		High (25 %)	Medium (20 %)	Low (10 %)		
Culture	100					
1. Training program	25		20		20	95
2. Creativity	25	25			25	
3. Innovation	25	25			25	
4. Entrepreneurship	25	25			25	
Policy	100					
1. Government role	25		20		20	70
2. Role of university	25		20		20	
3. Strategic focus	25		20		20	
4. Incubator funding	25			10	10	
Industry	100					
1. Incubators type	25	25			25	70
2. Incubators services	25	25			25	
3. Incubators size	25		20		20	
4. New product	25					
Economy	100					
1. Survival rate	25	25			25	70
2. Jobs creation	25		20		20	
3. Startup companies	25	25			25	
4. Number of patents	25					
Total	400					305
Average	100 %					76.25 %

4.2.8 Interview 8: The Petchey Center of Entrepreneurship, University of East London

Chart 4.15 shows the ratings for The Petchey Center of Entrepreneurship, University of East London. Two of the four categories-culture and policy-received high ratings, and the other two-industry and economy-received medium ratings. Only three indicators received low ratings-incubator funding, number of patents, and new products.

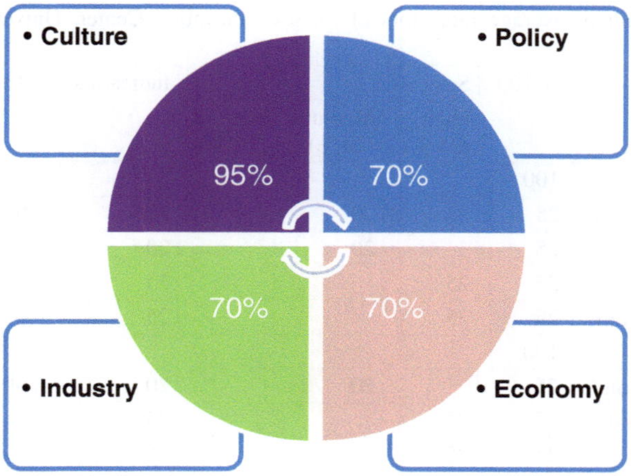

Chart 4.10 Percentage of total outcomes from radar chart

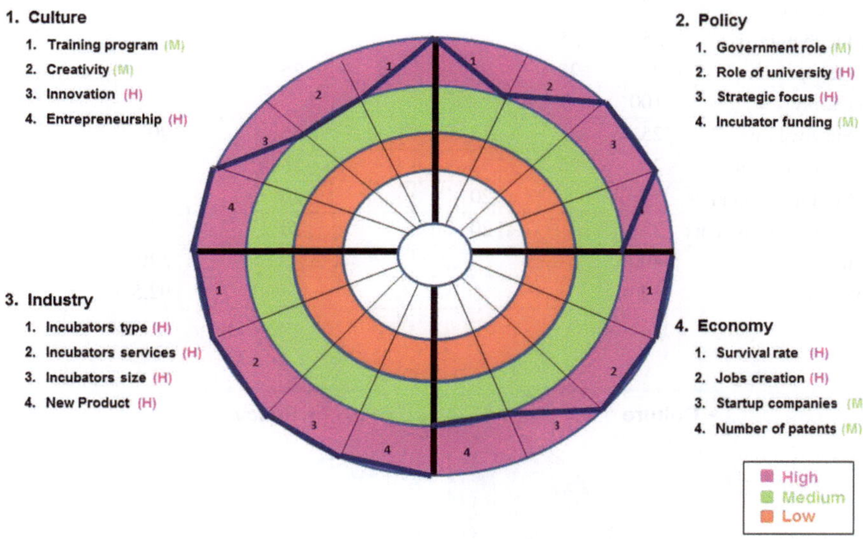

Chart 4.11 Radar chart of Sussex Innovation Center, University of Sussex, Brighton, UK

Table 4.12 presents the average of indicators as the result from the radar chart, at 78 %, which indicated an average outcome in the medium range. The percentage of each category from the radar chart indicated ratings for culture, policy, industry, and economy of 85, 85, 70, and 70 %, respectively. See Chart 4.16.

Table 4.10 Result of average indicators of Sussex Innovation Center, University of Sussex, Brighton, UK

	% 100	Scale			Indicators %	Total categories %
		High (25 %)	Medium (20 %)	Low (10 %)		
Culture	100					
1. Training program	25		20		20	90
2. Creativity	25		20		20	
3. Innovation	25	25			25	
4. Entrepreneurship	25	25			25	
Policy	100					
1. Government role	25		20		20	90
2. Role of university	25	25			25	
3. Strategic focus	25	25			25	
4. Incubator funding	25		20		20	
Industry	100					
1. Incubators type	25	25			25	100
2. Incubators services	25	25			25	
3. Incubators size	25	25			25	
4. New product	25	25			25	
Economy	100					
1. Survival rate	25	25			25	90
2. Jobs creation	25	25			25	
3. Startup companies	25		20		20	
4. Number of patents	25		20		20	
Total	400					370
Average	100 %					92.5 %

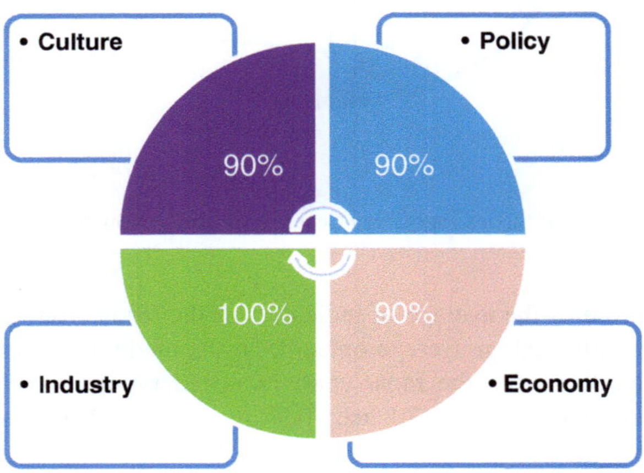

Chart 4.12 Percentage of total outcomes from radar chart

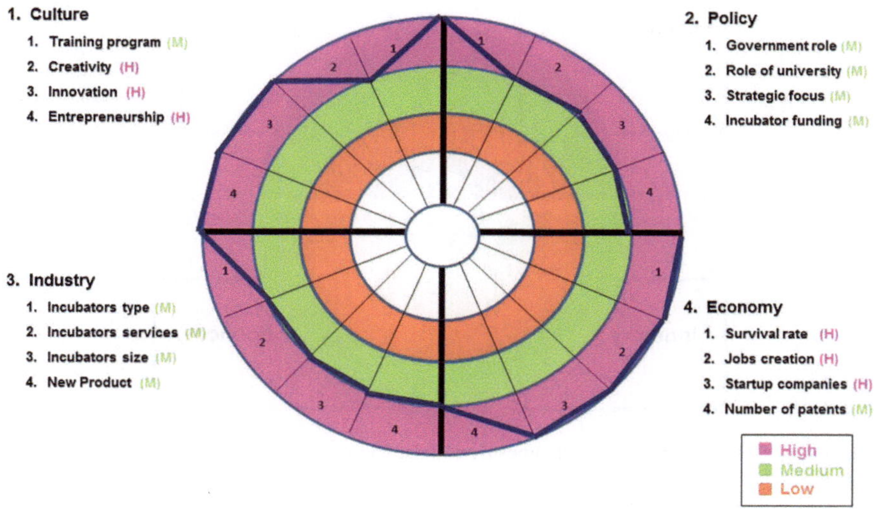

Chart 4.13 Radar chart of St. John's Innovation Centre

Table 4.11 Result of average indicators of St. John's Innovation Centre

	% 100	Scale			Indicators %	Total categories %
		High (25 %)	Medium (20 %)	Low (10 %)		
Culture	100					
1. Training program	25		20		20	95
2. Creativity	25	25			25	
3. Innovation	25	25			25	
4. Entrepreneurship	25	25			25	
Policy	100					
1. Government role	25		20		20	80
2. Role of university	25		20		20	
3. Strategic focus	25		20		20	
4. Incubator funding	25		20		20	
Industry	100					
1. Incubators type	25		20		20	80
2. Incubators services	25		20		20	
3. Incubators size	25		20		20	
4. New product	25		20		20	
Economy	100					
1. Survival rate	25	25			25	95
2. Jobs creation	25	25			25	
3. Startup companies	25	25			25	
4. Number of patents	25		20		20	
Total	400					350
Average	100 %					87.5 %

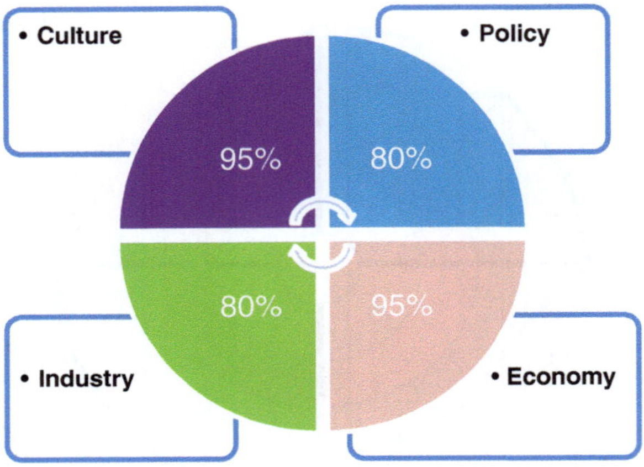

Chart 4.14 Percentage of total outcomes from radar chart

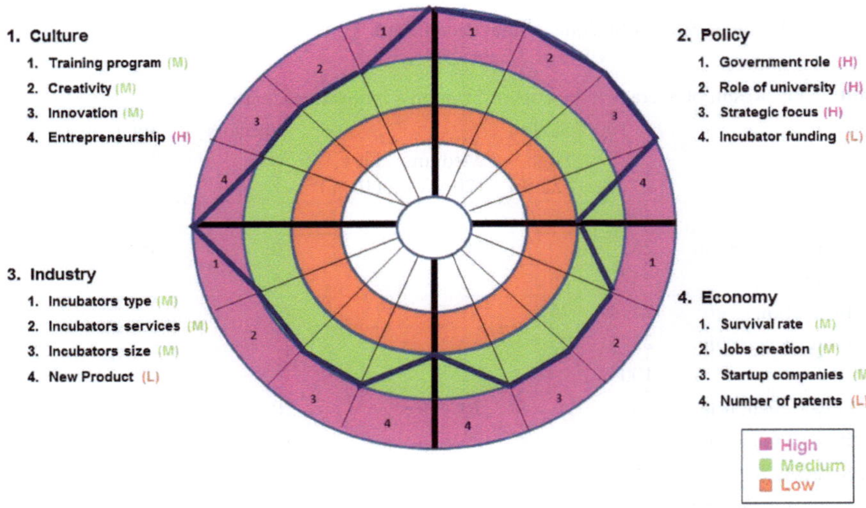

Chart 4.15 Radar chart of the Petchey Center of Entrepreneurship, University of East London

4.2.9 Interview 9: University of Birmingham, UK

Chart 4.17 shows the ratings for the University of Birmingham, UK. Two categories, industry and economy, received high ratings, the category of policy received a medium rating and the category of culture received a low rating. Three key performance indicators received low ratings-entrepreneurship, strategic focus, and creativity.

Table 4.12 Result of average indicators of The Petchey Center of Entrepreneurship, University of East London

	% 100	Scale			Indicators %	Total categories %
		High (25 %)	Medium (20 %)	Low (10 %)		
Culture	100					
1. Training program	25		20		20	85
2. Creativity	25		20		20	
3. Innovation	25		20		20	
4. Entrepreneurship	25	25			25	
Policy	100					
1. Government role	25	25			25	85
2. Role of university	25	25			25	
3. Strategic focus	25	25			25	
4. Incubator funding	25			10	10	
Industry	100					
1. Incubators type	25		20		20	70
2. Incubators services	25		20		20	
3. Incubators size	25		20		20	
4. New product	25			10	10	
Economy	100					
1. Survival rate	25		20		20	70
2. Jobs creation	25		20		20	
3. Startup companies	25		20		20	
4. Number of patents	25			10	10	
Total	400					310
Average	100 %					77.5 %

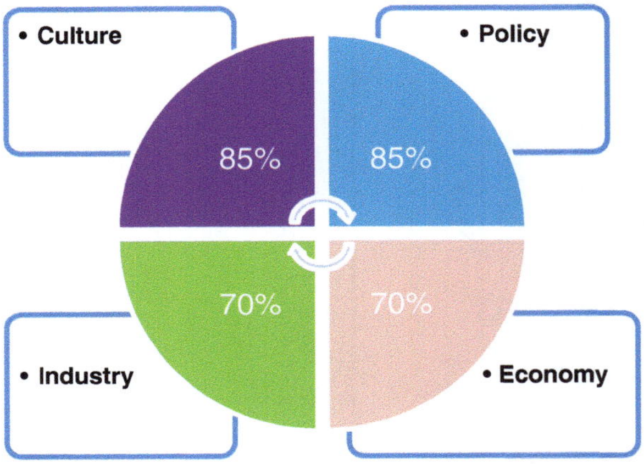

Chart 4.16 Percentage of total outcomes from radar chart

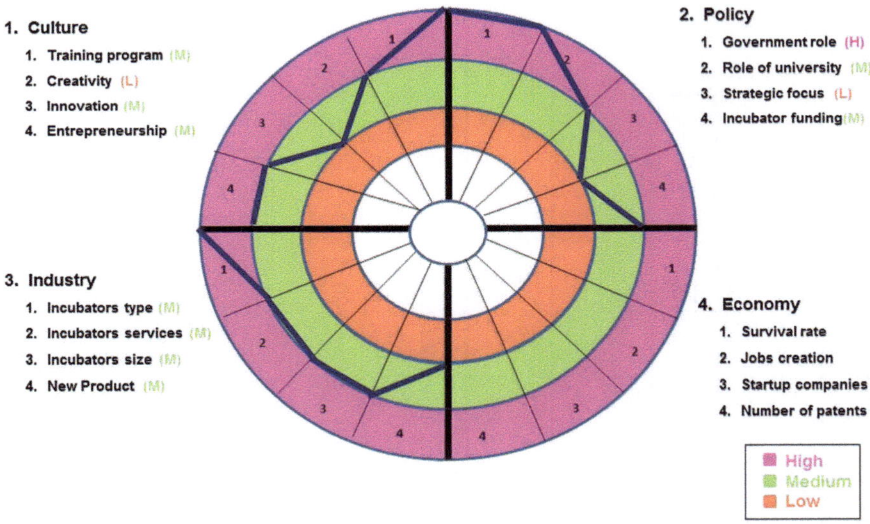

Chart 4.17 Radar chart of University of Birmingham, UK

Table 4.13 presents the ratings as the result from the radar chart, with an average of 79 %, which indicated an average outcome in the medium range. The economy category received a rating of 100 %. However, the policy, culture and industry categories received ratings of 75, 60, 80 %, respectively. See Chart 4.18.

Table 4.13 Result of average indicators of University of Birmingham, UK

	% 100	Scale			Indicators %	Total categories %
		High (25 %)	Medium (20 %)	Low (10 %)		
Culture	100					
1. Training program	25		20		20	60
2. Creativity	25			10	10	
3. Innovation	25		20		20	
4. Entrepreneurship	25			10	10	
Policy	100					
1. Government role	25	25			25	75
2. Role of university	25		20		20	
3. Strategic focus	25			10	10	
4. Incubator funding	25		20		20	
Industry	100					
1. Incubators type	25		20		20	80
2. Incubators services	25		20		20	
3. Incubators size	25		20		20	
4. New product	25		20		20	
Economy	100					
1. Survival rate	25	25				100
2. Jobs creation	25	25				
3. Startup companies	25	25				
4. Number of patents	25	25				
Total	400					215
Average	100 %					79

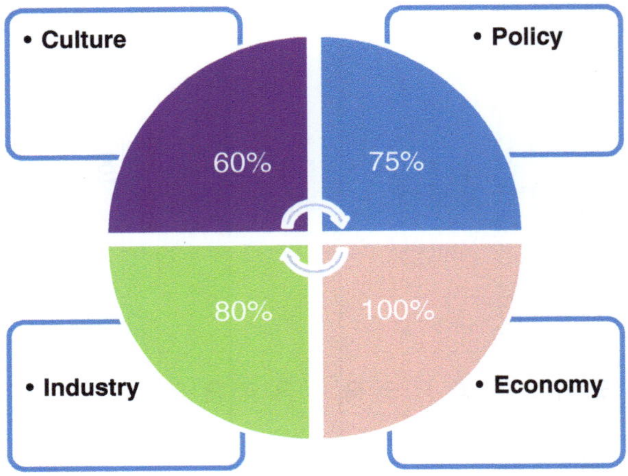

Chart 4.18 Percentage of total outcomes from radar chart

4.2.10 Interview 10: Southampton Management School, University of Southampton

Chart 4.19 shows the ratings for Southampton Management School, University of Southampton. The category of culture and policy received a high rating, while the category of industry and economy received low ratings. Very few described as low indicators. For example, entrepreneurship, incubators size, new product, jobs creation, incubator funding, and start-up companies.

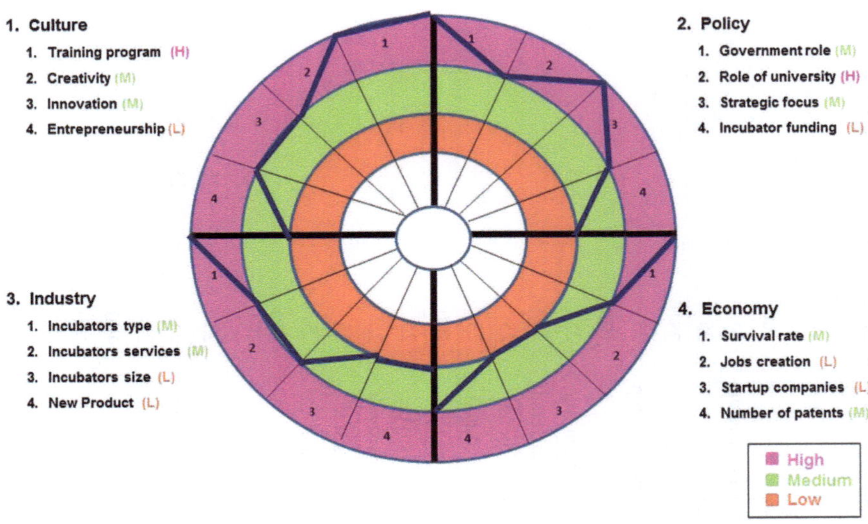

Chart 4.19 Radar chart of Southampton Management School, University of Southampton

Table 4.14 Result of average indicators of Southampton Management School, University of Southampton

	% 100	Scale			Indicators %	Total categories %
		High (25 %)	Medium (20 %)	Low (10 %)		
Culture	100					
1. Training program	25	25			25	85
2. Creativity	25		20		20	
3. Innovation	25		20		20	
4. Entrepreneurship	25		20		20	
Policy	100					
1. Government role	25		20		20	75
2. Role of university	25	25			25	
3. Strategic focus	25		20		20	
4. Incubator funding	25			10	10	
Industry	100					
1. Incubators type	25			20		60
2. Incubators services	25			20		
3. Incubators size	25				10	
4. New product	25				10	
Economy	100					
1. Survival rate	25		20		20	60
2. Jobs creation	25			10	10	
3. Startup companies	25			10	10	
4. Number of patents	25		20		20	
Total	400					280
Average	100 %					70 %

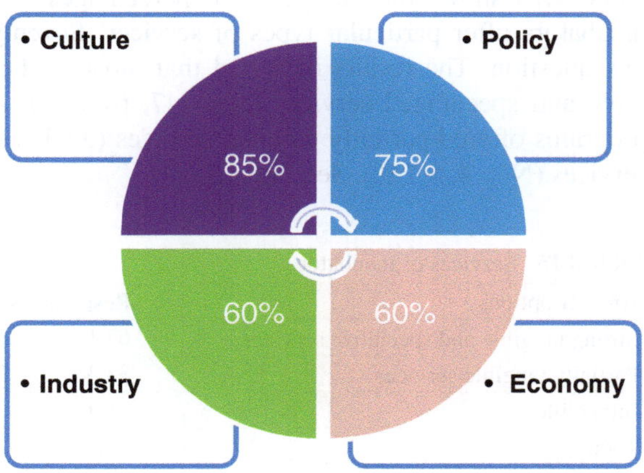

Chart 4.20 Percentage of total outcomes from radar chart

Table 4.14 presents the average of indicators as the result from the radar chart, with an average of 70 %, which indicated an average outcome in the medium range. The category of culture received a rating of 85 % and the policy category received a rating of 75 %. However, the economy and industry categories received ratings in the low range, 60 % each. See Chart 4.20.

4.3 Survey Results

Innovation is the driver of our future growth [34, 42, 43, 49, 75]. Innovation-based incubators are local economic development tools [3, 4, 5, 22, 43, 49], which favor the conditions for creation and growth of novel business activities and contribute actively to the development of the regions where they operate. In addition, innovation-based incubators support innovative business projects, provide new high-tech venture creation, technological entrepreneurship, commercialization, and transfer of technology [1, 6, 7, 8, 56, 57, 59, 67]. In order to find out the actual experiences of best practices of incubators and innovation programs, an internet-based survey was conducted with members of National Business Incubation Association (NBIA) representing worldwide countries including countries in the Middle East and the GCC, United Kingdom business incubation (UKBI) and United Kingdom Science Park Association (UKSPA).

Of the 124 survey invitations that were emailed to NBIA, UKBI, and UKSPA members via the Survey Monkey website, 50 were returned as undeliverable, leaving a sample frame of 74. The total number of survey responses was 74, representing a response rate (RR) of approximately 60 %. Each question was subjected to a descriptive analysis.

4.3.1 Services of Incubator

Table 4.15 shows the numbers and percentages of respondents whose business incubators offer particular types of services. Seventy-four respondents answered this question. The results indicated that most of the incubators offer strong tangible and specialized services (N = 47, 67.1 %). Also, twenty four incubation programs offered partially tangible services (34.3 %). Very few offered intangible services (N = 4, 5.7 %). See Fig. 4.1.

Table 4.15 Services of incubator

Answer options	Response percent (%)	Response count
Strong tangible and specialize services	67.1	47
Partially tangible services	34.3	24
Intangible	5.7	4
Total	100	

Fig. 4.1 Services of incubator

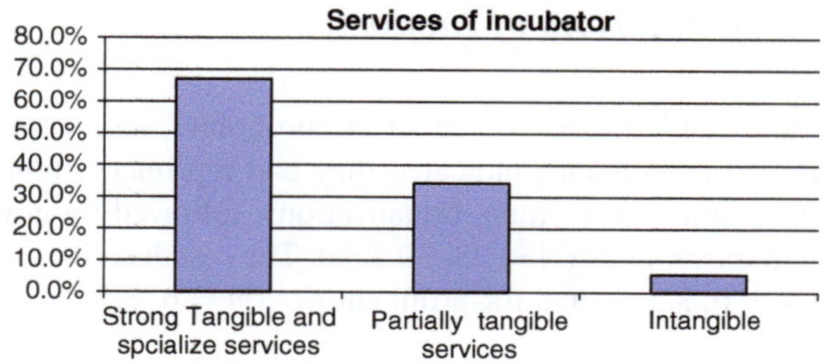

4.3.2 Goals of Incubator

Table 4.16 shows the distribution of respondents by their incubator program's goals. Nearly three-fourths (N = 52, 73.2 %) indicated that the primary goals of their incubators were an entrepreneurial climate, followed by innovation (N = 43, 60.6 %) and employment (N = 38, 53.5 %). To be consistent with the rest of the reports, I changed the reference here to the table instead of the figure. The rest described their programs as commercializing technologies (N = 35, 49.3 %) and diversifying local economies (N = 26, 36.6 %). See Fig. 4.2.

Table 4.16 Goals of incubator

Answer options	Response percent (%)	Response count
Employment	53.5	38
Entrepreneurial climate	73.2	52
Diversifying local economies	36.6	26
Commercializing technologies	49.3	35
Innovation	60.6	43
Total	100	

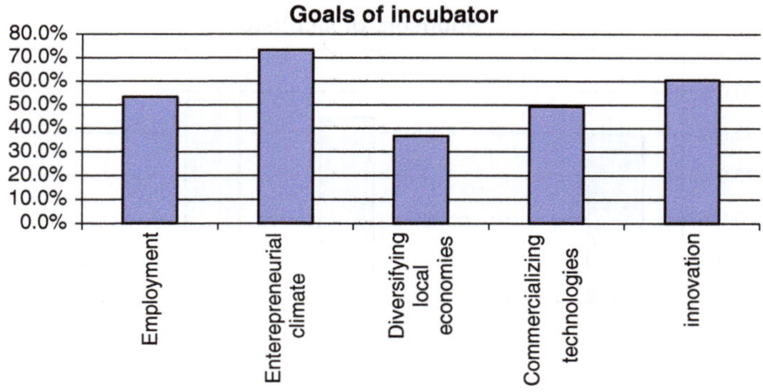

Fig. 4.2 Goals of incubator

4.3.3 Sponsors of Incubator

Table 4.17 shows data about the programs' sponsoring entities. Thirty-four out of the 74 respondents indicated they had a primary sponsoring entity with non-profit economic development organizations, followed by governments (N = 31, 43.7 %) and universities (N = 28, 39.4 %). The rest described hybrid sponsors of incubators (N = 6, 8.5 %) and for-profit entities (N = 6, 8.5 %). See Fig. 4.3.

4.3.4 Staff of Incubator

Table 4.18 reports the numbers and percentages of the staffing of the incubators. More than 50 respondents indicated that their staffs range in size from one to five persons (72.2 %). Less than one-fourth of 74 respondents indicated that their staffs range from six to twenty five persons (N = 13, 18.1 %). The rest described the staff as 26–50 persons (N = 5, 6.9 %) and more than 50 persons (N = 3, 4.2 %). See Fig. 4.4.

Table 4.17 Sponsors of incubator

Answer options	Response percent (%)	Response count
University	39.4	28
Hybrid	8.5	6
Government	43.7	31
Nonprofit economic development	47.9	34
For profit entity	8.5	6
Total	100	

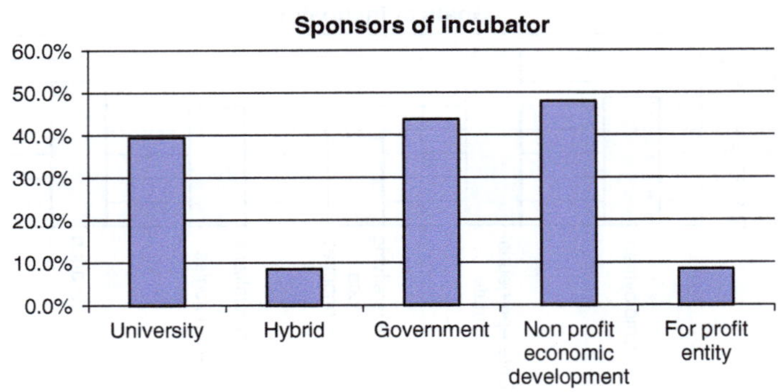

Fig. 4.3 Sponsors of incubator

Table 4.18 Staff of incubator

Answer options	Response percent (%)	Response count
1–5	72.2	52
6–25	18.1	13
26–50	6.9	5
>50	4.2	3
Total	100	

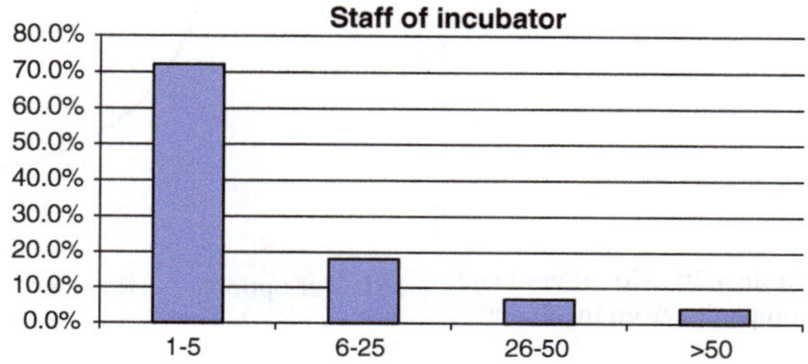

Fig. 4.4 Staff of incubator

Table 4.19 No. of jobs creation from the incubator

Answer options	Response percent (%)	Response count
1–5	9.9	7
6–25	14.1	10
25–50	16.9	12
>50	59.2	42
Total	100	

4.3.5 No. of Jobs Creation from the Incubator

Table 4.19 shows the number of jobs created from the innovation and incubator programs. Seventy-one respondents answered this question. The results indicated that the highest number of jobs created was more than 50 jobs (N = 42, 59.2 %). The second highest number of jobs created was between 25 to 50 jobs (N = 12, 16.9 %). The third was between 6 and 25 jobs created (N = 10, 14.1 %), and the last was 1–5 jobs (N = 7, 9.9 %). See Fig. 4.5.

4.3.6 No. of Graduate Companies from Incubator

Table 4.20 shows the number of graduate companies from the business incubators and innovation centers. Sixty-eight respondents answered this question. The results indicated that the highest response percentage of the number of graduate companies was between 6 and 25 companies (N = 29, 42.6), followed by those who indicated 1–5 graduated companies (N = 21, 30.9 %). The lowest response rate was those that reported more than 25 graduated companies (N = 18, 26.5 %). See Fig. 4.6.

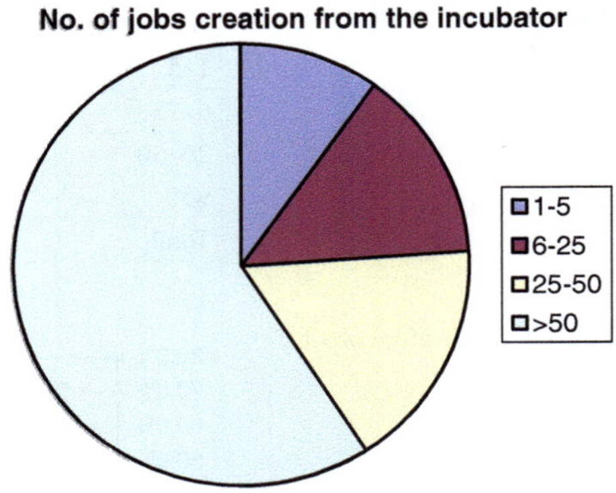

Fig. 4.5 No. of jobs created from the incubator

Table 4.20 No. of graduate companies from incubator

Answer options	Response percent (%)	Response count
1–5	30.9	21
6–25	42.6	29
>25	26.5	18
Total	100	

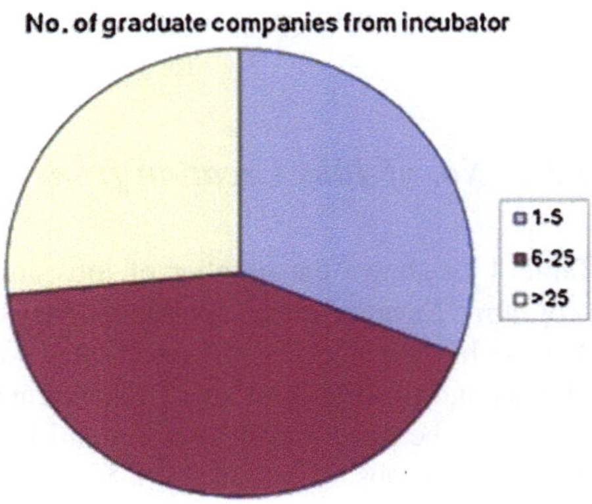

Fig. 4.6 No. of graduate companies from incubator

4.3.7 No. of Incubators Worldwide

Table 4.21 shows the number of incubators worldwide. The results indicated that most of respondents answered low increment of incubators worldwide (N = 29, 54.7 %), the other respondents answered high increment of incubators worldwide (N = 13, 24.5 %). Very few respondents answered medium increment (N = 12, 22.6 %). See Fig. 4.7.

Table 4.21 No. of incubators worldwide

Answer options	Response percent (%)	Response count
High increment	24.5	13
Medium increment	22.6	12
Low increment	54.7	29
Total	100	

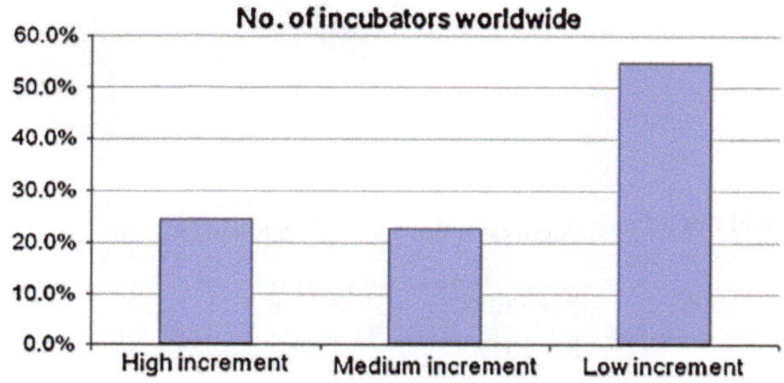

Fig. 4.7 No. of incubators worldwide

4.3.8 No. of Tenants Inside the Incubator

Table 4.22 shows the distribution of respondents by the number of tenants inside the incubators. Just over half (N = 35, 50.0 %) indicated that they have a number of tenants between 6 and 25, followed by those who have more than 25 tenants (N = 24, 34.3 %), and those with 1 to 5 tenants, respectively (N = 11, 15.7 %). See Fig. 4.8.

4.3.9 Survival Rate of Tenants

Table 4.23 presents the numbers and percentages of the tenants' survival rates. Survival rates in the mid-range of 81–90 % were reported by nearly half of the respondents (N = 34, 48.6 %), while lower tenant survival rates (less than 80 %)

Table 4.22 No. of tenants inside the incubator

Answer options	Response percent (%)	Response count
1–5	15.7	11
6–25	50.0	35
>25	34.3	24
Total	100	

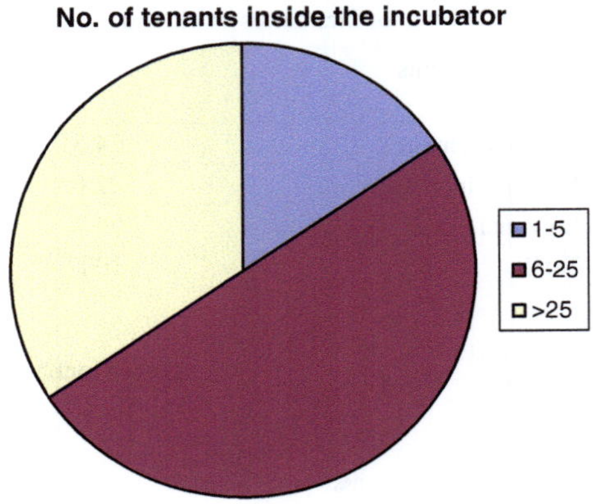

Fig. 4.8 No. of tenants inside the incubator

Table 4.23 Survival rate of tenants

Answer options	Response percent (%)	Response count
Less 80 %	27.1	19
81–90 %	48.6	34
>90 %	25.7	18
Total	100	

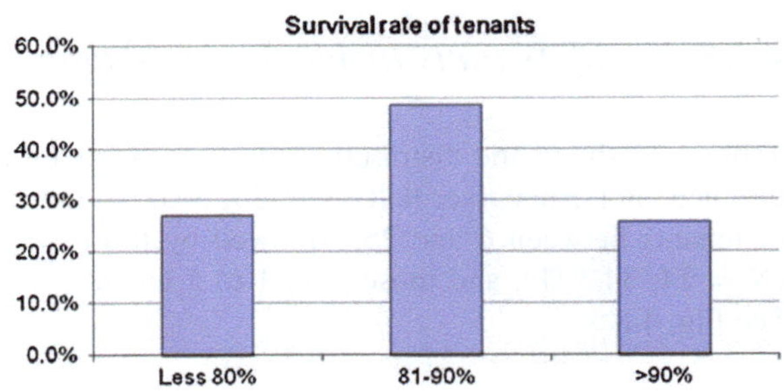

Fig. 4.9 Survival rate of tenants

were reported by approximately one fourth of the respondents (N = 19, 27.1 %), and higher survival rates (more than 90 %) by another fourth of the respondents (N = 18, 25.7 %) See Fig. 4.9.

4.3.10 Incubators Focus

Table 4.24 shows the distribution of respondents by their incubation program's focus. Just over half (N = 52, 73.2 %) indicated that the incubator's focus was on entrepreneurship, followed by economic development and jobs creation to describe the program respectively (N = 43, 60.6 %; N = 38, 53.5 %). The rest described their programs as technology transfer (N = 28, 39.4 %), create ventures (N = 24, 33.8 %), and high technology and patents (N = 18, 25.4 %). See Fig. 4.10.

Table 4.24 Incubators focus

Answer options	Response percent (%)	Response count
Economic development	60.6	43
Jobs creation	53.5	38
Technology transfer	39.4	28
High technology and patents	25.4	18
Entrepreneurship	73.2	52
Create ventures	33.8	24
Total	100	

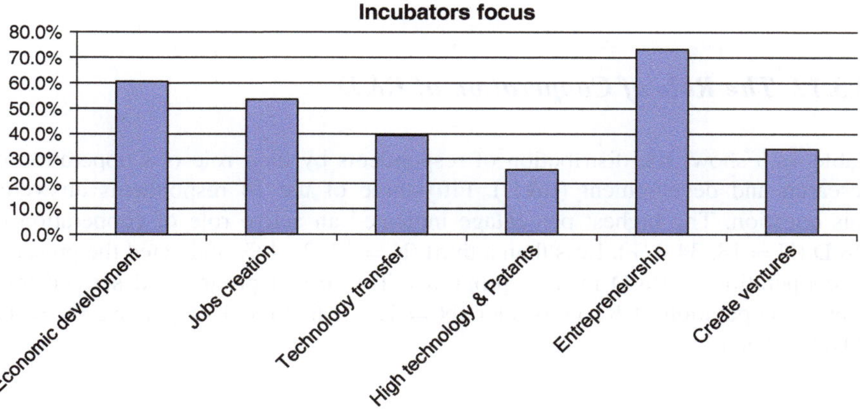

Fig. 4.10 Incubators focus

4.3.11 The Role of Science Park

Table 4.25 shows the distribution of respondents by their reports on the role of Science Park. Fifty-one of the 74 respondents answered this question. Just under half (N = 23, 45.1 %) indicated that the role of science park was poor. One-fourth of the respondents said (N = 13, 25.5 %) that indicated the presence of a science park in their programs. The rest of programs described their role of science park as very active (N = 8, 15.7 %) and active (N = 7, 13.7 %). See Fig. 4.11.

Table 4.25 The role of Science Park

Answer options	Response percent (%)	Response count
Very active	15.7	8
Present	25.5	13
Active	13.7	7
Poor	45.1	23
Total	100	

Fig. 4.11 The role of Science Park

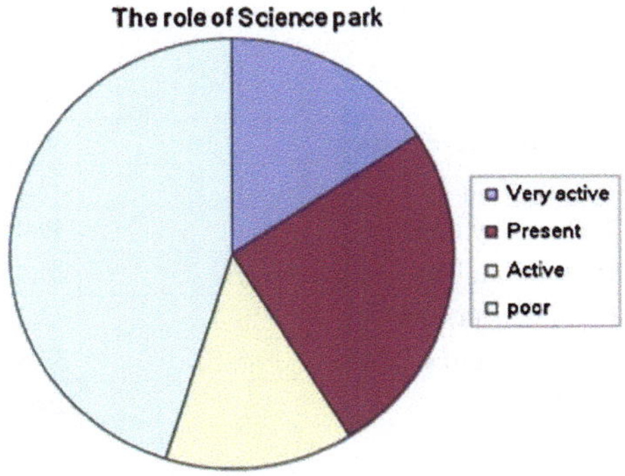

4.3.12 The Role of Cooperation of R&D

Table 4.26 shows the distribution of respondents by their role of Cooperation of research and development (R&D). Fifty-three of the 74 respondents answered this question. The highest percentage indicated an active role of cooperation of R&D (N = 18, 34.0 %). Less than a third (N = 14, 26.4 %) indicated the presence of cooperation of R&D in their programs. The rest of programs described their role of cooperation of R&D as poor (N = 12, 22.6 %) and very active (N = 11, 20.8 %). See Fig. 4.12.

Table 4.26 The role cooperation of R&D

Answer options	Response percent (%)	Response count
Very active	20.8	11
Present	26.4	14
Active	34.0	18
Poor	22.6	12
Total	100	

Fig. 4.12 The role of cooperation of R&D

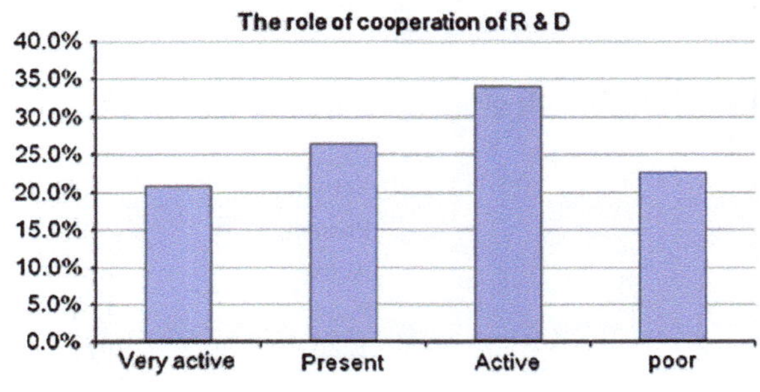

4.3.13 The Rate of Fostering Entrepreneurship

Table 4.27 shows the distribution of respondents by their role of fostering entrepreneurship. The highest percentage indicated their rate of fostering entrepreneurship at market rate (N = 46, 71.9 %). The lowest percentage indicated their rate of fostering entrepreneurship to be at a low rate (N = 8, 12.5 %). The rest of programs described their rate of fostering entrepreneurship to be below market rate (N = 10, 15.6 %). See Fig. 4.13.

4.3.14 The Role of Innovation

Table 4.28 presents the numbers and percentages of responses about the role of innovation in the programs. Approximately, 40 % of the respondents indicated that role of innovation was very active (N = 27, 39.7 %), followed by those who indicated that innovation had an active role (N = 22, 32.4 %). More than one-fourth (N = 18, 26.5 %) indicated a present role of innovation, followed by those who indicated a poor role of innovation (N = 3, 4.4 %). See Fig. 4.14.

Table 4.27 The rate of fostering entrepreneurship

Answer options	Response percent (%)	Response count
Below market rate	15.6	10
At market rate	71.9	46
Low rate	12.5	8
Total	100	

Fig. 4.13 The rate of fostering entrepreneurship

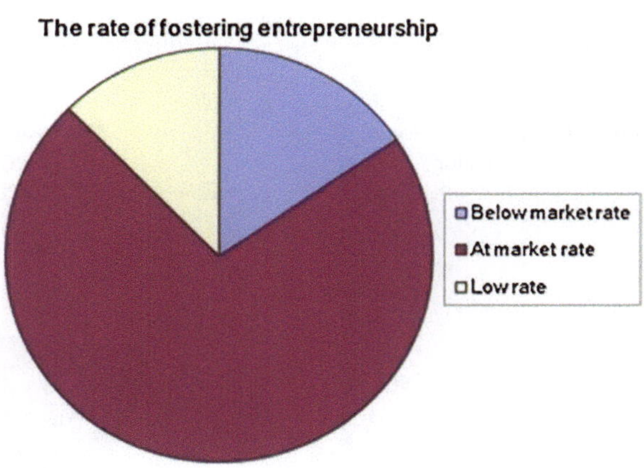

Table 4.28 The role of innovation

Answer options	Response percent (%)	Response count
Very active	39.7	27
Present	26.5	18
Active	32.4	22
Poor	4.4	3
Total	100	

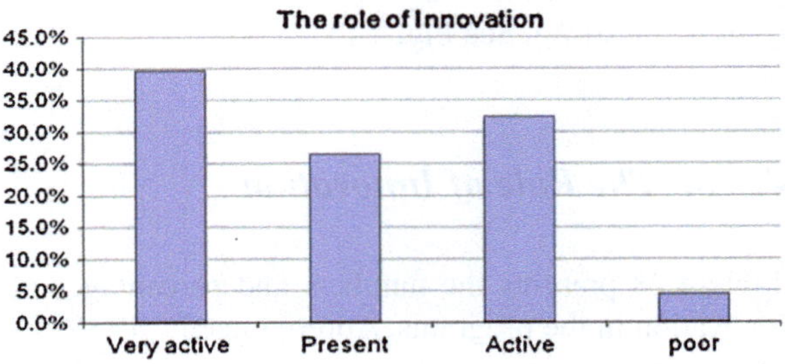

Fig. 4.14 The role of innovation

4.3.15 The Technology Transfer of Incubator

Table 4.29 shows the responses regarding the incubators' technology transfer. Nearly half of the survey respondents indicated a modest degree of technology transfer (N = 31, 47.7 %). The remaining respondents were evenly divided in their ratings of the degree of technology transfer in their programs, with some indicating a strong degree of technology transfer (N = 17, 26.2 %) and others indicated a weak degree (N = 17, 26.6 %). See Fig. 4.15.

Table 4.29 The technology

Answer options	Response percent (%)	Response count
Strong	26.2	17
Modest	47.7	31
Poor	26.2	17
Total	100	

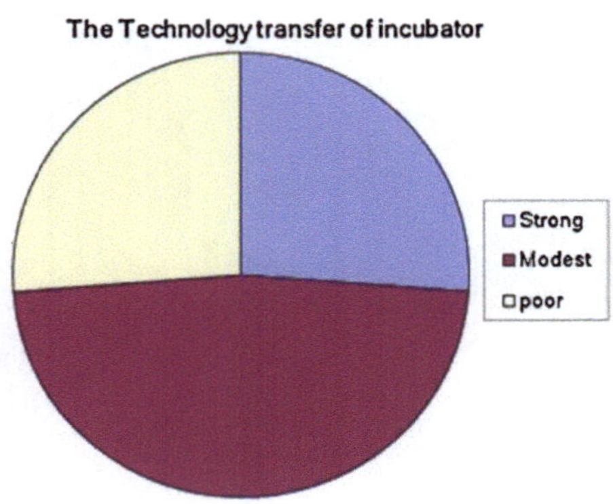

Fig. 4.15 The technology transfer of incubator

4.3.16 The Rate of Patents

Table 4.30 shows the distribution of respondents by the rate of patents from their innovation programs. Just under half (N = 32, 47.8 %) indicated a low rate of patents in their countries, followed by more than a third (N = 24, 35.8 %) that indicated a medium rate of patents. The rest described their innovation programs with a high rate of patents (N = 11, 16.4 %). See Fig. 4.16.

4.3.17 The Rate of Licensed Intellectual Property (IP)

Table 4.31 shows the distribution of respondents by their rate of licensed intellectual property (IP). Just under half (N = 31, 47.0 %) indicated a low rate of licensed IP in their countries, followed by more than a third (N = 25, 37.9 %) that indicated a medium rate of licensed IP. The rest described their innovation programs with a high rate of licensed IP (N = 10, 15.2 %). See Fig. 4.17.

Table 4.30 The rate of patents

Answer options	Response percent (%)	Response count
High rate	16.4	11
Medium rate	35.8	24
Low rate	47.8	32
Total	100	

Fig. 4.16 The rate of patents

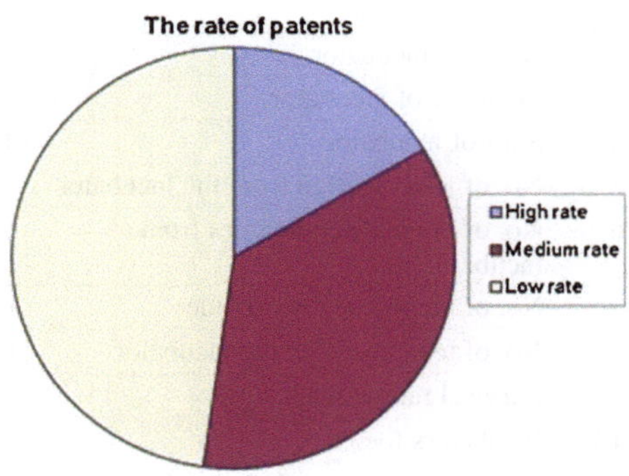

Table 4.31 The rate of licensed intellectual property (IP)

Answer options	Response percent (%)	Response count
High rate	15.2	10
Medium rate	37.9	25
Low rate	47.0	31
Total	100	

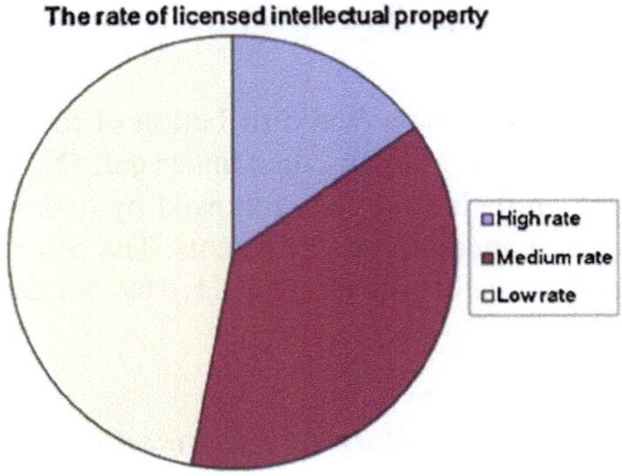

Fig. 4.17 The rate of licensed intellectual property (IP)

4.4 Summary

4.4.1 Summary of Surveys

Table 4.32 provides an overview of the responses to the 17 survey items by the 74 innovation centers and incubators in the survey sample. First, nearly three-fourths (73.0 %) indicated a goal of fostering an entrepreneurial climate. Second, most of the services offered by incubators focused on strong tangible and specialized

Table 4.32 Summary of surveys

No.	Survey questions	Percentage of high response	
1	Services of incubator	Strong tangible and specialize services	67.0 %
2	Goals of incubator	Entrepreneurial climate	73.0 %
3	Sponsors of incubator	Non-profit economic development	48.0 %
4	Staff of incubator	1–5	72.0 %
5	No. of jobs creation from the incubator	>50	59.0 %
6	No. of graduate companies from incubator	6–25	43.0 %
7	No. of incubators worldwide	Low increment	55.0 %
8	No. of tenants inside the incubator	6–25	50.0 %
9	Survival rate of tenants	81–90 %	49.0 %
10	Incubators focus	Entrepreneurship	73.0 %
11	The role of Science Park	Poor	45.0 %
12	The role cooperation of R&D	Active	34.0 %
13	The rate of fostering entrepreneurship	At market rate	72.0 %
14	The role of innovation	Very active	40.0 %
15	The technology transfer of incubator	Modest	48.0 %
16	The rate of patents	Low rate	48.0 %
17	The rate of licensed intellectual property (IP)	Low rate	47.0 %

services (67.0 %). Third, nearly half of the sponsors of incubators were non-profit economic development organizations (48.0 %). Fourth, the majority (73.0 %) of incubators focused on entrepreneurship. Fifth, nearly three-fourths (72.0 %) had staffs that ranged from one five persons to operate the incubators.

Sixth, most of the incubators created more than fifty jobs per incubator program (59.0 %). Seventh, less than half (43.0 %) had graduated in the range of six to twenty-five companies. Eighth, half of the programs reported that the number of tenants inside the incubation program ranged from six to twenty-five (50 %). Ninth, half (49.0 %) indicated that the survival rate for the companies ranged between eighty-one percent and ninety percent. Tenth, more than half (55.0 %) presented a low increment of the expansion of incubators worldwide.

Eleventh, one-third (34.0 %) of the innovation programs presented an active role of cooperation of R&D. Twelfth, a significant number of program indicated a very active role of innovation (40.0 %). Thirteen, nearly half of the programs indicated that the role of technology transfer from the incubators was at a modest level (48.0 %). Fourteen, the majority of programs responded that they had a moderate rate of fostering entrepreneurship (72.0 %). Fifteen, the rate of patents and the rate of intellectual property licenses (IP) were presented as a low rate by just under half of the respondents (48.0 and 47.0 % respectively). Finally, nearly half of the programs indicated a poor role of science parks in their programs (45.0 %).

4.4.2 Summary of Interviews

Table 4.33 presents the interview ranking of ten selected innovation centers. The Sussex Innovation Center presents the highest rank with an average of 93 %. Most of the case study averages indicated high outcomes for programs located in the US, UK and GCC. However, four case studies of programs in the UK indicated medium outcomes including Institute of Knowledge Transfer, Birmingham University, The Petchey Center of Entrepreneurship, and University of Southampton.

4.4.3 Summary of Case Studies

There are many ways to rank the case studies—for instance using descending or ascending order. The main purpose of descending order is to sort the case studies from higher to lower in any scale for the characteristics of case studies. Furthermore, the case studies consist of four characteristics including: the number of years of operation up till 2013, the ratio of jobs created over the years, the ratio of client companies over the years, and ratio of graduate companies over the years.

Table 4.33 Summary of interviews

Interview	Culture %	Policy %	Industry %	Economy %	Average %	Scale[a]	Rank
Sussex Innovation Center, University of Sussex, UK	90	90	100	90	93	High	1
Qatar Science and Technology Park, Doha, Qatar	85	80	95	100	90	High	2
St. John Innovation	95	80	80	95	88	High	3
South Side Innovation Center—"Inclusive Entrepreneurship", New York, US	100	75	75	85	84	High	4
High Tech Rochester Inc., West Henrietta, NY, US	95	65	65	95	80	High	5
Blue Valley Schools—Center for Advanced Professional Studies—"CAPS Incubator", Kansas, US	100	85	75	60	80	High	5
Birmingham University, UK	60	75	80	100	79	Medium	6
The Petchey Center of Entrepreneurship, University of East London, UK	85	85	70	70	78	Medium	7
Institute of Knowledge Transfer, London, UK	95	70	70	70	76	Medium	8
Southampton Management School, University of Southampton, UK	85	75	60	60	70	Medium	9

[a]Scale high ranged 81–100 %, scale medium ranged 41–80 %, scale low ranged less than 41 %

Table 4.34 gives the ranking of case studies based on the number of operation up till 2013. The Austria case study presents the highest ranking of 32 years; France presents a medium rank of 26 years, and Germany a low rank of 8 years.

Table 4.35 presents the case study ranking based on the average number of jobs created over the years. Austria presents the high rank with 25.00 jobs per year and Italy presents a medium rank with 14.85 jobs created per year.

Table 4.34 Ranking of years of operation

No.	Case study	Ratio of operation years
1	Austria	*32*
2	France	26
3	Italy	26
4	Luxemburg	26
5	Belgium	26
6	Spain	25
7	UK	19
8	N. Land	11
9	Germany	8

Table 4.35 Ranking the ratio of jobs created over the years

No.	Case study	Ratio of jobs created over the years
1	Austria	*25.00*
2	Italy	14.85
3	N. Land	8.82
4	UK	6.05
5	Spain	4.84
6	France	4.31
7	Germany	2
8	Belgium	1.31
9	Luxemburg	1

Table 4.36 Ranking the ratio of jobs created over the years

No.	Case study	Ratio of jobs created over the years
1	UK	*5.53*
2	Austria	5.31
3	N. Land	2.09
4	Belgium	1.73
5	Italy	1.62
6	Spain	1.56
7	Germany	1.25
8	Luxemburg	0.5
9	France	0.42

Table 4.36 presents the ranking of the ratios of client companies, with the UK showing the highest ratio at 5.53 per year, Austria demonstrating a medium rank with a ratio of 5.31 per year, and France with the lowest rank with a ratio of 0.42 per year.

Table 4.37 Ranking the ratio of graduate companies over the years

No.	Case study	Ratio of graduate companies over the years
1	Austria	*12.63*
2	Belgium	6.69
3	UK	5.84
4	N. Land	5.73
5	Spain	4.4
6	France	2.88
7	Italy	2.38
8	Germany	0.75
9	Luxemburg	0.46

Table 4.37 presents the ranking of the ratios of graduate companies, with Austria showing the highest rank at 12.63 graduate companies per year, Belgium demonstrating the moderate rank at 6.69 graduate companies per year, and Luxemburg presenting the lowest rank at 0.46 per year.

Chapter 5
Conclusion and Recommendation

Abstract This chapter provides the conclusion. The results imply that innovation centers and incubator programs that provide continuing tangible and intangible services are able to add value to the companies they support. This leads to a high number of tenants in incubator programs and a high number of start-up companies. The most common outcome from innovation is economic growth. In addition innovation systems in developed and developing countries lead to a platform for policy decisions and high technology derivatives. The results of the survey of a sample of seventy-four organizations, ten real interviews focused in the US, UK and GCC, and ten international multi-cases are reported. Key findings and recommendations for guidelines are also presented in this section.

Keywords Innovation · Incubators · Modern economy · Key findings · Recommendations

5.1 Overview

It has been widely acknowledged that the 21st century will be shaped by innovation and technology. Innovation will create new jobs and catalyze broadly shared economic growth. The growth of the economy and international competitiveness depends on our capacity to innovate. Furthermore, a business innovation centre is aimed at promoting economic development of its community by supporting start-up companies and their business development and offers services to support the establishment and development of new as well as existing small and medium companies. Also, incubators act as a positive active tool for economic growth by fostering entrepreneurship, innovation and technology commercialization.

The authors in this project have highlighted the importance of innovation and incubators as powerful tools for a modern economy. The project has been based on knowledge using quantitative and qualitative approaches to reach deeper insights into the phenomena under investigation. The study has included a survey of a

H.M. Al-Mubaraki et al., *Innovation and Entrepreneurship*,
SpringerBriefs in Business, DOI 10.1007/978-3-319-13683-7_5

sample of seventy-four organizations, ten real interviews focused in the US, UK and GCC, and ten international multi-case studies that investigated, addressed and explained the successful adoption of innovation centres and business incubators worldwide. Some case studies revealed better performance than others. Several reasons may contribute to this difference in performance. Key findings and recommendations for guidelines will be presented in section. Therefore, this project attempts to provide a new line of thinking and further scope for researchers in areas of entrepreneurship, incubators, innovation, and technologies. Finally, innovation is highlighted as the driver of the future growth that will shape the ideas and technologies suitable for building the 21st century.

5.2 Key Findings

Analysis of the survey, case studies, radar charts (international interview), and literature review resulted in the following key findings:

1. The case studies of innovation centers and incubation programs indicated that programs tend to become more successful with experience and maturity, making years in operation a very important metric.
2. Innovation centers and incubation programs that construct strategic relationships with an international organization such as the European Business and Innovation Centre Network (EBN) and governmental bodies are more successful in technology transfer.
3. Innovation centers and incubator programs that provide continuing tangible and intangible services are able to add value to the companies they support. This leads to a high number of tenants in incubator programs and a high number of start-up companies.
4. The most common outcome from innovation is economic growth.
5. Innovation systems in developed and developing countries lead to a platform for policy decisions and high technology derivatives.

Seventy-four incubators and innovation centers from both developed and developing countries participated in a survey. A descriptive analysis of the survey responses revealed the following:

1. Two-thirds (67 %) of the services offered by incubators were strong tangible and specialized services.
2. Three-fourths (73 %) of incubators' goals focused on the entrepreneurial climate.
3. The main sponsors of incubators were non-profit economic development (48.0 %).
4. Most incubators (72.0 %) have a staff of one to five persons.
5. The majority of incubators (59.0 %) reported creating more than 50 jobs.
6. A high percentage (43.0 %) reported having graduated between six and twenty-five companies.

7. A majority of respondents (55.0 %) indicated a low increment of incubators worldwide.
8. Half of the incubators surveyed (50.0 %) indicated having tenants in the range of six to twenty-five tenants.
9. Half of the incubators and innovation centers (49.0 %) indicated that the survival rate of their tenants ranged between 81–90 %.
10. Nearly half (45.0 %) of the incubators and innovation centers indicated a poor role of science parks.
11. One-third (34.0 %) of the programs indicated active cooperation with R&D.
12. A high percentage (72.0 %) of programs indicated fostering entrepreneurship at the market rate.
13. A significant percentage of programs (40.0 %) indicated a very active role for innovation.
14. Nearly half (48.0 %) of incubators and innovation centers indicated technology transfer at a modest level.
15. Most of incubators and innovation center present low rate of patents (48.0 %).
16. Most of the incubators and innovation centers present low rate of licensed intellectual property (IP) (47.0 %).

In addition, some of the leading rankings revealed in the study were these:

1. Among the ten programs located in the US, UK, and GCC that were interviewed based on selected best practices, Sussex Innovation Centre presented the highest outcomes (93 %) with respect to the other innovation centers.
2. An Austrian organization, Styrian Business Promotion Agency, SFG, was the oldest program studied, at 32 years.
3. The Austrian case study also presented the highest ratio of jobs creation, at 25.00 per year.
4. The UK case study demonstrated the highest ranking based on the ratio of client companies, at 6 per year.

5.3 Policy Recommendations

Based on the analysis of the surveys, case studies, and interviews we have identified many areas of opportunity for enhancing the work of innovation centers and business incubators in developed and developing countries. Following are the recommendations:

1. Construct an innovation data bank for best practice models including successful case studies and outcomes to share the knowledge worldwide.
2. Construct a worldwide database of innovation to upload annual reports with data for each country to facilitate networking and exchange of information.
3. Promote innovation as a tool to strengthen the modern economy based on the knowledge towards smart growth.

4. Provide incubator models with outcomes on innovation, entrepreneurship, and job creation.
5. Innovation centers and incubation programs should evaluate their results annually and analyze their strengths, weaknesses, opportunities, and threats for continuing planning and improvement.
6. Innovation programs should collaborate with universities, research institutes, and non-profit think tanks to maximize knowledge and technology transfer.
7. Provide funding organizations to support technology-based entrepreneurs, start-ups, and small business firms.
8. Develop communication platforms using software applications for worldwide innovation collaboration.
9. Focus on entrepreneurship as the most important element in generating innovation and economic growth including new firms, job growth, and small businesses.
10. Accelerate business innovation and business incubation toward smart, sustainable, inclusion and digital growth.
11. Strengthen education systems using innovation as a dynamic tool to adapt new education strategies and systems toward 21st century skills and create a world-class workforce.
12. Prepare governmental programs to promote innovation and competitiveness and sharpen understanding of the innovation process through the development of a research and innovation policy for economic development.

5.4 Kuwait Implications Policy

To accelerate the adoption of best practices for innovation in GCC as well as Kuwait the following priorities are suggested:

1. Governments at all levels should play major roles in providing innovation policy.
2. Research associations should play an active role in performing research about innovation as a driver for the 21st century.
3. Private enterprises and research institutes should play a vital role in supporting innovation activities financially to accelerate innovation in the country.
4. Universities and related institutions should provide appropriate resources in the form of knowledge, skills, and innovation infrastructure including laboratories, technology transfer offices, joint research institutes, patent offices, and training organizations.
5. Construct a collaborative innovation platform engaging all parties from academic institutions, government, policy makers, think tanks, and funding organizations to discuss strategies for success.
6. Include instruction about innovation in the curriculum of schools and colleges to give an earlier understanding about the role of innovation policy, practices, and strategies.

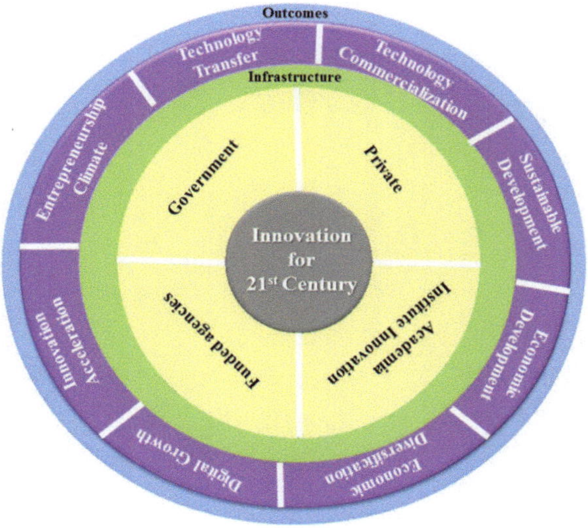

Fig. 5.1 The proposed model

7. Develop training programs at the government level for entrepreneurship, learning innovation processes, new technology, and creative thinking toward smart social growth.
8. Develop an annual award for the best innovation idea to increase the competitive climate among students, employees, and researchers.

5.5 Proposed Model

Figure 5.1 presents model of innovation strategy for 21st century with innovation infrastructure and innovation outcome, which contributed positively in developed and developing countries.

5.6 Future Studies

Future work can be continued from other regions such as the Middle East and South America which will fetch positive lessons that are valuable for future expansion and development of science park or innovation centre.

References

1. Al-Mubaraki H (2008) Procurement of International Business Incubation– Quantitative and Qualitative approaches. Melrose Books, United Kingdom. Available online: http://www.melrosebooks.com.
2. Al-Mubaraki H, Anderson B (2012) The gateway innovation center: exploring key elements of developing a business incubator. World J Entrepreneurship Manag Sustain Dev 8(4):208–216
3. Al-Mubaraki H, Busler M (2010a) Business incubators: findings from worldwide survey, and guidance for the G.C.C states. Glob Bus Rev 11(1):1–20
4. Al-Mubaraki H, Busler M (2010b) Business incubators models of the USA and UK: a SWOT analysis. World Assoc Sustain Dev 6(4):335–354
5. Al-Mubaraki H, Busler M (2010c) Sustainable development through the inclusion of incubator: A SWOT analysis. World Sustainable Development Outlook; pp 51-63. Available online: http://www.worldsustainable.org.
6. Al-Mubaraki H, Busler M (2011a) The development of entrepreneurial companies through business incubator programs. Int J Emerg Sci 1(2):95–107
7. Al-Mubaraki H, Busler M (2011b) Critical activity of successful business incubation. Int J Emerg Sci 1(3):455–464
8. Al-Mubaraki H, Busler M (2011c) Business incubation program for economic developments: the case study of G.C.C. Int J Arts Sci 4(22):431–444
9. Al-Mubaraki H, Busler M (2011d) Incubators: economic theories versus economic development. In The eighth international conference of global academy of business and economic research, Dubai, UAE, 22–23 Dec 2011
10. Al-Mubaraki H, Busler M (2011e) The incubators economic indicators: mixed approaches. J Case Res Bus Econ 4:1–12. http://www.aabri.com/manuscripts/11884.pdf
11. Al-Mubaraki H, Busler M (2011f) Innovation, entrepreneurship, and job creation, based on incubators: international experience. In: The Ninth biennial conference on entrepreneurship, Ahmedabad, India, 16–18 Feb 2011
12. Al-Mubaraki H, Busler M (2011g) University technology transfer through Innovation Incubator: a case study. In: World business, economics and finance conference, Bangkok, Thailand. http://www.wbiconpro.com/408-Hanadi.pdf. Accessed 26, 27 Sept 2011
13. Al-Mubaraki H, Busler M (2012a) Incubators landscapes in United States and Brazil: a comparison study. In: 4th International business and social science research conference, Dubai, UAE. http://www.wbiconpro.com/428-Hanadi.pdf. Accessed 5–7 Jan 2012
14. Al-Mubaraki H, Busler M (2012b) Quantitative and qualitative approaches of incubators as value-added: best practice model. J Am Acad Bus Camb 18(1):238–245

H.M. Al-Mubaraki et al., *Innovation and Entrepreneurship*,
SpringerBriefs in Business, DOI 10.1007/978-3-319-13683-7

15. Al-Mubaraki H, Busler M (2012c) Entrepreneurship spirit of Asia business incubation. Academic and business research, San Antonio, USA. http://www.aabri.com/SA12Manuscripts/SA12036.pdf. Accessed 22–24 March 2012
16. Al-Mubaraki H, Busler M (2012d) Innovation systems in European countries: a SWOT analysis. Eur J Bus Manag 4(16):106–117
17. Al-Mubaraki H, Busler M (2012e) Spark of business incubators roles: Latin American case studies. Glob Rev Bus Eco Res 9(1):17–23
18. Al-Mubaraki H, Busler M (2012f) Beyond incubators: youth entrepreneurship generation. Eur J Bus Manag 4(14):71–74
19. Al-Mubaraki H, Busler M (2013a) Towards a new vision for sustainability of incubator best practices model in the years to come. J Eco Sustain Dev 4(1):114–127
20. Al-Mubaraki H, Busler M (2013b) Fostering the innovation and entrepreneurship in developing countries. Int J Innovative Res Manag 1(1):67–73
21. Al-Mubaraki H, Hamad E (2012) Business incubation accelerator tool for 21st century. J Am Acad Bus 18(2):233–237
22. Al-Mubaraki H, Busler M (2009) Business incubators: findings from worldwide survey and guidance for the G.C.C states. World Sustainable Development Outlook, pp. 83–91
23. Al-Mubaraki H, Schrödl H (2011) Measuring the effectiveness of business incubators: a four dimensions approach from a gulf cooperation council perspective. J Enterprising Cult 19(4):435–452
24. Al-Mubaraki H, Schrödl H (2012) Incubating success towards gulf cooperation council (GCC). Int J Innov Knowl Manag Middle East North Afr 1(2):31–56
25. Al-Mubaraki H, Wong SF (2011) How valuable are business incubators? a case illustration of their performance indicators. In: European, mediterranean and middle eastern conference on information systems 2011. Athens, Greece. 30–31 May 2011
26. Al-Mubaraki H, Ahmed A, Al-Ajmei R (2014) Best practices of business incubators in developed and developing countries: the roadmap for the gulf cooperation council (GCC) countries. In: World association for sustainable development, University of Sussex, United Kingdom. http://www.amazon.co.uk
27. Al-Mubaraki H, Al-Karaghouli W, Busler M (2010) The Creation of business incubators in supporting economic developments. In: European, mediterranean and middle eastern conference on information systems (EMCIS2010). Abu Dhabi, 12–13 April 2010
28. Al-Mubaraki H, Al-Sharrah G, Aruna M (2012) Ranking business incubation: mathematical techniques. Int J Innov Reg Dev 4(6):466–487
29. Al-Mubaraki H, Busler M, Al-Ajmei R (2014) The impact of business incubation (BI) in developed and developing countries. Glob Rev Bus Eco Res 9(2):133–144
30. Archibugi D, Coco A (2004) A new indicator of technological capabilities for developed and developing countries (ArCo). World Dev 4(2004):629–654
31. Archibugi D, Coco A (2005) Measuring technological capabilities at the country level: a survey and a menu for choice. Res Policy 34:175–194
32. Bryman A, Bell E (2007) Business research methods, 3rd edn. Oxford University Press, Oxford
33. Eisenhardt K (1989) Building theories from case study research. Acad Manag Rev 14(4):532–550
34. European Business and Innovation Center Network (EBN) (2010) Case studies. http://www.ebn.eu/DisplayPage.aspx?pid=31
35. European Business and Innovation Centre Network (EBN) (2008) Business innovation centres (BIC) observatory facts and figures
36. European Business and Innovation Centre Network (EBN) (2009) BIC observatory facts and figures
37. European Business and Innovation Centre Network (EBN) (2012) BIC annual observatory. http://www.ebn.be/DisplayPage.aspx?pid=89
38. European Commission (EC) (2001) European Innovation Scoreboard 2001 (Ed.), SEC, Luxemburg, p. 1414

39. European Commission (EC) (2005) European Innovation Scoreboard 2005 (Ed.), Brussels
40. European Commission (EC) (2006) European Innovation Scoreboard 2006 (Ed.), Brussels
41. European Commission (EC) (2007) European Innovation Scoreboard 2007 (Ed.). Brussels
42. European Commission (EC) (2010) EUROPE 2020: A strategy for smart, sustainable and inclusive growth. http://ec.europa.eu/eu2020/pdf/COMPLET%20EN%20BARROSO%20%20%20007%20%20Europe%202020%20-%20EN%20version.pdf
43. European Union Regional Policy (EURP) (2010) The smart guide to innovation based incubators. http://www.ebn.eu/assets/assets/pdf/news/final_case-studies-nma-07042010.pdf
44. Freeman C, Soete L (2009) Developing science, technology and innovation indicators: what we can learn from the past. Res Policy 38:583–589
45. Godin B (2007) National innovation systems: the system approach in historical perspective, working paper No. 36, project on the history and sociology of STI statistics. Quebec, Montreal
46. Grupp H, Schubert T (2010) Review and new evidence on composite innovation indicators for evaluating national performance. Res Policy 39:67–78
47. Hart C (1998) Doing a literature review: releasing the social science research imagination. SAGE Publications Ltd, London
48. Jankowicz AD (1995) Business research projects. International Thomson Business Press, London
49. Eshun JP Jr (2009) Business incubation as strategy. Bus Strategy Ser 10(3):156–166. ABI/INFORM Global. (Document ID: 1882777971)
50. Katz JS (2006) Indicators for complex innovation systems. Res Policy 35:893–909
51. Katz JS (2005) Scale independent bibliometric indicators. Measurement: interdisciplinary. Res Perspect 3:24–28
52. Katz JS, Cothey V (2006) Web indicators for complex innovation systems. Res Eval 15(2):85–95
53. King N (1994) The qualitative research interview. In: Cassell C, Symon G (eds) Qualitative methods in organisational research. Sage, London
54. Kodama F (1987) A system approach to science indicators. In: Grupp H. (ed) Problems of measuring technological change. Verlag TÜV Rheinland, Köln, pp. 65–87
55. Lourenco MS (2004) Understanding communications network development and business incubation: an analysis of three incubators in Louisville, Kentucky. Doctoral dissertation, University of Louisville. Louisville, Kentucky
56. McAdam M, McAdam R (2008) High tech start-ups in university science park incubators: the relationship between the start-up s lifecycle progression and use of the incubator s resources. Technovation 28(5):277–290
57. Mian SA (1994) Are university technology incubators providing a milieu for technology-based entrepreneurship? Technol Manag 1:86–93
58. Mian SA (1996) Assessing value-added contributions of university technology business incubators to tenant firms. Res Policy 25(3):325–335
59. Mian SA (1997) Assessing and managing the university technology business incubator: an integrative framework. J Bus Ventur 12:251–285
60. Molnar L, Adkins D, Yolanda B, Grimes D, Sherman H, Tornatzky L (1997) Business incubation works. NBIA Publications, Athens
61. Monkman D (2010) Business incubators and their role in job creation. president and ceo national business incubation association (NBIA). Athens, Ohio. http://www.nbia.org. Accessed 30 Jan 2014
62. National Business Incubation Association (NBIA) (2010) Best practices in action: guidelines for implementing first-class business incubation programs. Revised 2nd Edition, Athens, OH
63. Niwa F, Tomizawa H (1996) A trial of general indicator of science and technology: methodological study of overall estimation of national S&T activities. Scientometrics 37(1996):245–265
64. Niwa F, Tomizawa H (1995) Integrated indicators: international comparisons of overall strengths in science and technology. In: NISTEP (ed) Science and technology indicators, 1994. Science and Technology Agency, Tokyo, pp. 345–365

65. Organisation for Economic Co-operation and Development (OECD) (2005) Science, Technology and Industry Scoreboard (Ed.), Paris
66. Organisation for Economic Co-operation and Development (OECD) (2010) Ministerial report on the OECD innovation strategy. Innovation to strengthen growth and address global and social challenges key findings. http://www.oecd.org/document/7/0,3343,en_2649_201185_44222919_1_1_1_1,00.html
67. Phillips RG (2002) Technology business incubators: how effective as technology transfer mechanism? Technol Soc 24:299–316
68. RTI International (2007) Study maryland incubator impact analysis and evaluation of incubator capacity, Maryland technology development corporation (TEDCO)
69. Sagasti F (2004) Knowledge and innovation for development: the sisyphus challenge of the 21st century. Edward Elgar, Cheltenham
70. Smith JM (1972) Interviewing in market and social research. Routledge, London
71. TEDCO (Maryland Technology Development Corporation) (2010) Annual report, fiscal year 2008, development. 13(3):454–468 ABI/INFORM Global. (Document ID: 1105634131).
72. Torrington D (1972) Face to face: techniques for handling the personal encounter at work. Gower Press, London
73. United Kingdom Business Incubation (UKBI) (2010) Business incubation. http://www.ukbi.co.uk/resources/business-incubation.aspx
74. United Kingdom Business Incubation (UKBI) (2011) The startup factories the rise of accelerator programmes to support new technology ventures. http://www.ukbi.co.uk/media/Download%20Docs/Incubation_for_Growth_-_NESTA_-_2011.pdf
75. White House (2010) A strategy for american innovation: driving towards sustainable growth and quality jobs. http://www.whitehouse.gov/assets/documents/SEPT_20_Innovation_Whitepaper_FINAL.pdf. Accessed 10 Jan 2014
76. Yin RK (1994) Case study research—design and methods, 1st edn. Sage Publications, Newbury Park
77. Yin RK (2004) The case study anthology, 2nd edn. Sage Publications, Newbury Park
78. Yin RK (2008) Case study research: design and methods, 4th edn. Sage publications, Newbury Park
79. Yin RK (2009) Case study research: design and methods, 5th edn. Sage publications, Newbury Park

MIX
Papier aus verantwortungsvollen Quellen
Paper from responsible sources
FSC® C105338

If you have any concerns about our products,
you can contact us on
ProductSafety@springernature.com

In case Publisher is established outside the EU,
the EU authorized representative is:
Springer Nature Customer Service Center GmbH
Europaplatz 3, 69115 Heidelberg, Germany

Printed by Libri Plureos GmbH
in Hamburg, Germany